$$\text{Skewness} = \frac{m_3}{m_2\sqrt{m_2}}$$

$$\text{Kurtosis} = g_2 = \frac{m_4}{m_2^2}$$

Pearson correlation coefficient

$$r_{xy} = \frac{n\sum XY - \left(\sum X\right)\left(\sum Y\right)}{\sqrt{\left[n\sum X^2 - \left(\sum X\right)^2\right]\left[n\sum Y^2 - \left(\sum Y\right)^2\right]}} \qquad (6.4)$$

For significance of r, use Table F, df $= n - 2$

Coefficient of determination $= r^2\ (100)$

Regression formulas for predicting Y values from X values

$$\hat{Y} = bX + a \qquad (6.6)$$

$$b_{yx} = r_{xy}(S_y/S_x) \qquad (6.7)$$

$$a_{yx} = M_y - b_{yx}(M_x) \qquad (6.11)$$

Standard error for predicting Y from X

$$\text{SE}_{yx} = S_y\sqrt{1 - r_{xy}^2} \qquad (6.13)$$

Partial correlation coefficient

$$r_{12.3} = \frac{r_{12} - (r_{13})(r_{23})}{\sqrt{(1 - r_{13}^2)(1 - r_{23}^2)}} \qquad (6.19)$$

Phi coefficient

$$\phi = \frac{(B)(C) - (A)(D)}{\sqrt{(A + B)(C + D)(A + C)(B + D)}} \qquad (6.21)$$

Spearman's correlation coefficient

$$\rho = 1 - \frac{6\left(\sum D^2\right)}{n(n^2 - 1)} \qquad (6.23)$$

PRACTICAL STATISTICS

for the

Physical Sciences

PRACTICAL STATISTICS for the Physical Sciences

Larry L. Havilcek

Ronald D. Crain

American Chemical Society
Washington, DC 1988

Library of Congress Cataloging-in-Publication Data

Havlicek, Larry L.
Practical statistics for the physical sciences.

(ACS professional reference book)
Bibliography: p.
Includes index.

1. Statistics. 2. Science—Statistical methods.

I. Crain, Ronald D., 1933– . II. Title.
III. Series.

QA276.12.H38 1988 519.5 88-10573
ISBN 0-8412-1453-0

PRINTED IN THE UNITED STATES OF AMERICA

About the Authors

LARRY L. HAVLICEK is currently Professor of Educational Research and Statistics at the University of Kansas. Since obtaining the B.M.E., M.M.E., and Ed. D. degrees from the University of Kansas, he has been director of several research studies and has 24 years of teaching, researching, and publishing experience in the areas of statistics and statistical applications. In 1985, he received a Fulbright award to lecture in Argentina and Uruguay on statistical applications in education. During the past 3 years, he has received three U.S. Information Agency grants to lecture, serve as a statistical consultant, and present seminars on statistical applications for educational institutions, government agencies, and private industries in Argentina, Paraguay, Peru, and Uruguay. Dr. Havlicek has published numerous articles in scholarly journals on theoretical and applied statistics and has made numerous presentations at national and international meetings, conferences, and seminars.

RONALD D. CRAIN was born in St. Joseph, MO, in 1938 and has an A.B. degree from William Jewell College (1954) and a Ph. D. in chemistry from Purdue University (1958). He worked in industry for the Ethyl Corporation and Farmland Industries and taught chemistry at McNeese State University and the University of Kansas. One of the first American Chemical Society (ACS) interactive computer courses in chemistry, “Structural Interpretation of Spectra”, was written by Dr. Crain for use on a network mainframe system. Along with Dr. Havlicek, Dr. Crain also wrote an introductory course on statistics for the ACS to be used on a microcomputer system. At the present time, Dr. Crain and his brother own a computer store in Bartlesville, OK.

Contents

Preface

Statistical procedures can make a major contribution to scientific research because these procedures enable the maximum amount of information to be drawn from the data. This book provides a comprehensive introduction to statistical procedures useful in the physical sciences.

This book was written with three types of readers in mind. The first type of reader is the user of research reports who has to interpret the results of statistical analyses and to make decisions on the basis of those analyses. An understanding of basic principles, assumptions, and statistical reasoning is necessary so proper and accurate interpretations of statistical analyses can be made. The second type of reader is the one who has to decide which analyses should be used for specific studies. These readers may have the responsibility for planning and conducting studies by using appropriate statistical analyses of the data. For these readers, this book will serve as a reference and guide for selecting and applying appropriate statistical analyses. The third type of reader is the technician who needs descriptions of actual applications of the analyses so that the analyses can be worked out. The worked-out examples included in this book can free these readers from having to translate textbook discussions into step-by-step computational procedures.

We have presented the concepts in a way that is understandable, practical to use, and easy to follow. More emphasis is placed on the concepts and assumptions underlying the use of statistics than on derivations of equations and mathematical computations. Emphasis has been given to the assumptions underlying each statistic and the limitations that should be considered when using and interpreting such analyses. This should provide the reader with the background necessary to design and conduct studies so that as much information as possible can be extracted from the results by using the most appropriate statistical analysis. Our main goal is to provide the reader with basic statistical principles that will result in thoughtful planning of statistical analyses and intelligent interpretation of the results.

The main emphasis is on the application of statistical analyses.

Each statistical procedure is illustrated by using a practical example similar to what would be encountered in chemistry or in related areas. Small whole numbers are used in the examples to avoid complications in arithmetic and to make it easier to follow. Readers do not need prior knowledge of statistics, and only a very basic knowledge of mathematics and algebra is required. The equations presented can be practically applied with hand or desk calculators or with computer programs, which are available for most of the analyses covered. References that provide additional information pertinent to the theoretical issues, derivations of formulas, and more complex analyses than those covered in this book are given. Included in the Appendix are tables of the distributions of the most commonly used statistics.

We thank the American Chemical Society Books Department staff for their interest, help, and encouragement in writing this book. We appreciate the assistance provided by the professional reviewers. We sincerely appreciate the permission to use tables given us by various authors, publishers, and associations. We are grateful to the Literary Executor of the late Sir Ronald A. Fisher, F.R.S.; to Dr. Frank Yates, F.R.S.; and the Longman Group Ltd, London, for permission to reprint Tables III, VII, VIII, and XXXIII from their book *Statistical Tables for Biological, Agricultural and Medical Research* (6th Edition, 1974).

LARRY L. HAVLICEK
Department of Educational Psychology and Research
University of Kansas
Lawrence, KA 66045–2338

RONALD D. CRAIN
MicroAge Computer Store
Bartlesville, OK 74006

PRACTICAL STATISTICS for the Physical Sciences

Chapter 1

Basic Statistical Concepts

Statistics is generally defined as a branch of mathematics concerned with the collection, analysis, presentation, and interpretation of data. The methods that are used range from simple counting and percentages to complex multivariate analyses designed to extract the maximum amount of information from a given set of data. Thus, *statistics* is a methodology for presenting and summarizing data, extracting information and meaning from that data with a specified degree of certainty, and making decisions from that data.

Statistics can be grouped into two categories: *Descriptive statistics* are used to describe the findings and results of studies, whereas *inferential statistics* enable inferences to be made beyond the data collected. Descriptive statistics will be presented in Chapters 3–6, and inferential statistics will be presented in Chapters 7–15.

1.1 Concepts Pertinent to Scientific Studies

This book is concerned only with statistical designs applicable to samples of data for which only one measurement is taken for each unit or sample element in a given study. These analyses are referred to as *univariate analyses*, regardless of the number of independent variables. The analyses covered in this book are applicable for experimental designs in which one or more treatment variables, each at two or more levels, can be analyzed. Thus, the designs covered in this book are applicable for studies in which one or more independent variables can be selected by the researcher. For example, many factors are known to have an effect on a chemical reaction, such as reaction time, temperature, pressure, and concentration level.

1453–0/88/0001 $06.00/1

The relative effect of each selected factor can be determined as a unique effect as well as an interactive effect with the other selected factors. However, the analyses covered in this book are appropriate only for studies having only one dependent variable (e.g., yield of a chemical or reaction time).

If two or more intercorrelated measurements are taken for each unit or sample element, then *multivariate statistical analyses* would be appropriate. Although multivariate analyses are beyond the scope of this book, an overview of these analyses and where they would be applicable is presented in Chapter 16.

Throughout this textbook, two important concepts pertinent to chemical studies will be emphasized. The first is *repeatability*, which is the closeness of agreement between individual results obtained with the same method on identical test material under the same conditions (i.e., same operator, apparatus, laboratory, and within a short period of time). This concept is closely related to the general concept of reliability, which will be discussed in Chapter 2. Reliability deals with the variability within groups or laboratories in chemical studies. The second concept is *reproducibility*, which deals with the closeness of agreement between results obtained under different conditions such as different settings, procedures, operators, apparatus, or laboratories. Both of these terms are defined in other statistics textbooks (*1–3*) and are closely related to the concepts of within and between variance, which are basic to the statistical analysis known as analysis of variance.

For the chemist who is engaged in production or research, there are as many uses of statistical methods as there are problems and needs for data in these areas. The extent of the varied uses of statistics in chemical applications is indicated by the list of studies under the subject heading "Statistics and Statistical Analysis" in *Chemical Abstracts* (*4*).

For example, statistics can be used to determine the most efficacious mixture of an insecticide dust with an inert powder for spreading over a given area, or to provide the best estimate of the percentage of insects killed after such a mixture is used. Statistics can be used to determine the optimum conditions for the manufacture of an organic chemical by taking into consideration various blends of the compound, temperature ranges, types of catalysts, and the quality control for the production of that chemical. Or possibly, the optimum combination of time of reaction and concentration of a mixture in the manufacture of a given chemical is needed.

1.2 Basic Questions That Statistics Answers

Sections 1.2.1–1.2.10 contain a discussion of some of the basic questions that can be answered through the use of statistics. For each question, the appropriate statistics are mentioned and then the specific chapters are given where each will be described in detail along with the procedures and examples for using each.

1.2.1 Organizing and Presenting Results

What is the general pattern of data and how can the data be presented? One of the basic uses of statistics is to describe the results of a study in a clear and meaningful manner. Various statistical methods can be used to answer this question. These statistics range from frequency distributions to various types of charts and graphs, as well as pictorial forms. These methods are presented in Chapter 3 along with guidelines for their use.

1.2.2 Summarizing Data: Central Tendency

What one central value best typifies the data, or what is the central tendency of the data? In addition to tabulations and pictorial presentations of data, statistical methods can be used to summarize the results by use of one or two values. Such methods are important when several sets of data are to be presented and compared. One type of summarizing statistic answers the question, "What is the average value or most typical value?" For example, what is the typical production output per hour, day, week, month, or year? What is the average concentration level of a chemical over several production runs? What is the average percentage of impurities in a specific chemical? Three statistics can be used to answer these questions: the mean, median, or mode, depending on the type of data and how it is distributed. Statistics dealing with measures of central tendency are presented in Chapter 4.

1.2.3 Summarizing Data: Variability

How variable are the data, or how closely do the values cluster around the measure of central tendency? Is there a high degree of consistency in a product, or is there a wide range between low and high points? Are products produced within specified tolerance limits? Are products produced by one production procedure more consistent than another procedure? These and similar questions can be answered by analyzing the variation of products, procedures, or just about any type of data. Several types of summary statistics can be used to communicate the variation in data, which in statistical terminology is known as *variance*. The most common of these summary statistics are the range, standard deviation, and the variance. Each of these will be described in Chapter 5 as well as the methods to communicate individual values within a set of values, which can be used to indicate the relative position of one data point in a set of values.

1.2.4 Relationships among Variables

How are two or more variables related to each other and how can one estimate or predict future events? Often a researcher or someone involved with productivity is interested in how two or more variables vary together (i.e., as one variable varies, is there also a similar or inverse variation in the other variable?). For example, a researcher may be interested in whether or not variations in temperature are associated with speed of reaction for a specific catalyst. A number of statistical measures describe the relationship, or *correlation* as statisticians prefer, between two variables, and these are presented in Chapter 6. These measures range from simple correlation coefficients between two variables to multiple regression analysis when dealing with one variable possibly being related in various degrees with several predictor variables.

1.2.5 Probability

How likely is it that a specific event will happen (e.g., that a specific production procedure will result in a specified percentage of defective products)? *Probability* deals with the likelihood that a specific event will happen. Any possible event has some probability of happening, and statistical procedures can express the likelihood of the occurrence of that particular event. The mathematical theory of probability provides the basis for inferential statistics and is thus important for this reason as well as for determining the specific likelihood that an event will happen. Various theoretical and empirical probability distributions are used by statisticians for both descriptive and inferential statistics. The basic principles of probability and a description of the basic theoretical distributions are presented in Chapter 7.

1.2.6 Hypothesis Testing

A *hypothesis* is defined as an assertion or supposition about an event that is unproven but can be empirically tested to determine whether it should be rejected or retained. Examples of where hypotheses are appropriate are given in Sections 1.2.8 and 1.2.9. In testing hypotheses, a number of principles must be adhered to, otherwise the conclusions may not be justified, and an incorrect decision may be made. The basic principles of hypothesis testing are presented in Chapter 8.

1.2.7 Confidence Limits

How much error is involved in the data, or how accurate are the data? Often a researcher or someone involved in quality control in production is interested in whether there is a deviation from a given or theoretical standard. For example, are current production runs producing more defective products than have been experienced before? The principle of confidence limits and probability is presented in Chapter 9, which deals with testing hypotheses for single sets of data.

1.2.8 Testing Hypotheses for Two Data Sets

How do two sets of data differ? Is one production schedule more effective than another? Is one type of catalyst more effective than another? Does one product contain more impurities than another? Several statistical analyses can be used to answer these types of questions by comparing two sets of data. Generally, these analyses involve comparing mean levels of productivity effectiveness, or some other characteristic, but many tests determine differences between other statistics that can be calculated for each group. These methods are presented in Chapter 10.

1.2.9 Testing Hypotheses Involving More Than Two Data Sets

The questions in Section 1.2.1–1.2.8 were concerned with two sets of data. But what about when more than two production schedules, catalysts, or products are to be compared? For example, there may be three or four types of production schedules, four or five catalysts, or six or more products to be compared to determine which are the most effective or most advantageous to use. Several statistical analyses are used for comparing more than two data sets. These analyses range from comparisons among the groups on just one independent variable to complex designs studying many different factors at the same time. One-way designs are presented in Chapter 11, and factorial designs involving simultaneously studying several factors at the same time are presented in Chapters 12 and 13.

1.2.10 Statistical Selection Guide

What statistic should be computed for a given situation? A guide summarizing the statistics that should be used for various types of situations and problems is presented in Chapter 16. This guide should serve as a quick reference to identify which statistic should be used. Once the statistic is identified, then the person needs to review the principles and procedures for correctly using that statistic. The guide will cover all of the statistics covered in this book. References will be given for each of these analyses so that the reader interested in these analyses can obtain additional information for either running these analyses or for interpreting the results of studies that have used these analyses.

1.3 Basic Statistical Principles

As with many concepts in scientific research, there are legitimate uses and, conversely, misuses of the concept that can be due to either lack of knowledge or a deliberate deception. As Huff (*5*) pointed out, statistics can be used to "sensationalize, inflate, confuse, and oversimplify". The purpose of this book is to emphasize legitimate applications of statistics so that honest and correct interpretations can be obtained from the data of a study. Both those applying statistics and the user of statistical reports must be familiar with the basic concepts of statistics so that valid interpretations and correct decisions can be made from the analyses of the data for a given study. Thus, some overall principles should be stated in using statistics.

First, statistics are merely tools to be used for evaluating scientific evidence. These tools must be used according to the rules and procedures associated with each statistic in order to arrive at sound decisions. Each statistic is based on both mathematical and logical assumptions, which must be followed to yield results that are meaningful and isomorphic to reality. As pointed out by a number of statisticians, statistical analyses can be computed on any set of numbers. However, the results may not be interpretable. Thus, the basic assumptions and rules for using each statistic must be met.

Closely associated to this idea is the concept that many statistical analyses, especially the complex factorial designs, are associated with a specific experimental design, and the results of such analyses can only be interpreted contingent upon how the design was implemented. The basic principles of randomization, blocking, and confounding of variables have to be taken into consideration in the design of any study. To the extent that the design was not implemented in a valid way, the results of the statistical analyses will be suspect.

Because statistical analyses are based on data of some sort, the basic data or raw material upon which statistics are computed must be valid and reliable. That is, the data must accurately reflect what exists in reality. High-speed computers can do many complex statistical analyses in a fraction of a second. However, if the data going into the computer are not valid or are biased, the results will also not be valid or will also be biased. "Garbage in–garbage out" is a concept that has evolved from zealously searching for meaning from invalid or biased data by running every possible statistic on such data and hoping to arrive at something that appears to be valid.

Not only is consideration of the basic data used for analyses important, but consideration must also be given to how the data were collected and recorded. As indicated, statistical analyses are often closely related to specific experimental designs used in experiments, and interpretation of such analyses is always based upon the design specifications. A statistic merely represents the calculations done on a given set of data. What the analyses imply or what kind of inferences can be drawn can be made only after consideration of how the basic data were collected. Statistics merely indicate, they do not per se show causes or reasons for their existence.

For inferences to be valid, the information or data upon which statistical analyses are based must adequately represent all cases that will be included in the inferences. In other words, the sampling of data must be representative of the population that is inferred. Methods of sampling, how the data were collected, how missing data were handled, and sample size all have to be considered when making inferences. The importance of random sampling cannot be overemphasized.

One must always consider that the results of any single study can never be considered as conclusive evidence because improbable outcomes can occur by chance. Thus, the concept of statistical error must be considered, and all results must be interpreted in terms of probability statements. This concept also demonstrates the necessity for replications of a study as one method of providing conclusive evidence.

Finally, the principle of reasonableness must always be applied. In interpreting statistical analyses, one must always rely heavily upon their knowledge of the situation, past experience, common sense, and the basic principles of applied statistics. One needs to consider all possible factors that might be influencing the statistical analyses and exercise caution before making definitive conclusions regarding any data analysis.

1.4 Parametric and Nonparametric Statistics

Most of the statistical analyses that are applicable for studies in chemistry and related sciences fall into the area of *parametric statistics*, which rely on the central limit theorem or on sampling distributions that are normally distributed. These statistics are usually applicable for interval and ratio measurements and for large-size samples, and often are valid for sample sizes as small as five cases per data set. However, situations may arise in which the basic data collected for a study are represented as *nominal* or *ordinal measurement*, where the distributions of these data are not normal, or where sample sizes are very small. For these situations, some statisticians recommend *nonparametric* or *distribution-free statistical methods*, which are valid regardless of the shape of the population distribution for a specific variable and can be used with nominal or ordinal measurements. Although these methods are generally not as powerful as parametric statistics, they are recommended when the basic data collected for a study are represented as nominal or ordinal measurement. These methods are described in several textbooks (*6–12*).

1.5 Statistical versus Practical Significance

Two types of decisions can be made regarding statistical analyses used for testing hypotheses. One type deals solely with the probability that the obtained event could or could not be due merely to chance, and the other type is whether the difference is large enough to be considered in a practical way. Statistical significance must be obtained in order to consider the practical significance because if the result could have happened by chance, no need arises for further consideration of the result. However, no generally accepted criteria is available for deciding how a significant difference should be considered from a practical point of view. Some statisticians recommend that the difference between any two groups should be at least one-third of a standard deviation to be of practical value. Others consider the basic measurement scale and consider practical differences on the basis of how the values are distributed. Regardless of the procedure selected, the interpretation of the results should include how practical differences were determined and the criteria used. This type of problem becomes acute when dealing with very large sample sizes for which minute differences could produce highly significant differences in terms of probability.

1.6 Computer Statistical Packages

Although most of the descriptive and many of the simpler inferential statistical analyses can be computed fairly quickly with a hand or desk calculator, the amount of arithmetic involved increases dramatically with complex analyses and increasing numbers of variables included in a study. Thus, complex analyses should be run with an electronic computer with either a package of statistical programs or programs written specifically for a particular analysis. The economics of using the computer with regard to cost and time are well documented. Additional advantages include reducing the possibility of computational errors and increasing the thoroughness of the study because many different analyses can be run once the data are in the computer and complex analyses that would be virtually impossible to do by hand can be run.

However, all of the analyses in this book have been done with a desk calculator to emphasize that these analyses can be done without a computer and to show the computational procedures.

A number of statistical packages are available for both mainframe and microcomputers. Some of the more common packages are widely available and are listed in Chapter 16. Many of these packages are reviewed periodically in various computer journals.

Many of these packages of programs include many other programs besides basic statistics. For example, the BMDP package, includes programs for detecting outliers, random case selection, and analysis patterns for missing data and checking for unacceptable values as well as articles pertinent to the statistical analyses included. Also, a statistical package should be tested by using examples from this or some other statistics textbook so that the user can verify that the program yields the results desired and that the results can be interpreted in a valid and meaningful way. Often programs provide several statistical analyses in addition to those desired and are formatted in a way that is different than expected. Thus, by running analyses on data that has been analyzed, the computer program can be verified as to accuracy, and the format of the results will become evident.

1.7 Planning for Statistical Analyses

Many reasons can be found for planning what statistical analyses are to be run on the data from the beginning or even at the conceptual stages of a study. First, planning for statistical analyses requires one to be definite and exact with regard to types of data necessary for a study, how the data are to be collected, what instruments are needed for collecting the data, the sampling required for the data to be representative, the sample size needed, and how the data are to be categorized for future partitioning into subgroups. Often, the planning for statistical analyses is done after a study is begun or sometimes even after all of the data have been collected. Often in such situations, running desired analyses is not possible because the data were not identified appropriately or not enough cases for a subgrouping were obtained. Thus, planning for statistical analyses from the conceptual stages of a study will ensure that all of the statistical analyses desired will potentially be possible. Such planning will make it possible to legitimately run a priori or planned comparisons rather than having to run unplanned or post hoc comparisons. Thus, choosing the type of analyses at the conception stage of a study is wise. The advantages of a priori and post hoc analyses will be discussed in Chapter 11.

As suggested by Caulcutt and Boddy (*1*), planning of extensive chemical studies should be done by a panel of experts familiar with the chemical processes, procedures, and products. This panel should include professional specialists, operators, supervisors, an executive officer, and a statistical research design expert. This panel should be involved in all stages of the research study from planning, development, and implementation through interpretation and presentation of the results. If a professional statistician is included as a panel member, a major chemical study of a complex process could possibly be designed in a way that would provide the most information with the least effort and number of replications. This design could result in considerable savings of personnel time, research effort, and amount of money expended for the research. However, smaller studies can be designed and conducted by individuals.

Many considerations have to be taken into account and decisions have to be made at the planning stage, such as conceptualizing the problem in terms that are feasible to carry out. One of the basic questions to be considered is whether the problem is potentially researchable or solvable. If the problem is researchable, then decisions have to be made with regard to the specific goals of the study, workforce and facility requirements, budget requirements, timetables, the number of replicates necessary, the number of levels of treatments, and many other aspects pertinent to each specific study and situation.

The Appendix at the end of this chapter is a summary of the sequence of activities and the types of questions that should be asked in the planning and the implementation of any study involving statistical analyses. Often in the planning of extensive studies, planning and conducting preliminary experiments is advisable before a large-scale experiment is run. These experiments may reveal problem areas or areas where procedures should be modified.

1.8 Guide for Interpreting Statistical Reports

What does a person interpreting statistical reports look for to determine if the reports validly and reliably communicate actual effects? This section will serve as a guide as to what types of questions should be asked as statistical reports are interpreted. This guide follows four basic questions that should be sequentially applied to any statistical report.

The first basic question deals with statistical conclusion validity. This question is "Was there an effect?" or "Are there results that should be considered?" Essentially, these questions lead to the following: "What evidence is presented to support the conclusions drawn for that study?" To answer this basic question, the following set of questions should be answered sequentially.

1. Were the results statistically significant? If so, at what level of significance?
2. What statistical analysis was used? Was this analysis appropriate for the hypothesis that was tested and the research design that was used? Were the results correctly interpreted?

3. Were all controlled sources of variance identified and accounted for? Was the error or residual variance examined to determine which uncontrolled extraneous variables might be operating, such as time drifts, recording errors, or impurities in the chemicals used?
4. What data were used for the analyses? Were the observations and measures used to collect the data valid and reliable? Were appropriate observations made? Were the data recorded and handled in a proper way to ensure few errors?
5. Were sampling procedures used to identify and select cases for the study done in an appropriate way? Were sample sizes adequate?
6. On the basis of the researcher's knowledge, experience, previous research, and theoretical and practical considerations, are the results believable? In other words, can the results be considered feasible on a rational basis? Are the results consistent with previous findings or what results are possible for such studies?
7. Are there any inconsistencies in the data? Do the results presented in or among tables and in the text agree? Do the conclusions logically follow what would be expected from these data?

The second basic question deals with the reason for the results or why the results happened. In research terminology, this concept is known as ***internal validity***, and answers the question "How likely is it that the observed results are due to the process or intervention and not to other factors?" Some of the basic questions that should be answered include the following:

1. Were all extraneous factors that possibly could have had an influence on the results taken into account?
2. Which of these extraneous factors were controlled? If factors were not controlled, what effect might these factors have on the results of the study?
3. What procedures were used in the study to control for these extraneous factors? Did these procedures provide satisfactory control of these variables?

4. Can plausible alternative explanations or hypotheses be considered as to why the results might have happened?

The third basic question deals with how the study was conceptualized and how conditions were defined. This concept is known as *construct validity* in research terminology.

1. Did the procedures used to conduct the study follow standard, accepted practices?
2. Were the ways of defining variables and conditions acceptable and standard?
3. Were the measures or observations used to collect data standard and acceptable? Were the instruments for obtaining this data reliable, standard, and acceptable?
4. Were proper conventions followed with regard to formulas, laboratory procedures, production schedules, and other operational conditions?

The final question deals with whether the results of the study can be generalized to another place, time, or condition, which is known as *external validity*.

1. Are there unique factors operating in the study situation so that the same result may not occur if these same unique factors are not present? For example, altitude may affect atmospheric pressure so that what works at a location of high altitude may not work in the same way at sea level.
2. Are there equipment or facility factors that might influence the results? What produces results in one factory may not produce the same results in a factory with different equipment, facilities, or personnel.
3. Do the results have practical or useful application? In other words, do the results justify the additional expense to implement the concept? Are the results cost-effective?

These four basic questions to be considered from a user's point of view can also serve as a guide for those doing research and statistical analyses. That is, a study should be planned and analyses done so that answers to all of these questions are provided when the results are presented. One approach that could be used in planning and conducting research is to change one's role from a doer of research to that of a consumer of the research to ensure that valid answers are being provided for each question that a consumer might have regarding the research. Another procedure is to have a colleague review the study by using this list of concerns and questions.

References

1. Caulcutt, R; Boddy, R. *Statistics for Analytical Chemists;* Chapman and Hall: London, 1983.
2. Kateman, G; Pijpers, F. W. *Quality Control in Analytical Chemistry;* Wiley: New York, 1981.
3. Miller, J. C.; Miller, J. N. *Statistics for Analytical Chemistry;* Ellis Horwood: Chichester, England, 1984.
4. *Chemical Abstracts;* Chemical Abstracts Service: Columbus, OH.
5. Huff, D. *How to Lie With Statistics;* Norton: New York, 1954.
6. Bradley, J. V. *Distribution-Free Statistical Tests;* Prentice Hall: Englewood Cliffs, NJ, 1968.
7. Conover, W. J. *Practical Non-Parametric Statistics;* Wiley: New York, 1971.
8. Daniel, W. W. *Applied Nonparametric Statistics;* Houghton Mifflin: Boston, 1978.
9. Hollander, M.; Wolf, D. A. *Nonparametric Statistical Methods;* Wiley: New York, 1973.
10. Siegel, S. *Nonparametric Statistics;* McGraw–Hill: New York, 1956.
11. Sprent, P. *Quick Statistics;* Penguin Books: New York, 1981.
12. Walsh, J. E. *Handbook of Nonparametric Statististics;* Van Nostrand: Princeton, NJ, 1965.

Appendix. Planning and Implementation of a Statistical Study

1. State the goals and objectives of your study. These can be in the form of questions to be answered or hypotheses to be tested. Both should be conceptualized in terms of the types of decisions or evaluations that the study is to provide. A good approach to formulating these goals and objectives is to ask the question, "When the study is completed, what information will be provided, what decisions will it be possible to make from the results of the study, and what kind of generalizations can be made for the results?"

2. Review existing knowledge on your goals and objectives. Such knowledge helps to formulate procedures that could be followed as well as procedures to avoid, identify additional variables to consider, formulate additional hypotheses, select instruments for data collection, identify statistical analyses that might be used, and serve as a basis for comparing the results.
3. Define the variables that your study will deal with: independent, dependent, and extraneous variables that can be controlled, and extraneous variables that cannot be controlled.
4. Decide how each variable is to be measured, including types of data for each and the instrument that will be used to measure each variable. These decisions are known as the *operational definitions* of your identified variables.
5. Determine the design that will be used to investigate your problem and to provide data for your goals and objectives (e.g., observational, time sampling, or experimental design).
6. Plan statistical analyses for each of the goals and objectives. Often these will be based upon the design that you selected in step 5.
7. Set your levels of significance or confidence limits for each hypothesis that you will be testing.
8. Determine your sampling procedure and determine sizes of samples needed so that each question can be answered with the degree of precision desired and so that each hypothesis can be tested with an adequate degree of power.
9. Collect data in the manner specified in your plan with the instrumentation specified and following the data collection procedure.
10. Analyze your data according to the specifications for each question to be answered or each hypothesis to be tested.
11. Interpret the results and draw conclusions. Compare your results with previous research. Consider all of the factors that might influence your study, and inter-

pret your results rationally. Be familiar with the limitations and advantages of the procedures that you have used, the assumptions upon which they are based, and how your results might have been influenced by these assumptions in your study.

12. Present your results in an effective, clear, and efficient way. Know how to communicate your results in understandable ways so that the user will be able to comprehend the meaning of your results. Always consider the level of the user. One suggestion is to present your data by using methods that the user is accustomed to, which can be determined by reviewing similar types of studies and reports that your particular audience has been exposed to in the past. Often this procedure can be done by reviewing the presentations found in journals, research reports, or textbooks commonly used by those for whom your results are intended.

Chapter 2

Basic Data Concepts

2.1 Principles of Numerical Data

Numerical data are the raw materials for statistical analyses. Statistical analyses cannot be done until the data for a study are gathered and organized in a useful manner. Such data collected for a particular study will be referred to as a *data set* that is made up of a collection of elements or individual cases. Because statistical analyses depend upon the basic data upon which they are computed, all studies must have valid and reliable data that also have a high degree of sensitivity to measure the properties of interest in a study. If the basic data are meaningless, any analysis computed on that data will also be meaningless.

Basically, as originally stated by Campbell (*1*), *measurement* is the assignment of numbers to objects or events according to rules. Because numbers are the basis of measurement and the ingredients for statistical analyses, consideration of how numbers are assigned to events and the implications of this assignment are imperative for statistical analyses. Most statistics assume specific measurement properties for the numbers used for their computation. These properties are meaningful only if the basic data represent numbers that can be legitimately used for mathematical functions.

2.2 Mathematical Scales

Broadly considered, three properties of numbers are important from a statistical point of view: identity, rank order, and additivity. *Identity* means that a number is used to identify an object or a thing as a substitute for a name. *Rank order* implies that numbers can be placed

1453–0/88/0017$06.00/1

in an indisputable order along a linear scale and represent the rank order of the objects or items being considered. *Additivity* implies that numbers can be added together in a meaningful way so that the results are consistent with the rational number system. If additivity is valid, then the fundamental operations of addition, subtraction, division, and multiplication will also result in consistent and valid results.

Several classifications of measurement have been developed and are often referred to in statistics as *measurement scaling*. Four commonly accepted types of scales are used for statistical analyses. From the lowest to the highest levels, these types are the

- nominal scale,
- ordinal scale,
- interval scale, and
- ratio scale.

Each is based on a set of rules for assigning numbers to objects and events, and, as stated by Nunnally (*2*), the intended use of those numbers. The higher levels of measurement require more restrictive rules, but at the same time, communicate more meaning and permit more mathematical computations.

2.2.1 Nominal Measurement

The lowest level of measurement scale is the *nominal scale*, and the rule for the use of numbers for this scale is that the number merely serves as a name or label for a class or category of objects, and thus only has the identity property of numbers. For nominal scales, each object or event must be capable of being classified into one and only one category or class, and thus the events are regarded as being equal or equivalent with regard to the classification system. For example, sex is a nominal variable because we usually assume all persons can be classified as male or female, and that no person can be classified as both. Other examples are graduates and nongraduates of a specific college, marital status, specific production plants, or a chemical compound with or without a specific element. Because nominal numbers are merely used to represent the class or category, arithmetic operations performed with these numbers are meaningless. Nominal scales are discrete in that the classification results in a set of unique categories of data into which all objects or events are classified, and an event cannot be classified between two categories. Thus, the properties of a nominal scale are that the categories are *mutually exclusive and exhaustive.*

The main use of nominal data in research and statistics is to designate subgroups for differential experimental treatments and statistical analyses. For example, a chemist may want to study the effects of three different catalysts on the yield of a chemical, and these could be labeled batch I, II, and III for statistical analysis. Several statistical operations can be done by using nominal data, such as counting to obtain frequencies and percentages, modes to indicate most populous classes, and statistics such as phi (ϕ) or the contingency coefficient (C) to determine relationships or independence. Data obtained by using nominal scales are known as *categorical data* because data are tabulated into categories, or *qualitative data* because each class or category has unique characteristics.

2.2.2 Ordinal Measurement

Measurements on an ordinal scale use the property of rank order (i.e., one category will be thought of as being higher or lower than an adjacent category). Although the number indicates the rank order from lowest to highest, or highest to lowest, nothing is specified with regard to the magnitude of the distance between any two ranks. Thus, quantity is not indicated in the absolute sense because the ranks are relative. For example, being ranked tallest in a group of midgets is quite different from being ranked the tallest basketball player. Rank ordering is basic to all higher forms of measurement, but because it only indicates the relative rank order, little information is conveyed about the attributes being measured. Ordinal scales are often used when more precise measurements are not possible, such as for ranking color preferences, tastes, or the psychological merit of administrative procedures. As will be pointed out in this chapter, most of the data used in chemical studies are considered interval data. Thus, except for preferences or administrative types of studies, ordinal measurements are used very little in chemical studies.

Statistics that are possible with ordinal measures include those that are possible with nominal measures plus nonparametric statistics such as the median, centile ranks, rank-order coefficients of correlation, and various tests for differences in rank orders.

2.2.3 Interval Measurement

Interval measures have the properties of nominal and ordinal data plus the important property of equal-unit distances along the entire scale. Thus, interval measures have the property of additivity. Examples of interval scales are a ruler, a thermometer, calendar time, aptitude measures of human performance, and pressure gauges. Interval measures provide information regarding exact and mathematical distances, and because the units are equal, most arithmetic operations on the numbers are permissible. Thus, nearly all of the common statistical analyses can be applied to interval measures. Most of the data commonly encountered in chemical studies can be classified as interval data. However, because an arbitrary zero or reference point may exist, it is not possible to say that, for example, 60 °F represents twice as much heat as 30 °F, because the zero reference point does not coincide with absolutely no heat. Such comparisons are only possible with ratio scales.

2.2.4 Ratio Measurement

Ratio measures have all of the properties of the other three basic scales plus a true zero reference point, which represents neither more nor less than none of the property being considered. Measures of age, height, weight, sales data, costs, market share, and number of purchases are usually considered ratio data. Because a meaningful zero point exists, meaningful ratios can be computed, and one object can represent 2 or 3 times as much as another object. Often this mathematical function is not important because interval data are sufficient for nearly all common statistical analyses for which differences in amounts are important.

2.3 Basic Scale Considerations

Many discussions have occurred among statisticians concerning the basic scale properties needed for and the influence of scale properties on statistical analyses. Many have adhered to the basic assumption that a true equal-interval metric scale is required for any parametric statistic. Recent research indicated that scales that retain the basic properties of identity, rank order, and additivity but that may not have exactly equal distances between points have very little influence on statistical results (*3–9*). Most parametric statistics are robust to scale variations. The most important scale considerations are as follows: (1) the intent of the scale coincides with the properties of a particular scale (e.g., if a classification is desired, a scale is used that is based on the rule that objects will be classified into the categories provided) and (2) the scale is isomorphic to reality (i.e., the scale is susceptible to the required mathematical operations required by the statistic being applied). As pointed out previously, all fundamental operations of algebra are permissible with intervals along an interval scale, and for statistical purposes, this property is sufficient for nearly all parametric statistics. Violations of the assumption of true equal distances have very little effect on inferential statistics applied to that data (*2*).

For interpreting the results of statistical analyses, interval and ratio scales are considered as continuous data, and an infinite number of meaningful points could be contained along the continuum from the lowest to the highest point. Thus, if a gauge is used that is divided into 10 scale values from the lowest to the highest point, this scale could be further refined to 100 or even 1000 discernible values from lowest to highest points. Thus, a mean value computed on the original 10-point scale of 6.8 is interpretable because the scale can represent any point or fraction of a point along the continuum, and the value of 6.8 has inherent meaning because this value would represent a value of 68 on a 100-point scale. However, with nominal data, for example, with three discrete categories (1, 2, and 3) into which all objects are categorized, then a mean value of 1.6 is difficult and often impossible to interpret, because this value falls between categories. Statistics such as the mean are often used with nominal data, such as the average family has 1.4 children. It is not possible to have 1.4 children, but because statistics have often been used this way, the implied meaning is communicated. However, the mode would have been a more appropriate statistic to use. This situation should serve as a simple example of why consideration of the basic scale properties is important when using statistics.

Scale values along a continuum are observed to their nearest unit, which could be inches, feet, or miles (e.g., 7 in.). However, in measurement theory, these values are considered as occupying a distance or interval from half a unit below to half a unit above the value (e.g., the value of 7 represents the continuum distance from 6.5 to 7.5). These numbers are known as the *real limits* of each value and provide the basis for meaningful interpretation of statistical analyses using interval or ratio scales. For decimal values, 4.58 would represent a distance from 4.575 to 4.585. The only notable exception to this general rule is when values are reported to the last event in a series, such as chronological age being reported as the last birthday. Here a 10-year-old child is anywhere from 10 years to 11 years of age. Consideration of the real limits of values is important in presenting data in charts and in the interpretation of statistical analyses based on those values.

2.4 Characteristics of Measurements

Three basic characteristics of measurement must be considered for any statistical study:

- validity,
- reliability, and
- sensitivity.

Validity refers to the degree to which a value accurately describes what it purports to measure. The values used for statistical analyses in chemistry—pressure, temperature, speed of reaction, chemical components, and density—usually have well-accepted meanings. However, for production, sales, or administrative data, measures are often used that are more indirect and for which validity of the values may have to be substantiated. References for determining validity of such measures (*2, 10–12*) should be consulted.

Reliability concerns consistency of measures and deals with the amount of error involved with the measured values. In measurement theory, reliability is the proportion of true variance in a set of values compared to the error component, which is made up of a number of factors that could include the instrument or device itself; human errors in reading or recording each value; temporary variations due to differences in temperature, humidity, pressure, or inert materials; or impurities in the chemicals. Errors can be grouped into two types: (1) constant errors (e.g., a gauge may consistently read higher than the actual value), and (2) random errors, which may cause the observed value to be higher, lower, or at the true value of the object being measured. Measurement theory and statistical analyses deal with random error.

In chemical applications, reliability can be determined by obtaining two sets of values obtained, for example, by having two independent persons taking measures on the same object at the same time, which in measurement theory is known as *parallel-forms reliability*. Another way is to have the same person take two measures within a reasonable time period, which is known as the *test–retest form of reliability*. For each method, a correlation coefficient is computed between the two sets of measures obtained to determine how closely related the two sets of data are. Such reliability coefficients usually vary from 0.85 to 0.99 depending on the type of object being measured and the type of measure being taken. The same references cited for validity can be consulted for determining the reliability of measures used in chemical studies.

The last characteristic is *sensitivity*, which is the degree to which the measure is able to sense minute differences. Sensitivity is closely related to accuracy. For some measures, a percentage change of 5% might provide enough accuracy; for others, parts-per-million data might be necessary. The sensitivity and degree of accuracy of a measure have to be considered and selected on the basis of acceptable standards.

Closely related to accuracy is the determination of how many decimal places or significant figures would be retained in calculations and in reporting statistical results. The general guideline is to retain and report values to one more decimal place than was available in the original data. However, for a large sample size (n) (e.g., n = 1000), two or three decimal places can be reported if the original units were observed to the whole number. However, with a small sample, reporting such values to one decimal place would be all that could be justified. Reporting to several additional decimal places may be either unwarranted or unnecessary, and may imply a spurious accuracy of the data. Most computer analyses are done by using floating-point arithmetic operations; thus, several decimal places are retained. Statistics calculated by hand would probably yield results that are different in the second, third, or fourth decimal place. Thus, consideration of significant figures is important in interpreting such analyses, and results should be reported only to what could be considered meaningful in terms of the number of decimal places.

Closely related to the number of significant figures to retain is how to round numbers. Rounding of numbers is necessary in all analytical areas, and if done incorrectly or prematurely, rounding can affect the final results of statistical analyses. The general rule for rounding numbers is that if the figure following those to be retained is greater than 5, the last retained figure is raised by 1. If the original measurement justifies only two significant figures, then the result of arithmetic calculations should be rounded to only two significant figures. For example, the number 26.837 would be rounded to 26.84. If the figure following those to be retained is less than 5, the last figure is dropped and the retained figures are kept unchanged. For example, 26.384 would be rounded to 26.38. If the figure following the last figure to be retained is 5, then the general rule is that if the last significant figure is an odd number, it is increased by 1, and if the last figure to be retained is an even number, it is kept unchanged. For example, 26.835 would be rounded to 26.84, whereas 26.845 would be rounded to 26.84.

When a series of arithmetic operations is done, the general rule is to complete the operations with all digits carried through the operation, and then to round the final result. For example, when two numbers are to be multiplied or divided, the arithmetic operation is carried out by using all of the digits in the multiplier, divisor, or dividend that has the fewer significant digits. For most statistical operations, the recommended procedure is to carry several extra figures through all calculations and then to round the final answer to the proper number of significant figures.

2.5 Basic Statistical Terminology

This section is a brief description of some of the basic concepts and terminology used in research and statistical studies. Unambiguous communication is impossible without clear definition, understanding, and expression of the concepts and terminology used in statistics. Statisticians use terms and symbols in a very precise way that is often different from everyday usage of those terms. Therefore, becoming familiar with the vocabulary used in statistics is important.

In statistical work, the *population* is all of the objects, events, organisms, or whatever things that are specified in a precise way. Population membership must clearly be defined so that it will be explicit whether or not something can be considered as belonging to the defined population to which the results of statistical analyses are inferred. Such a population is often referred to as the *target population*. Kish (*13*) stated that the target population must be defined in terms of (1) content, (2) units, (3) extent, and (4) time. For example, in a study to determine the purity of a manufactured chemical, the population might be defined as (1) the specific chemical ammonium chloride, (2) packaged in 55-gal. barrels, (3) produced at a specified plant, and (4) during a specified time period.

The elements of a population are the individual units or cases for which information is sought. They are the units used for analyses, and thus it is very important to specify how they are identified and selected for study. Most populations are so large that it is impractical, if not impossible, to study every element in the population. Thus, a *sample*, which is part of a population, is usually used for nearly all statistical analyses. Many ways are available to select a sample of elements, and these ways will be described in Chapter 8.

When dealing with populations and samples, two terms and their corresponding symbols are used by statisticians to refer to either population or sample values. When referring to a population value, the term *parameter* is used, and when referring to a sample, the term *statistic* is used. Statistical convention traditionally uses Greek letters to denote population values or parameters, and English letters to represent sample values. For example, the mean value calculated by using all elements in a defined population is symbolized by μ, the Greek letter mu. The mean calculated on a sample from that population would be symbolized as $\bar{X}$, (read "bar X"), where X would represent an entity such as amount of sodium chloride present in a given solution.

The elements, entities, or attributes that are the focus of a study are called variables or factors in research and statistical analyses. A *variable* is any observable or measurable property of an entity that could vary from one unit to another. Weight, density, reaction time, tolerance levels, pressure, and volume are examples of variables in chemistry. When a characteristic cannot vary, that characteristic is referred to as a *constant*. For example, in the formula for the circumference of a circle, $C = 2\pi r$, 2 and π are constants, and r is a variable.

Variables are defined in a number of different ways. Statisticians deal with both quantitative variables (i.e., those variables that can be measured by a number that reflects the amount of the property in question) and qualitative variables (i.e., those variables that vary with regard to quality or type, such as classifying people into two categories, male and female). In experimental research, qualitative variables are usually known as *independent variables* because the researcher selects, manipulates, and classifies the variables that will be studied. The independent variable can be the circumstances or conditions that exist or are set up to be studied and are usually at discrete levels. However, the independent variable can also be quantitative, and if it is, random levels of that variable can be set up and studied. The result of these circumstances or conditions, which is the entity that is measured as the outcome, is known as the *dependent variable*. Variables that might have an influence on either the independent or dependent variables are known as extraneous, nuisance, confounding, or intervening variables. These are the variables that need to be controlled in a study either directly through manipulation or indirectly through the research design or through statistical control. For example, in a study to determine whether different blends of acetanilide yield different percentages of loss during the manufacture of chlorosulfonamide, the different blends would be the independent variable, the percentage loss would be the dependent variable, and extraneous variables to control for or to take into account could be temperature, amount of impurities in the elements, and pressure.

The selection of independent variables for any study is determined by the specific interests of those conducting the study and is usually based on need, previous studies, or theoretical considerations. Selection of dependent variables should be based on a consideration of the validity, reliability, sensitivity, and practicality of potential available measures. On the basis of experience, previous research, and a failure–avoidance analysis of potential hazards to research, a list should be made of the extraneous or nuisance variables to be controlled or considered, and then a research design should be selected so as to minimize the influence of these variables to be controlled or considered. Because research usually represents a sizable investment of time and cost, careful consideration of these three types of variables is very important. And, considering the complexity of most cause–effect relationships and the research cost involved, multivariate statistical procedures should be considered over univariate procedures (i.e., study the effects of several variables at the same time rather than one single factor or variable). With present computer facilities and programs, multivariate analyses can be run as efficiently as simpler analyses, and often such analyses are possible with minimal extra research effort in terms of time and cost. Thus, planning for statistical analyses at the conceptual stages of a research study is essential so that analyses can be planned to yield the most information from any given study.

2.6 Other Considerations

2.6.1 Missing Data

Often in studies being conducted, unexpected circumstances happen either during the ongoing study or during the collection of data. For example, if one aspect of a study is not completed because of accidents or careless implementation of the conditions, so that data from part of the study are missing, or if the data were collected and then lost, or errors were made in collecting some data, then the statistician has to deal with missing data. Depending on the type of study, the research design, and the amount of data being collected, the planned analyses might be carried out without that particular data or set of observations, or all comparisons could be made except those involving missing data. Often a study is designed in such a way that if some data are missing, the planned statistical analyses are limited because some analyses require equal or proportional numbers of cases in each subgroup or condition. In such situations, methods are available to estimate what that data would be if the data had been obtained. For basic descriptive statistics, one way to estimate is to compute the mean value for that data set and then use the mean value in place of the missing values. If fewer than 10% of the values are missing and the sample size is 20 or more, this method is usually satisfactory. For most inferential statistical analyses, specific formulas are used for estimating missing values, and these formulas will be presented with each of these analyses.

2.6.2 Outliers

In some instances, one or a few values that are observed and recorded for a given data set deviate markedly from the majority of the other values. These extreme values are *outliers*. There are two concerns regarding outliers. First, how deviant must a score be before it is judged to be an outlier, and if it is an outlier, how does one deal with the deviant value? The problem is of particular import in studies in which the observations are costly in both time and money.

If a value appears to be an outlier, the researcher should first check the observation for obvious errors such as a misread dial, a recording error, a misplaced decimal, a misfunction of the data collection device, a simple computational error, or any special circumstances that may have caused the outlying observation (e.g., pressure or temperature fluctuations). If this kind of error is obvious, then this data would be discarded. If none of these errors are evident, then the statistician could determine if the suspected outlier is as deviant or extreme as it appears. One method is to transform the extreme value into a z score by using the formula

$$z = (X - M)/S \tag{2.1}$$

where X is the value being considered, M is the mean for that distribution, and S is the standard deviation. The z obtained could then be compared to the tabled values expected by chance alone at the 0.05 or 0.01 level of significance, which are, respectively, 1.96 and 2.58. (These values can be found in a table of the normal curve, which is discussed in Chapter 7.) If the obtained z is greater than the preferred of these two values, then that observation could be defined as an outlier.

Once an outlier has been verified, then the statistician has to decide whether to discard or retain this value for future calculations. Although no standard guidelines are available for making this decision, one possible procedure, advocated by Barnett and Toby (*14*), is to check with other findings to see if such values have been observed in other studies. Following the logic of probability theory, it is possible to have legitimate extreme values (i.e., the values expected to occur by chance out of every 100 values included in a data set). If such extreme values have a rational existence and no obvious observational or recording errors are found, then the decision may be made to retain these extreme values. Regardless of whether extreme values are retained for further analyses, ethically the statistician should describe not only the extreme values noted, but also how they were handled in statistical analyses. This information should then be included in any technical report of the study. The user of the report will then be aware of such values and how they were considered in the statistical analyses.

2.6.3 Adherence to the Basic Assumptions of Parametric Statistics

In the past, much emphasis was placed on meeting the basic assumptions of parametric statistics and the consequences of violations of those basic assumptions. As presented in statistics textbooks (*8, 15–17*), these basic assumptions are that

1. the observations must be independent,
2. the observations must be drawn from normally distributed populations,
3. the populations must have the same variance (homogeneity of variance), and
4. the variables involved must have been measured by at least an interval scale.

Bradley (*18*), Senders (*19*), Siegel (*20*), and Stevens (*21*) are among those who advocated strict adherence to the basic assumptions underlying parametric tests and argued that the results of parametric statistics depend on the validity of all the assumptions. Anastasi (*10*), Eisenhart (*17*), Senders (*19*), Stevens (*21*) and Siegel (*20*) stated that parametric statistics can be used only with interval data and indicated that analyses would be in error to the extent that the successive intervals are unequal in size. These and other authors stated that one cannot be certain that all of these assumptions have been met in a study, and thus parametric tests should be regarded as approximate rather than exact when all of the assumptions have not been met.

Opposing those who advocated strict adherence to the basic assumptions of parametric statistics, Lord (*9*) and Havlicek and Peterson (*3, 4*) argued that for most studies, the effects of violations of the basic assumptions are not sufficiently great to invalidate the statistical test. Dixon and Massey (*22*), Ferguson (*23*), Hays (*16*), Burke (*24*), and Lord (*9*) pointed out that the validity of a parametric test does not depend on the type of measuring scale used. The empirical studies by Boneau (*6*) and Havlicek and Peterson (*3, 4*) indicated that (1) unequal interval scales have little effect on parametric tests; (2) nonnormal distributions have little effect on such tests; and (3) if an equal number of cases are compared, unequal variances have little effect on such tests. However, when two or more samples being compared have unequal numbers of cases and unequal variances, the obtained results will vary considerably from the theoretical and thus yield spurious results. Our point of view as presented in this book is based upon the results of these studies (*2–4, 9, 22–24*) for most parametric statistical analyses, and these results will be considered as each of these analyses is presented.

2.7 Data Collection Procedures

One of the most important, if not the most important, aspect of any study is how the data are collected. For any study, the data collection procedures must follow standard procedures so that the study can be replicated to verify the results. This necessity is especially true for pharmacological studies, which are typically replicated several times before a product is put on the market. Thus, the study's design plan should be followed, especially with regard to data collection. The following concerns should be followed in any study.

The overall basic procedure for collecting data should be delineated in detail, and such a procedure should be formulated at the conceptual stage of a study before any data are collected. All aspects of the data collection procedure should be specified in detail. Some of the major concerns are as follows:

- What measuring devices or instruments were used to collect the data? For some studies, specifics with regard to manufacturer, model number, calibration, and date of manufacture of the measuring instrument should be reported.
- What type of forms were used to record the data? Or, if direct computer input was performed, what type of input was used? If optical-scan forms were used, these should be specified and described in detail.
- Who collected the data? Were these people competent and qualified for the type of data dealt with?
- Are all data labeled correctly, so that specific data can be identified? Are all data properly identified so that the data can be categorized for future analyses?
- Is a verification procedure available to detect and correct errors in data collection and management? If so, these procedures should be delineated so that such procedures are known.
- What are the time spans for data collection? If data are collected over a long time span, are there any implications for the results of the study?
- If data collection procedures use instruments other than direct observation, have these been administered following the prescribed procedure? The standard administration procedure for collecting all data should be delineated.

The planning for statistical analyses should include data collection procedures to be followed during the conduct of the study. These procedures should include how the data will be handled from observations through final data analyses. The general rule is to reduce the handling and transposition of data to an absolute minimum so that errors can be reduced as much as possible. Direct measurement and computer input of that data is the most reliable system of data handling.

Except for studies using direct computer input of the raw data, most data collection procedures can be facilitated through the use of data recording forms that can be optically scanned by machine to those from which a person reads the data for hand analyses or keys the data for computer input. Usually, data recording forms use columns to represent identifying data and each variable being studied. Rows usually are used for each case or element included in the study. Figure 2.1 presents a typical data collection form for hand analyses or keying into a computer. The types of identifying data will vary from study to study, and the variables included will be determined by the purpose of the study and the number of desired subgroup analyses.

Use of such a form will reduce not only errors in recording or misplacing data but will also reduce errors in working with the data. Several points should be noted for such forms. Provision should be made on the form for the largest possible values. To aid keying of such data, each column should provide for the number of digits possible, and leading zeros should be written in. Provision should be made on the form for decimal points, and these decimal points should always be in the same column. However, decimal points are usually not keyed in, but are taken care of by format statements in most computer programs. Often, indicating the keying position for each column is a wise step. Instructions and codes should be printed on each form to remind users of how these forms are to be completed. Finally, data verification procedures should be used, and these should be included in the description of the data collection procedures in technical reports.

Many types of report forms can be found, and these should be selected on the basis of the details necessary for the study or quality control procedures. Most will be of the loose-sheets type, which can be retained in ring binders. Such loose sheets permit easy addition of new sheets, replacement of older data, and handling of specific data segments. However, loose sheets make it possible to lose or misplace data, or to get data out of sequence. To reduce these problems, bound books can be obtained for data recording. Bound books can be used for chronologically inserting data. Such books include book and page numbers, spaces for titles of projects, project numbers, researchers' signatures, witness' signatures, and dates. Such books are recommended for extensive long-term studies, especially if conducted by different researchers over an extended period of time.

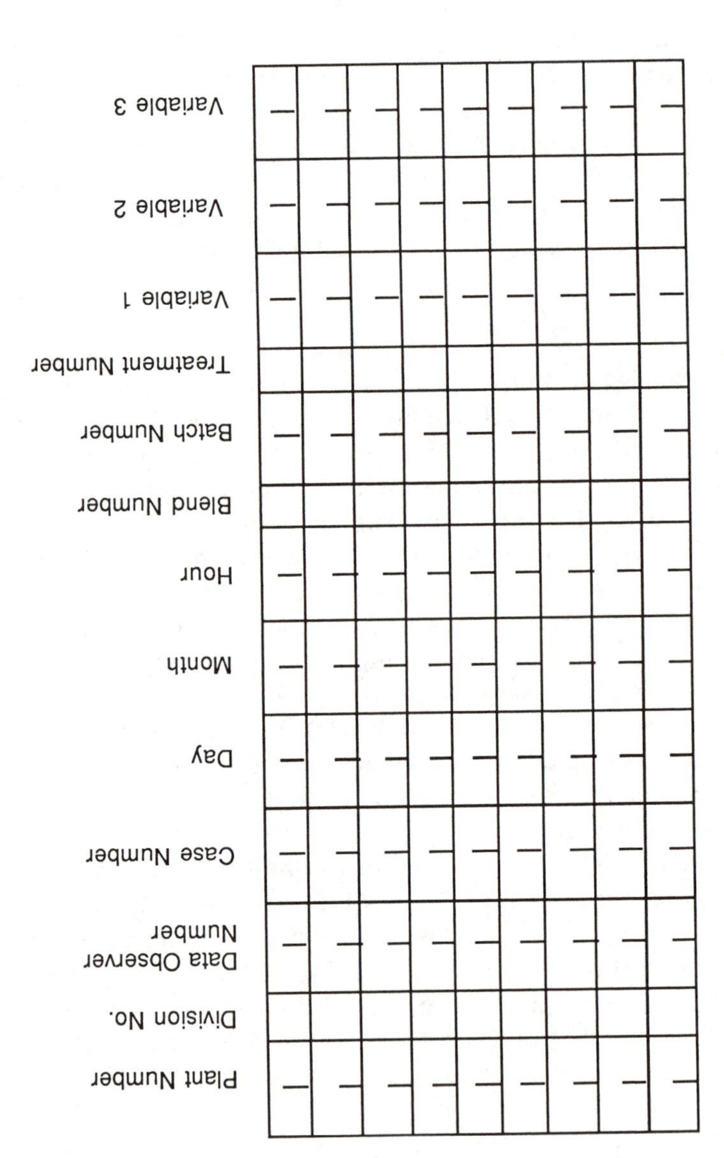

Name of Study________ Study #________ Sequence#________ Page________

Plant Number	Division No.	Data Observer Number	Case Number	Day	Month	Hour	Blend Number	Batch Number	Treatment Number	Variable 1	Variable 2	Variable 3

Figure 2.1. A typical data collection form.

The researcher should make use of automated data recording and collection instruments when these are appropriate. For example, instrumental analyses such as the autoanalyzer, the pH meter, the selective electrode meter, and the atomic absorption spectrophotometer provide digital readout of concentrations that can then be recorded directly onto data sheets or fed directly into on-line computer systems. The on-line systems reduce data handling errors because the basic data are fed directly into the computer, which then does the calculations for statistical analyses or performs quality control functions automatically.

2.8 Notation

Throughout this book, the symbols and notations used for each statistic will be defined. However, the following are some general descriptions of the common terms and symbols used.

The Greek capital letter sigma, Σ, is the summation sign and means to add the values corresponding to the letter following the symbol that represents the attribute being studied. The index of summation (i) is usually $i = 1$ and is often presented under the sigma letter. However, because this is the usual increment, this will not be included unless the increment is other than $i = 1$. The limits of summation is usually presented above the letter. For example, $\sum_{i=1}^{n} X$ would be interpreted as the sum of the X values from the first through the nth value in that data set. If values are to be summed over specific factors, then the identifying letter for that factor will be used.

Capital English letters, usually X or Y, will be used to denote the value of an element or entity. A single subscript i will be used to denote the ith value in a set, whereas two subscripts will be used to denote the ith row and the jth column in a bivariate matrix. The usual notation is to let the first subscript denote the row and the second subscript denote the column. When dealing with some statistical analyses (e.g., multiple regression), the subscript will denote the series of variables used in the equation.

The results of statistical analyses based upon or referring to population values use Greek letters (e.g., mu (μ) and sigma (σ) to represent the mean and standard deviation, respectively). For samples, the symbols $\bar{X}$ and S.D. (or S) will be used to represent the mean and standard deviation, respectively.

References

1. Campbell, N. R. *Advancement of Science* **1940,** *No. 2,* 331–349.
2. Nunnally, J. C. *Psychometric Theory;* McGraw–Hill: New York, 1978.
3. Havlicek, L. L.; Peterson, N. L. *Psych. Reports* **1974,** *34,* 1074–1114.
4. Havlicek, L. L.; Peterson, N. L. *Psych. Bull.* **1977,** *84,* 373–377.
5. Anderson, N. H. *Psych. Bull.* **1961,** *58,* 305–316.
6. Boneau, C. A. *Psych. Bull.* **1959,** *57,* 49–64.
7. Cochran, W. G. *Biometrics* **1947,** *3,* 22–38.
8. Lindquist, E. F. *Design and Analysis of Experiments in Psychology and Education;* Houghton–Mifflin: Boston, 1953.
9. Lord, F. M. *Am. Psych.* **1953,** *8,* 750–751.
10. Anastasi, A. *Psychological Testing;* Macmillan: New York, 1976.
11. Thorndike, R. L.; Hagen, E. *Measurement and Evaluation in Psychology and Education;* Wiley: New York, 1977.
12. Sax, G. *Principles of Educational and Psychological Measurement and Evaluation;* Wadsworth: Belmont, CA, 1980.
13. Kish, L. *Survey Sampling;* Wiley: New York, 1965.
14. Barnett, V.; Toby, L. *Outliers in Statistical Data;* Wiley: New York, 1978.
15. Guilford, J. P. *Fundmental Statistics in Psychology and Education;* McGraw–Hill: New York, 1965.
16. Hays, W. L. *Statistics;* Holt, Rinehart & Winston: New York, 1963.
17. Eisenhart, C. *Biometrics* **1947,** *3,* 1–21.
18. Bradley, J. V. *Distribution Free Statistical Tests;* Prentice–Hall: Englewood Cliffs, NJ, 1968.
19. Senders, V. L. *Measurement and Statistics;* Oxford: London, 1958.
20. Siegel, S. *Nonparametric Statistics;* McGraw–Hill: New York, 1956.
21. Stevens, S. S. *Handbook of Experimental Psychology;* Wiley: New York, 1951.
22. Dixon, W. J.; Massey, E. J. *Introduction to Statistical Analysis;* McGraw–Hill: New York, 1957.
23. Ferguson, G. A. *Statistical Analysis in Psychology and Education;* McGraw–Hill: New York, 1981.
24. Burke, C. *J. Psych. Review* **1953,** *60,* 73–75.

Chapter 3

Organizing and Presenting Results

Descriptive statistics deal with all of the procedures that can be used to organize, summarize, display, and communicate the findings of a study or to compare the results of several studies. Description of the findings is usually the first priority in any study for a number of reasons. First, the researcher is given a good idea as to what the data look like, and this data description helps to both plan for future analyses and interpret those analyses. For example, if the data are not normally distributed, then some data transformation might be considered to normalize the distribution of values. Second, statistics can be used to communicate findings in clear, meaningful, and efficient ways. Third, the researcher becomes familiar with the details of the data and develops a feeling for individual observations as well as the overall distribution of data. Finally, most studies deal with data sets that are so large that the data must be organized so that analyses can begin. This chapter focuses on *frequency distributions* and their graphic representation. As with most statistics, the procedures presented in this chapter deal with studies involving many observations. If a study deals with very few observations, for example, fewer than five, creating a frequency distribution may not be the most economical way of presenting the data unless such distributions were used for comparisons with other results.

For many studies, direct input of the observations or basic data is made into a computer. In such situations, computer programs are available or can be developed that will organize and compute statistics and display the results in a number of different ways. However, the procedures presented in this chapter follow those procedures done by hand tabulations.

1453–0/88/0035$06.00/1

As with all statistics, the main reason for use is to provide as much information as possible from the data. Three guidelines with regard to using descriptive statistics are that the procedures used should be (1) as useful, clear, and understandable as possible; (2) an accurate representation of reality; and (3) as effective and economical as possible.

3.1 Frequency Distributions

Usually, the first step in presenting findings is to set up a frequency distribution, which is a way of organizing data in a systematic order. Some of the basic uses of information communicated through frequency distributions are to indicate

- what the data look like,
- the high and low values observed,
- how the values are distributed,
- what the most frequent values are, and
- how the values accumulate going from low to high.

Frequency distributions are of two types: ungrouped, which uses each data point or value, and grouped, which uses intervals of the values or classes of values, each of which contains several scale values. The type to use depends upon the range of values to be tabulated and the number of cases or observations included in the study.

3.1.1 Ungrouped Frequency Distributions

We will consider first the *ungrouped frequency distribution*, which is an arrangement that lists all possible values in a data set and shows the frequency of occurrence of each value. Usually, these values are arranged from lowest to highest, where the highest value is at the top of the distribution. Such an arrangement shows the range of values, whether the values are evenly distributed or if they tend to bunch up or are concentrated at the low or high end of the scale.

Consider 60 determinations of the sulfur content of randomly drawn samples of kerosene from a given refinery, obtained by using the lamp sulfur method. The values presented in the list on page 37 were recorded in the order that the samples were drawn, and the desired result is to communicate how these values are distributed.

Kerosene Sulfur Content for 60 Random Samples

38	28	37	41	40
42	36	30	37	32
34	40	36	33	36
30	33	29	35	29
33	38	33	39	36
37	34	35	36	34
32	36	33	31	39
35	31	42	28	35
39	27	37	37	31
34	36	34	35	37
36	30	32	41	34
38	35	38	35	36

The values are tabulated into the frequency distribution presented in Table 3.1, which shows the values of observed sulfur content in the first column, the tallies in the second column (omitted in final presentations for reports), and the frequency, *f*, of each value in the third column.

Often, additional information can be shown in a frequency distribution, and such information is presented in the last three columns. Column four shows each frequency reported as a proportion of the total number of observations. These data are known as *relative fre-*

Table 3.1 Frequency Distribution of Sulfur Content for the 60 Kerosene Random Samples

X	Tallies	*f*	rf	cf	cp
42	|	1	0.017	60	1.000
41	∟	2	0.033	59	0.983
40	⊔	3	0.050	57	0.950
39	⊔	3	0.050	54	0.900
38	□	4	0.067	51	0.850
37	⧄|	6	0.100	47	0.783
36	⧄□	9	0.150	41	0.683
35	⧄∟	7	0.117	32	0.533
34	⧄|	6	0.100	25	0.417
33	⧄	5	0.083	19	0.316
32	⊔	3	0.050	14	0.233
31	⊔	3	0.050	11	0.183
30	⊔	3	0.050	8	0.133
29	∟	2	0.033	5	0.083
28	∟	2	0.033	3	0.050
27	|	1	0.017	1	0.017

quency, rf, and are calculated by simply dividing each frequency by the total number of observations. Relative frequencies are important when comparing one study with another study when the number of cases or elements is different. Comparisons can then be made on a common basis (i.e., 5 out of 50 would represent the same proportion as 10 out of 100). As a check, all relative frequencies added together in the distribution must equal 1.00. If desired, percentages can be used in place of proportions by multiplying each proportion by 100.

Column five shows the *cumulative frequency*, cf, for the 60 values and is obtained by adding the frequency of observations at each data point to the frequencies of preceding values. Thus, the frequencies accumulate as one reads up the listing of data values and can be displayed in a chart as a cumulative frequency curve, or ogive. The last column shows the cumulative frequency expressed as cumulative proportions of the observations. This column shows the proportion of scores in the distribution that fall at and below that specific value. This column, cp, will be referred to again when dealing with percentile points and is an important concept in statistical inference. An important aspect to consider is that both the cumulative frequency and cumulative proportion columns indicate the frequency or proportion of cases that fall below the upper real limits of each value, which will be explained when dealing with percentile points.

What information is communicated by Table 3.1? First of all, the range of values is from a high content of 42 to a low content of 27. Second, most of the values are concentrated from 33 to 38, with a few lower and a few higher values fairly evenly distributed; nine observations are at content level 36, seven at 35, etc. The proportion of cases at the content value of 36 is 0.150, or 15%, whereas only 0.017, or 1.7%, of the values are at content level 42. Reading from low to high, one can see that 0.850, or 85%, of the values are accounted for from the lowest reading of 27 to the reading of 38. Through subtraction, one can also figure that only 15% of the values are above the data point 38, only 23.3% below the data point 32, and 61.7% of the values fall between 33 and 38. Once the data are organized as presented in Table 3.1, additional statistical analyses are easily computed and can be interpreted by referring to this distribution of values.

Data can be tabulated in several ways. Two hand methods are to use slashes or squares, depending on the statistician's preference. Most statistical software packages have programs to create such distributions and then to display the data by using several types of charts and graphs. For such programs, the data would be entered into the computer as presented in the list, or as it is recorded and keyed from the data collection forms. Regardless of how frequency distributions are tabulated, this first step in taking a look at data is highly recommended.

3.1.2 When To Use Grouped Frequency Distributions

If a data set has a wide range of values, for example, from a low of 20 to a high of 100, an ungrouped frequency distribution showing each of the 81 values would be cumbersome, and comprehending the distribution of values would be difficult, especially if many values had zero frequency. A common solution to this problem is to group observed values together into groupings called *class intervals*. The size of the class interval is the number of values within it, and class interval boundaries must be specified with a degree of precision so that every observation can be unambiguously placed into only one interval. This specification usually means that the class interval limits must be expressed with refinements equal to or smaller than those of original measures.

Some generally accepted conventions are used in the selection of class intervals for setting up a grouped frequency distribution. These conventions are as follows:

- The class interval containing the highest value should be placed at the top.
- All intervals should be the same size.
- Class intervals should be mutually exclusive.
- Intervals should be continuous throughout the distribution (i.e., no gaps should occur).
- Most distributions should have from 10 to 20 class intervals.
- A convenient interval width should be selected. Usually, an odd number should be used because then a whole number will be the midpoint. However, 10 is often used for ease of tabulation.
- The lowest interval should be a multiple of the interval width so that the lowest value will be included in this interval. The intervals should continue until reaching the top interval, which will contain the highest value.

In constructing a grouped frequency distribution, the number and size of class intervals should first be estimated in a rational way. If between 10 and 20 class intervals are desired, the range of values would be divided from highest to lowest by numbers between 10 and 20 to determine what interval width would be appropriate. Then, a convenient width, which is usually an odd number, would be chosen. The lowest class interval should include the lowest value, and class intervals should be marked off until the class interval containing the highest value is reached. The values would then be tallied into these class intervals just like values are tallied into an ungrouped frequency distribution.

As an example of where a grouped frequency distribution would be appropriate, consider a study in which 200 measures of the carbon content of a mixed powder is taken over a period of 2 months. The total range of these measures was from 3.62% to 4.84%, measured to two decimal places. To determine the interval size, the range of values, 1.22, is divided by numbers ranging from 10 to 20. The resulting interval widths range from 0.122 to 0.061. For ease of tabulation and to arrive at an appropriate number of class intervals, an interval width (i = 0.10) could be considered that would result in a total of 13 class intervals to handle all observed values. By using i = 0.10, class intervals could be set up starting with 3.60–3.69 to include the lowest observed value. The completed grouped frequency distribution for these 200 observed values is displayed in Table 3.2.

Table 3.2 Grouped Frequency Distribution of Measures of Percentage Carbon in a Mixed Powder

Values	Tallies	Midpoints	f	rf (%)	cf	Cumulative (%)
4.80–4.89	𝍷𝍷	4.845	2	1.0	200	100.0
4.70–4.79	𝍷𝍷𝍷𝍷	4.745	4	2.0	198	99.0
4.60–4.69	𝍸 𝍷𝍷𝍷	4.645	8	4.0	194	97.0
4.50–4.59	𝍸 𝍸 𝍷	4.545	11	5.5	186	93.0
4.40–4.49	𝍸 𝍸 𝍸 𝍷𝍷	4.445	17	8.5	175	87.5
4.30–4.39	𝍸 𝍸 𝍸 𝍸	4.345	20	10.0	158	79.0
4.20–4.29	𝍸 𝍸 𝍸 𝍸 𝍸 𝍷𝍷𝍷𝍷	4.245	29	14.5	138	69.0
4.10–4.19	𝍸 𝍸 𝍸 𝍸 𝍸 𝍸 𝍷𝍷𝍷𝍷	4.145	34	17.0	109	54.5
4.00–4.09	𝍸 𝍸 𝍸 𝍸 𝍸 𝍷	4.045	26	13.0	75	37.5
3.90–3.99	𝍸 𝍸 𝍸 𝍸 𝍷𝍷𝍷	3.945	23	11.5	49	24.5
3.80–3.89	𝍸 𝍸 𝍸	3.845	15	7.5	26	13.0
3.70–3.79	𝍸 𝍷𝍷	3.745	7	3.5	11	5.5
3.60–3.69	𝍸	3.645	4	2.0	4	2.0

For both grouped and ungrouped frequency distributions, each interval is defined in terms of value limits and in terms of real limits. Value limits are used for tabulation purposes, and real limits for statistical computations. *Value limits* are the highest and lowest values used in setting up a class interval, whereas *real limits* are taken as one-half a unit below to one-half a unit above the value limits for each interval. For the ungrouped frequency distribution in Table 3.1, the value limits are each of the values listed in the table, and the real limits for each value are 26.5–27.49, 27.5–28.49, 28.5–29.49, etc. For the class intervals in Table 3.2, the value limits would be 3.60–3.69, 3.70–3.79, etc. The real limits would be 3.595–3.6949, 3.695–3.7949, 3.795–3.8949, etc.

3.1.3 Conventions Regarding Grouped Frequency Distributions

Three points should be remembered regarding grouped frequency distributions. First, there is some loss of the original precision of the values observed. This loss is dependent upon the size of the class interval. For example, in Table 3.2, all original values that could range from 3.60 to 3.69 are tallied into this same class interval. All of the values could have been 3.69, or 3.60, or any other value between these value limits. Once the original values are tabulated, they loose their unique values and are considered by using the two conventions discussed in the next two paragraphs.

The first convention is that the midpoint of the class interval represents all the values within that interval and is used in drawing frequency polygons and in the calculation of statistics such as the mean and standard deviation. The midpoint is calculated by dividing the size of the class interval by 2 and then adding this resulting value to the lower real limits for each interval. For Table 3.2, $i = 10$; thus, 0.05 is added to each lower real limit to obtain the midpoints shown in Table 3.2.

The second convention is that the values are considered evenly distributed over the real limits of each class interval. This assumption is made when displaying the results in a histogram and in the calculations of statistics such as the median and percentile points.

3.1.4 Bivariate Frequency Distributions

The concept of frequency distributions, or *univariate scatterplots* as they are sometimes referred to, can be extended to situations in which two or more measures are taken for each element or case, and these pairs of values can also be tabulated into a single table. Such tables are known as ***bivariate frequency distributions*** or as a ***contingency table*** when two measures are involved, or as ***cross-tabulations*** when two or more measures are involved. Bivariate frequency distributions are important in correlational analyses to indicate if the two variables represent a linear or some other form of relationship. Such distributions will be considered in Chapter 6.

3.2 Graphic Presentation of Data

Graphs and *charts* can be thought of as pictorial representations of data and are often useful in communicating research findings to others, especially to general audiences. The major purpose of a graph is to present a clear picture of what the data look like, and they are useful in comprehending the essential features of a frequency distribution and in comparing one frequency distribution with another. Often, the visual display of data helps in interpreting the results of complex statistical analyses (e.g., correlation and higher order factorial analyses).

Much has been written on graphic representation, especially recently with computer graphics. Many options are available for constructing graphs, ranging from color-coded displays to three-dimensional arrangements. Although much has been written on graphic representation, especially with regard to computer graphics, we shall consider only the basic aspects of developing graphs and charts.

3.2.1 Basic Recommendations for Charts and Graphs

Some recommendations are commonly accepted in the construction of graphs and charts, and these will be followed for all charts and graphs presented in this book.

- The horizontal axis (X axis, or the abscissa) usually represents the observed values, and the vertical axis (Y axis, or ordinate) represents frequency or relative frequency.
- The intersection of the two axes should be zero on both scales, and if this is not possible, preferably on the vertical scale. If the data are such that the vertical axis would be out of proportion, then designate the point of intersection as the zero point and make a small break in the vertical axis.

- Units along each scale should be chosen to represent the data being graphed, and limits of class intervals or interval midpoints usually are used along the horizontal axis. Frequencies or relative frequencies are usually plotted on the vertical axis. The low numbers on the horizontal scale should be on the left, and toward the bottom on the vertical scale.
- The height of the chart is usually from 60% to 75% of the total width. This procedure results in both a standardization for charts and for a general aesthetic appeal.
- Every graph should have an appropriate descriptive title, as well as labels for both axes.

Many variations can be used in the construction of graphs and charts. For example, color coding can be used very effectively. Different types of lines can also be used.

3.2.2 Bar Graphs

For qualitative data (i.e., nominal data), the bar graph is appropriate because the space between the bars shows the discontinuity among the categories. For example, if the number of rejects produced by each of three plants is to be displayed, a bar graph would be the preferred method. Figure 3.1 shows such a graph, and from reading this graph, an observer can tell that the greatest number of rejects is from the third plant.

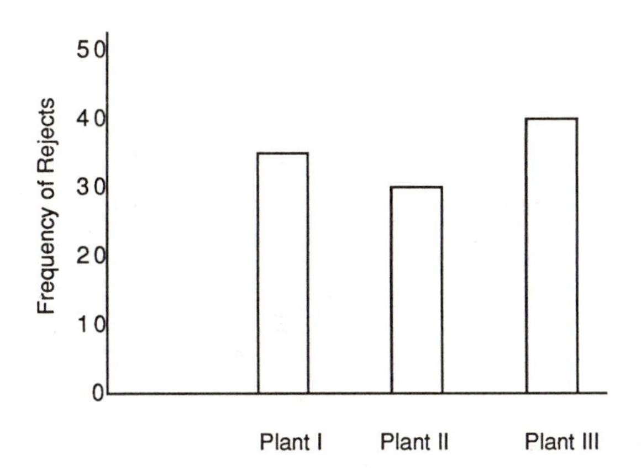

Figure 3.1. Number of rejects for three plants.

3.2.3 Histograms

A *histogram* is similar to a bar graph and is used to graph quantitative variables. It is constructed in a similar way, but the vertical bars are erected contiguously over the real limits of the values or the class intervals; thus, the continuity of the values is shown. Also, in using a histogram, all the cases within a class interval are assumed to be evenly distributed over the range of the interval. This even distribution is what the histogram indicates.

A histogram of the data presented in Table 3.2 is shown in Figure 3.2. The frequency column starts at zero frequency, whereas the values start with the lowest observed value. Usually, a good idea is to make some break in the horizontal line near the zero point to indicate that these values do not begin at the zero point. Although the class limits are used for the horizontal points, the midpoint values for each class interval could have been used. Also, if data from other studies are to be compared with the graphed data and if the number of cases in each study varies, more meaning would be derived if relative frequency is used rather than the actual frequency. In this

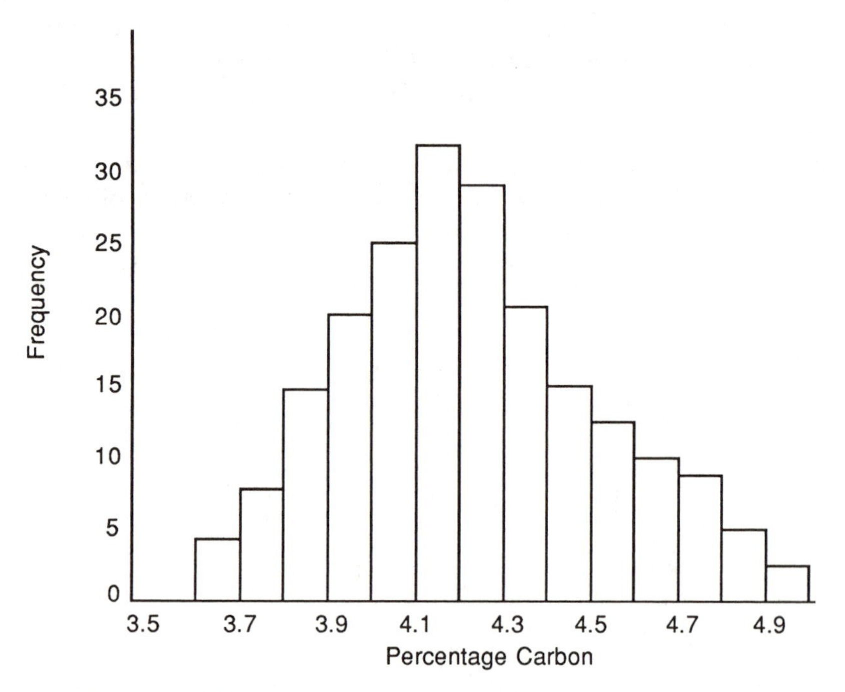

Figure 3.2. Histogram of percentage carbon in mixed powder.

way, both histograms would be based on the same reference scale (i.e., proportions or percentages).

3.2.4 Polygons

A frequency polygon is constructed by placing a dot at the appropriate frequency above the midpoint of each observed value and then connecting the dots by straight lines. It is also customary to show an additional midpoint at each end of the horizontal scale and to indicate zero frequency for these two points; thus, starting and ending points are given to the graph. In the frequency polygon, the midpoint is assumed to represent all of the cases within a class interval, and the line is assumed to represent the way actual values are distributed within the interval. That is, as values proceed from low to high, more frequencies are expected within each interval at the higher values within each interval, and fewer frequencies of higher values are expected within each interval as the midpoint of the graph is passed. As with the histogram, class limits could have been used along the horizontal axis as well as the midpoints. Relative frequency could also have been plotted rather than actual frequency. Figure 3.3 is a polygon of the data displayed in Table 3.2.

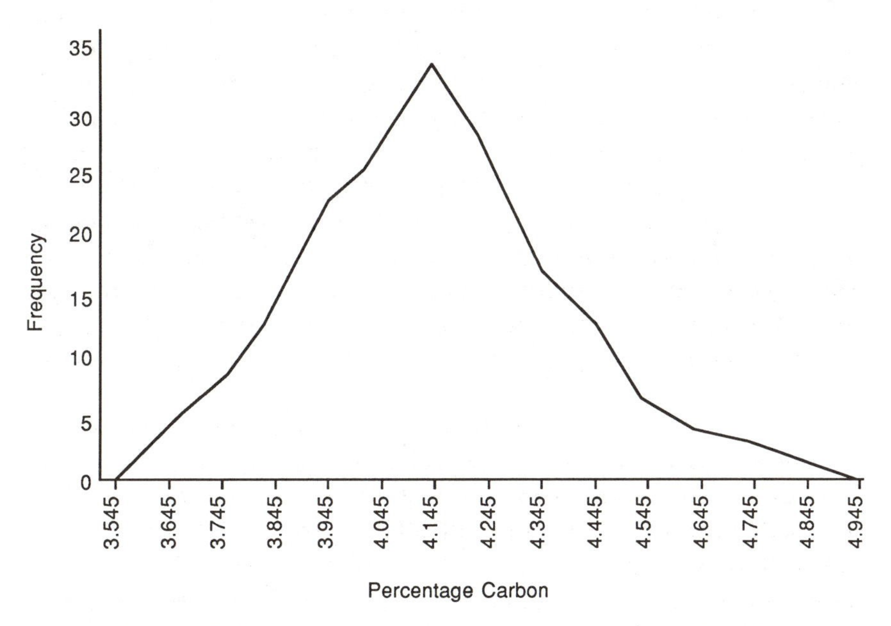

Figure 3.3. Polygon of percentage carbon in mixed powder.

Advantages and disadvantages can be found for both types of graphs. The frequency polygon gives a better conception of the shape of the distribution and the transition from one interval to another. It also usually represents how the actual cases are distributed within each class interval because the cases on both sides of the middle point are usually more frequent on the side near the middle except for some inversions. Also, if two sets of data are to be plotted on the same graph, the frequency polygon is easier to construct and indicates the two or more sets of data more clearly than a histogram. However, the histogram may show the number of cases more clearly in each interval because each case represents the same amount of space.

3.2.5 Cumulative Polygons

Often, statisticians desire graphs to show how the values accumulate from low to high values so that the frequency or percentage of the observed values falling below each interval can readily be seen. For example, such graphs would be important for production schedules over a period of time. *Cumulative frequency polygons* can also be used to estimate centiles and percentile ranks, which will be presented in Chapter 5.

In constructing a cumulative polygon, the vertical scale can represent the cumulative frequency, cumulative proportions, or cumulative percentages, and the horizontal scale can represent observed values, or for grouped data, the scale limits for each class interval. The dots in a cumulative polygon are placed above the upper limit of each class interval because the graph is used to show the frequency, proportion, or percentage of the cases falling at and below each interval. For ungrouped data, the dot is placed above each observed value. The cumulative polygon for the data in Table 3.2 is shown in Figure 3.4. As with all graphs, lined paper or graph paper can be used so that the values on both axes can be read more accurately.

If the distribution is symmetrical, the cumulative distribution is usually S-shaped, and thus is often referred to as an *ogive*. Such graphs can also be drawn on normal probability paper, and if the data are normally distributed, the dots will fall close to a straight line. Normal probability paper uses a vertical scale corresponding to the percentile points characteristic of the normal curve, which will be covered in Chapter 5.

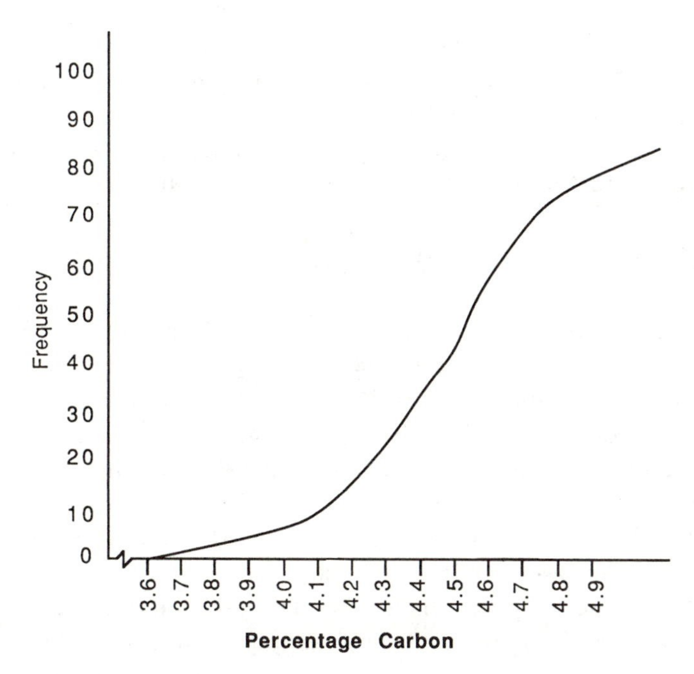

Figure 3.4. Cumulative percentage polygon for percentage carbon in mixed powder.

3.2.6 Other Charts and Graphs

For some data (e.g., production costs and distribution of personnel), or whenever a total amount of something can be divided into parts, a pie chart might be useful to display data. Corporate reports often use two pie charts to show (1) sources and (2) distribution of income.

If both negative and positive values are dealt with, then these might be plotted by using a *Cartesian coordinate system*. This system uses two number scales placed at right angles to each other with the zero point or point of origin at the place where the two lines meet. Negative and positive values are then scaled by using the same recommendations as for other charts. Such charts are often useful in plotting the relationship between two variables, which will be discussed in Chapter 6.

To emphasize specific items, *pictorial charts* could be used. In such charts, pictures of the item of interest are drawn to different scales to represent relative frequency. However, care must be exercised so that the pictures in the chart are not out of proportion to the actual frequencies that they represent. For example, if one corporation sells twice as many television sets as another corporation, this information should be displayed as in Figure 3.6 but not as in Figure 3.5. Although the height showing the frequency in Figure 3.5 is appropriate, the additional width makes it appear that corporation B has sold four times as many sets as corporation A.

3.3 Shapes of Distributions

One of the reasons for displaying a distribution of values is to get an idea as to how the values are distributed. Knowing how the values are distributed is important in decisions regarding the appropriateness of statistical methods to be used for further analyses and in the interpretation of those analyses. Some of the more common shapes of distributions found in research and production data are shown in Figure 3.7. Each shape will be briefly described and will be referenced later during discussions of various statistical analyses. Figure 3.7 presents the frequency polygons for the data presented in Table 3.3.

Frequency distributions have four important properties:

- central location,
- variation,
- skewness, and
- kurtosis.

These four properties are used to describe the frequency distribution

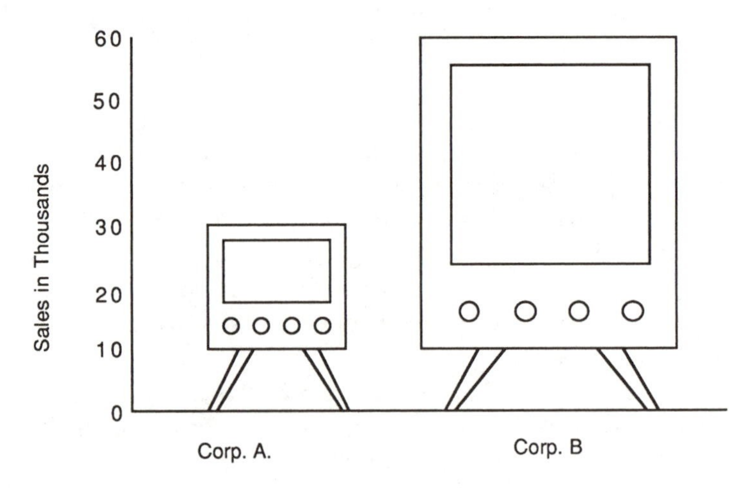

Figure 3.5. A misleading pictorial chart of sales of television sets by corporation.

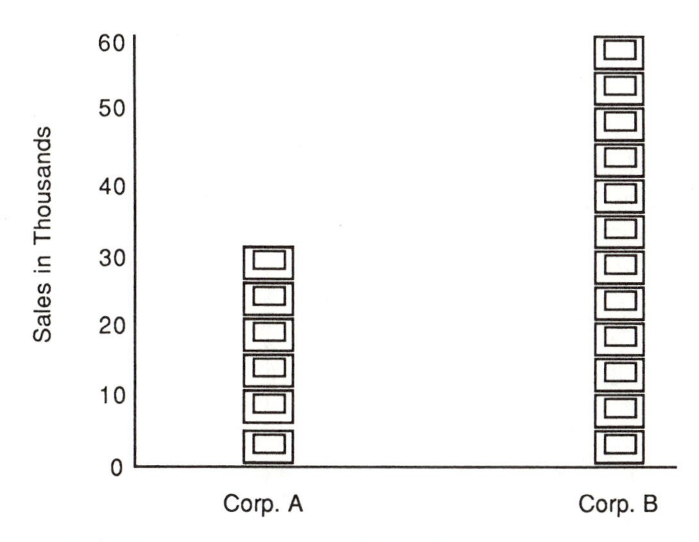

Figure 3.6. An accurate pictorial of sales of television sets by corporation.

of observed values in either tabular or graphic form. *Central location* refers to the value of the measures being considered near the center of the distribution or the middle point, which is referred to as the average. *Variation* refers to how close or scattered the values are from this central location and is usually measured by the range of values or by a standardized measure known as the standard deviation. *Skewness* is a measure indicating whether the distribution is symmetrical or asymmetrical. A skewed distribution is one in which most of the values fall at one end of the curve and only a few values fall at the opposite end. If most of the values are low but a few of the values are high, a pulling-out of the curve toward the high end of the scale results; this situation is referred to as a *positive skew* or skewed to the right (Figure 3.7D). If the values tend to be distributed in the opposite way with many high values and a few low values, then the distribution is said to be *negatively skewed* or skewed to the left (Figure 3.7E).

Figure 3.7A approximates the shape of the symmetrical normal curve, which is also referred to as the *normal distribution*, the bell-shaped curve, or the Gaussian curve. This standard curve is used for many statistical analyses and will be referred to very often. Mathematically, the normal curve is the limit of the binomial distribution where the probability of success p is equal to the probability of failure q and when the number of trials n approaches infinity.

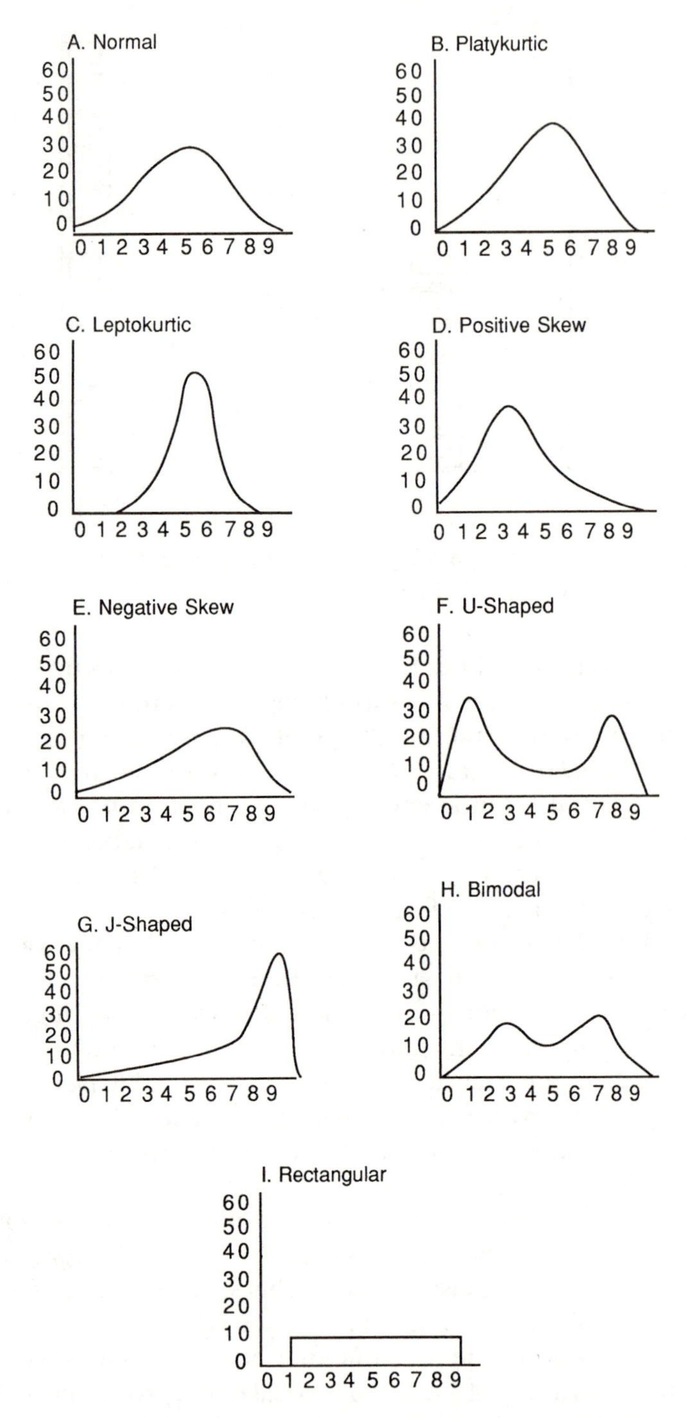

Figure 3.7. Frequency polygons of data from Table 3.3.

Table 3.3 Hypothetical Data Showing Frequency Distribution of Different Shapes

Values	A Normal	B Platy-kurtic	C Lepto-kurtic	D Positive Skew	E Negative Skew	F U-Shaped	G J-Shaped	H Bimodal	I Rec.
9	2	2	0	1	8	27	60	5	11
8	5	7	0	2	27	10	17	10	11
7	11	14	3	3	27	7	7	20	11
6	18	17	27	4	20	4	5	11	11
5	28	20	40	8	8	4	4	8	11
4	18	17	27	20	4	4	3	11	11
3	11	14	3	27	3	7	2	20	11
2	5	7	0	27	2	10	1	10	11
1	2	2	0	8	1	27	1	5	11

Kurtosis is a measure of whether the distribution of values is relatively normal, flat, or peaked. If the majority of the values cluster around the center, the distribution of values will peak at the center; the resulting distribution is *leptokurtic* (Figure 3.7C). If the values tend to scatter away from the center, then the distribution is *platykurtic* (Figure 3.7B). The normal distribution is the reference point for determining kurtosis and is referred to as being *mesokurtic*. The leptokurtic and platykurtic distributions are symmetrical in Figure 3.7. However, this does not always have to be the case, for such distributions could also be asymmetrical.

Other shapes of distributions are common. For example, if the extreme values have the highest frequencies and fewer values are at the center, a U-shaped distribution results (Figure 3.7F). Plotting the value of an automobile over 50 years might result in a U-shaped distribution in that the dollar value depreciates in the first few years, then remains about the same for a few years, and then rises as the car becomes an antique. U-shaped distributions can also occur as an inverted U-shaped distribution, where many high values are near the center of the distribution and very few extreme values occur.

If the skewness is extreme, as in Figure 3.7G, the resulting curve is termed J-shaped. If two peaks are found in a distribution, the distribution is *bimodal* (Figure 3.7H). If more than two peaks occur, the distribution is *multimodal*. Finally, if all values have the same frequency of occurrence, the result is a rectangular distribution (Figure 3.7I). If values are converted to percentiles and then plotted, the result is a rectangular distribution.

Formulas for computing indices of skewness and kurtosis will be presented in Chapter 5. In a research study, it is often not only important to show how the values were distributed, but also how to use an index to measure the extent of the deviation so that comparisons with other studies are possible.

3.4 Final Considerations

Except for ungrouped frequency distributions and their respective ungrouped graph, the same data can be presented in many tabular or graphic forms. For example, different sizes of class intervals or different starting places could change the shape of the frequency distribution, especially for samples with a small number of cases.

For graphs that deal with very large values for which it would not be possible to plot all values on the vertical scale or the horizontal scale, the graph might present a picture that either spuriously inflates or conceals a trend. For example, suppose the increase in sales for the past year is to be graphed. The months could be plotted along the horizontal axis and sales (in millions of dollars) along the vertical axis. Figure 3.8 shows that for 12 months, a 10% increase in sales occurred. The same data are displayed in both Figure 3.8A and 3.8B. However, in Figure 3.8B, the vertical axis has been changed in two ways: (1) the vertical scale does not begin at the zero point, and (2) the scale has been changed so that now each vertical unit represents only one-tenth the amount as in Figure 3.8A. Figure 3.8A follows the recommendations for constructing graphs, and pictures the yearly 10% growth in sales in a realistic but not impressive way. This graph takes into consideration the high level of sales. Figure 3.8B also pictures the same 10% growth in sales, but because the starting point and the vertical scale units have been changed, the growth seems very dramatic.

As another example of how changing either the vertical or horizontal scale, or both, could change the appearance of a graph, the normal curve, as usually conceptualized, has the form shown in Figure 3.7A. However, changes in the scales used for either or both axes could result in the normal distribution being depicted as in Figures 3.7B or 3.7C. Thus, the recommendation was made that the scales selected for both axes be such that the height is approximately 60%–75% of the width.

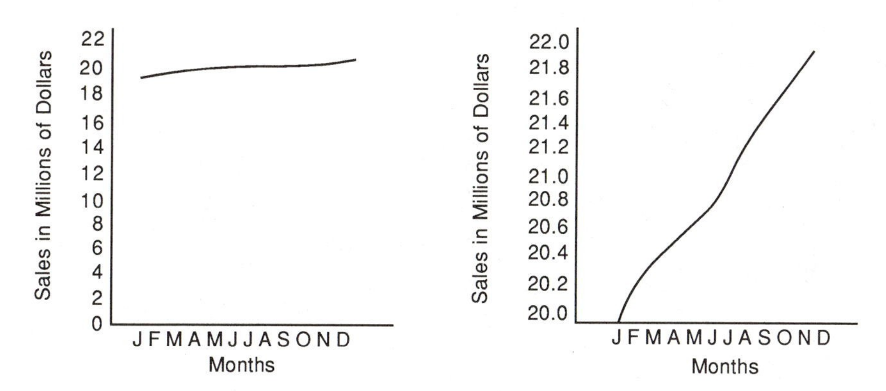

Figure 3.8. Graphs of monthly changes in sales.

Huff (*1*) gave other examples of how graphs can distort the data to make it appear as the preparer might want it to appear or to emphasize or conceal some finding. He also pointed out that often graphs and statistics can mislead the reader of reports, so readers must recognize how results could be presented in a misleading way so that the results can be more accurately interpreted.

Reference

1. Huff, D. *How to Lie With Statistics;* W. W. Norton: New York, 1954.

Chapter 4

Measures of Central Tendency

*F*igures and graphs are useful for retaining and communicating a high level of detail in data, especially for mass communication. However, further analyses for prediction or inferences are generally not possible from figures or graphs. Although these items are useful to communicate the findings for a study, they tend to be inefficient because they require considerable amounts of space to present, and they are an end in themselves. Thus, a need arises for ways to describe and communicate the findings of a study that will be efficient, provide the basis for further statistical analyses, and at the same time provide an accurate index of the findings.

Statistics most commonly used to represent and communicate the properties of a set of data fall into three major categories:

1. measures of central tendency, giving the location of the central or typical value;
2. measures of dispersion, showing the degree to which the values vary or scatter; and
3. measures describing the shape of the distribution, which deals with symmetry and peakness.

This chapter presents the commonly used numerical measures of central tendency, generally referred to as the average value, or what is typical, usual, normal, or representative for a set of data. However, because these terms have different connotations, statisticians prefer more precise terms to refer to the measures used to represent central location. The three main measures of central location presented in

1453–0/88/0055$06.00/1

this chapter are the mode, the median, and the means: arithmetic, geometric, and harmonic. Each will be discussed in terms of its use, stability, representativeness, and utility for further analyses.

4.1 The Mode

The *mode* is the most frequently occurring value, and is most applicable with nominal data. The mode can be viewed as a typical or average value only in that it communicates which value occurred most frequently. For example, the mode could be at either extreme in a data set, as for example, with a highly skewed distribution.

The mode is easily obtained by inspecting the data to see which value occurred most frequently. This inspection is facilitated by constructing a frequency distribution. When only one mode is found, the distribution is called *unimodal*. If two nonadjacent values are found with equally high frequencies, the distribution is *bimodal*. If the two highest frequency values are adjacent to each other, then the mode is the midpoint between these two values. When data are grouped, the mode is the midpoint of the interval containing the highest frequency. If the distribution of scores is rectangular, there is no mode because all values would have the same frequency.

The mode is the least preferred measure of central tendency (except for nominal data, for which it is the only legitimate measure that should be used). The mode tends to be unstable because small shifts in frequency result in changes in the mode. Because the mode is not derived mathematically, it cannot be used for further analyses. For distributions with two or more modes or extremely skewed distributions, the modes obtained are not measures of central location.

For the frequency distribution in Table 3.1, the mode is 36 because more frequencies occur for this value than for any other value. The mode for the data in Table 3.2 is 4.145, the midpoint of the interval with the most frequencies.

4.2 The Median and Centile Values

The *median*, often denoted Mdn, is the point in an ordered frequency distribution that cuts the distribution in half: half of the observations fall above this point and half below it. As such, it is also known as the 50th percentile, or P_{50}, because 50% of the distribution falls below this point. The median requires at least ordinal measurement.

The median is computed for any frequency distribution by first determining $n/2$, which is one-half the number of cases in a sample. For an ungrouped frequency distribution, one then counts from the bottom of the distribution to the point at which half of the cases fall below and half of the cases fall above. If n is an odd number, the median is easily found. Consider the following nine values:

X	4	6	7	8	9	10	10	11	13
cf	1	2	3	4	5	6	7	8	9

Reading across the cumulative frequency, $X = 9$ is the median because this is the middle point; four values are below and four values are above this point. When n is an even number, interpolation between the two middle numbers is made to compute the median. Consider the following 10 values:

X	4	6	7	8	9	11	12	12	13	14
cf	1	2	3	4	5	6	7	8	9	10

the median falls between the values 9 and 11. Thus, interpolation is made to determine that the median is 10, which is the central point between these two values. For more complex frequency distributions such as in Tables 3.1 and 3.2, the procedure is as follows:

1. First, set up a cumulative frequency distribution.
2. Determine $0.5n$, which is one-half the number of cases.
3. Find the value or class interval in which the median case falls.
4. Determine the exact lower limits of this interval.
5. Interpolate to find the value along the scale above and below which one-half of the cases fall.

To accomplish step 5, use the equation

$$\text{Mdn} = \text{LL} + \frac{(0.5n - n_b)}{n_i}\, i \qquad (4.1)$$

where LL is the lower limit of the interval containing the median, n is the number of cases, n_b is the cumulative frequency below the LL of the interval, n_i is the frequency of cases within the interval, and i is the size of the class interval.

For the data in Table 3.1, $0.5n = 30$; thus, the median will fall somewhere in the interval occupied by the value 35. For this example, LL = 34.5, $n = 60$, $n_b = 25$, $n_i = 7$, and $i = 1$. Therefore,

$$\text{Mdn} = 34.5 + \frac{[0.5(60) - 25]}{7}(1) = 35.21$$

The median falls at exactly five-sevenths of the distance up from the lower limit, or at the value $34.5 + 0.71 = 35.21$.

For the data in Table 3.2, $0.5n = 100$. Thus, the median will fall in the class interval whose lower limit is 4.095. For this example, the median is

$$\text{Mdn} = 4.095 + \frac{[0.5(200) - 75]}{34}(0.10) = 4.169$$

In computing the median, the values are assumed to be evenly distributed throughout the interval, and the value is derived from counting frequencies, not the actual observed values. As such, the median is not influenced by extreme observed values and is thus appropriate for skewed distributions. The median is somewhat more stable than the mode, but like the mode it does not use the observed values in its calculation. Thus, the median cannot be used for most further analyses. The median can be estimated from an ogive that uses percentages for the vertical scale.

Another application of the median concept is in determining the half-life of a radioactive element as a measure of its activity and as a measure of the toxicity of insecticides. In both of these uses, the *half-life* is the time at which half of the atoms are disintegrated or half of the insects die, which in each case, is often expressed as the median value of time.

Although not in the same category as averages, centile points other than the median, which is the 50th centile point, are computed in the same way as the median. The only difference is that the point of interest is other than the center of the distribution. Such centile points are useful for constructing control charts, for determining cutoff points in quality control or production, and in determining measures of variation.

A centile point, percentile point, or simply *percentile*, is the value in a distribution below which a given percent of the cases fall. For example, if in a distribution of values, 80% of the cases fall below the value of 64, then 64 is the 80th percentile point. Another term closely related is *percentile rank*, which indicates the percentage of cases falling below that scale value. In this example, the percentile rank for the value of 64 is 80. Thus, centile points refer to the scale values, whereas percentile rank indicates what percent of the distribution falls below the midpoint of the value identified.

To compute centile points other than the median, the only change in the equation for computing the median is to substitute the percentage n required in place of $0.5n$. For example, if the point below which 10% of the cases would fall is desired, then $0.10n$ would be used in equation 4.1. For the data in Table 3.1, $0.10n = 6$. Thus, the 10th percentile point, or first decile, will fall somewhere in the real limits for the value 30, or

$$P_{10} = 29.5 + \frac{(6 - 5)}{3}(1)$$

$$P_{10} = 29.5 + 0.33 = 29.83$$

Thus, $P_{10} = 29.83$, below which exactly 10% of the cases fall. For the same data, to find the value below which 25% or 15 of the cases fall, calculate as follows:

$$P_{25} = 32.5 + \frac{(15 - 14)}{5}(1)$$

$$P_{25} = 32.5 + 0.2 = 32.7$$

To find the value below which 75% of the cases fall, the calculation would be

$$P_{75} = 36.5 + \frac{(45 - 41)}{6}(1)$$

$$P_{75} = 36.5 + 0.67 = 37.17$$

and the calculation to find the value below which 90% of the cases fall would be

$$P_{90} = 38.5 + \frac{(54 - 51)}{3}(1)$$

$$P_{90} = 38.5 + 1 = 39.50$$

These percentile points can be checked by looking at the cumulative percent column. For example, the cumulative frequency column indicates that 90% of the cases fall below the upper limit of the value 39, or 39.5, and this is exactly what the computed value is. These four percentile points—P_{10}, P_{25}, P_{75}, P_{90}—are often used to compute measures of variability that will be discussed in Chapter 5. Other percentile points can be computed in the same manner by using the desired percentage of cases.

4.3 The Arithmetic Mean

The most widely used measure of central tendency is the arithmetic mean. The *arithmetic mean* is simply the sum of all observed values divided by the number of cases in the set. The common equation is

$$\text{mean} = \overline{X} = \frac{\sum X}{n} \tag{4.2}$$

where X represents an observed value, and n represents the number of cases. The capital Greek letter sigma, Σ, indicates that all of the values through the nth case are to be summed. Several symbols are used for the mean, depending upon whether one is referring to a population mean for which the Greek letter mu (μ) is used, or to a sample mean for which $\overline{x}$, $\overline{X}$ (read "bar X"), or M is used. For most studies, sample means are used to estimate the population mean, and as shown in Chapter 9, this estimate becomes more precise as the number of cases increases.

The mean can be computed from data in any form. However, computation by hand is facilitated by having the data in a frequency distribution. For example, the raw data in the list on page 37 could be used, and each value could be added to the total sum as each value is considered. However, a more convenient method would be to tabulate the raw data into a frequency distribution as in Table 3.1. Then the mean can be computed by using the equation

$$M = \frac{\sum fX}{n} \tag{4.3}$$

where f is the frequency of each value, X is the observed value or the midpoint of the interval for grouped data, and n is the total number of cases. For the data in Table 3.1, the mean sulfur content is 34.88 according to equation 4.3.

The *mean* is the point in a distribution of values such that the algebraic sum of the differences of all values from that point is zero. In later analyses, this property will be called the sum of the least-squared values, which is calculated by using the squared differences from the mean. Because the mean is computed by using the observed values, it is sensitive to extreme low or high values. Thus, for skewed distributions, the mean may not give an accurate representation of the set of data. However, the mean is generally the most stable measure of central tendency, and because it is a mathematical function of all values, it is used in most of the statistical analyses that will be covered in this book.

4.4 Selecting a Measure of Central Tendency

Three factors should be considered when selecting a measure of central tendency. The first factor is the type of data. The mode is recommended for nominal data but can also be used with ordinal and interval data; the median is recommended for ordinal data but can also be used with interval data; and the mean should only be used with interval data or data that has a theoretical continuum. Ratio data for statistical analyses are treated the same as interval data.

The second factor is the shape of the distribution. Consider the relative positions of the three measures of central tendency for the three shapes of distributions displayed in Figure 4.1. For normal distributions, the three measures are identical. For skewed distributions, the mean is pulled out in the direction of the skew. Thus, if one wanted to communicate the impression that the values were high for a positively skewed distribution, the mean would be chosen. However, the mean may not accurately represent the data. For skewed distributions, all three values of central tendency should be reported.

The third factor is the reason for computing the average. If the reason is primarily for descriptive purposes, then the measure that is most descriptive should be used. If inferences are to be made, then the mean would be chosen because other statistical analyses used to make inferences are also computed by using each observed value as is the mean.

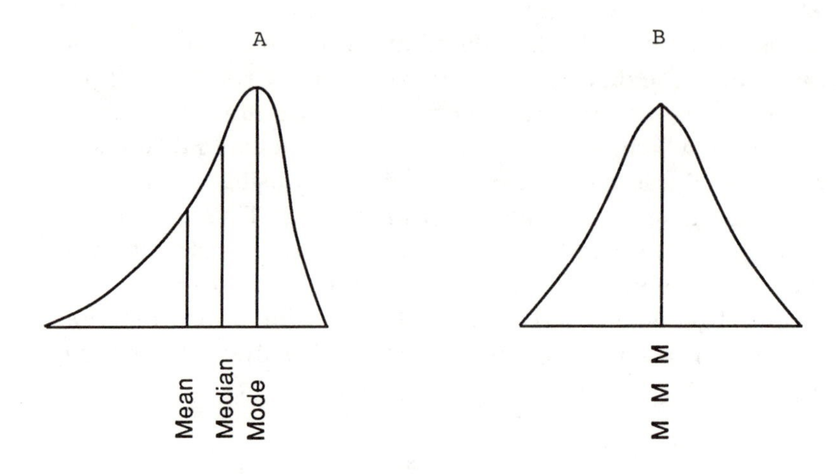

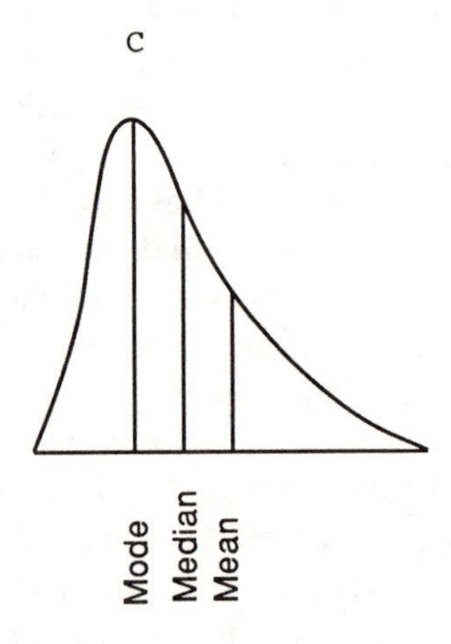

Figure 4.1. Relative positions of the mean, median, and mode for three shapes of distributions: (A) negatively skewed, (B) normal, and (C) positively skewed.

4.5 The Mean of Combined Data Sets

Often the means for different data sets are available, and the mean for all cases in all data sets is needed. If the number of cases in each data set is the same, then the mean of the means could be computed. If $n = 10$ for each of the three groups for which the means are 15, 18, and 20, the mean would simply be $(15 + 18 + 20)/3 = 53/3 = 17.67$. However, if the number of cases varies from one data set to another, then the means would have different weightings because they are based on differing numbers of cases. To adjust for this, the weighted mean ($\overline{X}_w$) is calculated as

$$\overline{X}_w = \frac{n_a\overline{X}_a + n_b\overline{X}_b + n_c\overline{X}_c + \ldots + n_k\overline{X}_k}{n_a + n_b + n_c + \ldots + n_k} \tag{4.4}$$

where $n_a, n_b, \ldots n_k$ are the sample sizes for each group, and $\overline{X}$ is the mean for each group.

For this same example of the three means of 15, 18, and 20, suppose the values of n were, respectively, 14, 12, and 8. Computing the weighted mean would give

$$\overline{X}_w = \frac{14(15) + 12(18) + 8(20)}{14 + 12 + 8} = 17.24$$

which is lower than the simple average of the three means because the smallest mean has the greatest frequency and thus the greatest weight. Thus, each mean is weighted by its number of cases. Essentially, the sum of scores for each data set is recomputed, these sums are added together and divided by the total combined n.

When averaging a series of percentages or proportions, this weighting procedure must be used unless the number of cases in each data set is the same. The weighted mean for percentages or proportions is

$$\overline{X}_w = \frac{n_aP_a + n_bP_b + n_cP_c + \ldots + n_kP_k}{n_a + n_b + n_c + \ldots + n_k} \tag{4.5}$$

where n is the number of cases in each group, and P is the percentage or proportion.

4.6 The Geometric Mean

The arithmetic mean, median, and mode are the most widely used measures of central tendency. However, under certain conditions other measures of central tendency may be useful, because none of the three common measures may be appropriate. For example, the geometric mean would be used when dealing with data in the form of ratios and when the average rates of change are desired, such as in production sales from one year to another. The geometric mean is useful in studies of any type of change and in any situation in which interpolation or extrapolation is needed (i.e., estimation of intermediate or projected values is desired in a series, in which a linear constant change can be assumed).

The *geometric mean*, GM, is the nth root of the product of the items, or

$$\mathrm{GM} = \sqrt[n]{(X_1)(X_2)(X_3)\ldots(X_n)} \tag{4.6}$$

Thus, for four values 6, 9, 10, and 12, the geometric mean is

$$\mathrm{GM} = \sqrt[4]{(6)(9)(10)(12)} = \sqrt[4]{6480} = 8.972$$

whereas the arithmetic mean would be 9.25. For any series of positive values that are different, the geometric mean is always smaller than the arithmetic mean. The geometric mean is not appropriate if any of the values are zero or negative. One useful property of the geometric mean is that taking its nth power (i.e., multiplying together n numbers all equal to the geometric mean) gives the same result as multiplying together all of the original observed values. This property is useful in predicting future or intermediate values.

For actual computations, the arithmetic mean of the logarithms is computed, and the antilogarithm of this mean is taken. Consider the following example of sales of a chemical over a 4-year period where obtaining the average percentage change over the 4 years is desired.

Sales Units	1981	1982	1983	1984
Tons	720	700	820	860
Percentage of previous year		97.2	117.1	104.9

The arithmetic mean of the three percentage changes is 106.4, which implies an average rate of growth of 6.4%. However, applied to the data, if sales had increased each year by 6.4% , by 1984 sales would have been 867.3 tons (720 × 1.064 × 1.064 × 1.064). The arithmetic mean percentage change is obviously not correct. The geometric mean of the percentage change will provide the correct average percentage change. Equation 4.7 can be used for such data.

$$\log \mathrm{GM} = \frac{\log X_1 + \log X_2 + \log X_3 + \ldots + \log X_n}{n} \tag{4.7}$$

For the sales data,

$$\log \mathrm{GM} = \frac{\log 97.2 + \log 117.1 + \log 104.9}{3} = 2.0257$$

$$\mathrm{GM} = \text{antilog } 2.0257 = 106.1$$

Thus, the geometric mean indicates an average yearly increase of 6.1%. To check this out, starting in year 1981 with 720 tons, by 1984 sales would have been 720 × 1.061 × 1.061 × 1.061 = 859.96, or 860 tons. If other factors are constant, this product can be multiplied again by 1.061 to estimate the tons that will be sold in 1985, which would be 912 tons.

4.7 *The Harmonic Mean*

The harmonic mean (HM) is useful in connection with problems involving averages of rates of work and when data are given in terms of productivity per hour, distance covered per time span, units purchased per dollar, or to estimate total productivity based on different rates of production for a given group of workers. For rating of work, the data are usually expressed in either of two forms: (1) time required per unit amount of productivity, or (2) productivity per unit of time. When rates of productivity are expressed in terms of time per unit of productivity, the harmonic mean of the rates is equivalent to the weighted arithmetic mean of those rates. Thus, the harmonic mean provides the same information as the weighted average mean. However, the harmonic mean might be more convenient to use when someone dealing with a long series of rates or prices in one form (e.g., unit per time) wants to know what the arithmetic mean would be if the series were expressed in another form (e.g., total productivity). As with the geometric mean, the harmonic mean is meaningless for any series containing zero or negative values and is smaller than the geometric mean or the unweighted arithmetic mean.

The harmonic mean of a data set is the reciprocal of the arithmetic mean of the reciprocals of the individual values.

$$\mathrm{HM} = \frac{n}{(1/X_1) + (1/X_2) + (1/X_3) + \ldots + (1/X_n)} \qquad (4.8)$$

where X is the individual value, and n is the number of values.

As an application, suppose that for five different conditions, the number of minutes required to produce one unit for each condition were 6, 9, 12, 12, and 15 min., respectively, and the average time required per unit is needed for all five conditions. By using equation 4.8, the harmonic mean would be

$$\mathrm{HM} = \frac{5}{(1/6) + (1/9) + (1/12) + (1/12) + (1/15)} = 9.78$$

Thus, the average number of minutes to produce each unit would be 9.78 min. If all five conditions were operating, a total of $5(60/9.78) = 30.67$ units would be produced per hour. The number of units per hour could also be computed for each condition, and these units could be averaged to come out with the same result. The harmonic mean may be more convenient to use than the weighted mean, especially if the weights to be applied are in the numerator of the equation, as in the previous example, where the unit produced was in the numerator and the amount of time in the denominator.

4.8 Uses and Cautions

Selection of a measure of central tendency depends on (1) the shape of the distribution, (2) the reasons for using the statistic, (3) the type of measurement scale used, and (4) further use of the statistic. If the purpose is purely descriptive, the measure that best represents the data and that the statistician wants to communicate should be used.

If the desired outcome is to show where the greatest concentration of values fall, then the mode would be appropriate. However, except for nominal data or for merely showing where the greatest frequency of values is, the mode has little utility as an average in applied statistics.

If the desired outcome is to indicate a point at which half of the cases fall below and half above, then the median would be appropriate. If a set of data contains a few extreme scores at either the high or low end of the distribution (e.g., a skewed distribution), then the median might best represent the data. For example, for a sample of 10 persons, if the average annual salary in thousands of dollars for each person is 16, 17, 18, 18, 19, 20, 22, 23, 23, and 80, then the median salary of 19.5 would be the most appropriate measure of central tendency because two modes (18 and 23) occur, and the one extreme salary pulls the mean up to 25.6, which really does not represent most of the salaries. Other situations in which the median might be preferred are when grouped data and open-ended or unspecified intervals occur at either the lower or upper extreme, if there is a question about the equalness of the intervals along the scale values, or if the values are described and interpreted in terms of percentiles.

The arithmetic mean is the most widely used and useful measure of central tendency because it is the most reliable and is simply and clearly defined. Being an algebraic quantity, the mean is the measure of central tendency most often used in further analyses. Thus, the mean should be used as the measure of central tendency unless there is special reason for not using it (e.g., skewed distributions).

The geometric mean is useful when dealing with change or growth data, especially when in the form of ratios, and when the values suggest a geometric series. The harmonic mean results in the same average as the weighted mean but is useful when the data are given as averages of rates of work, time, cost, or distance.

The best advice that can be given with regard to which measure of central tendency should be used is that the user should become thoroughly acquainted with the limitations and advantages of each and then select the most appropriate measure on the basis of which best suits the purposes for using that measure. The user of statistical reports should be acquainted with how each average can be affected by the type of data and how it is distributed so that a valid picture of the actual results has been communicated rather than what the author of the report wanted to communicate. In general, the arithmetic mean is most often used in research studies.

Finally, averages are not perfect, and variation will occur from one sample to another. This concept will be discussed in Chapter 8, which deals with confidence limits for various statistics.

Problems

1. Find the mode, median, and mean for frequency distributions A, E, and G in Table 3.3 and compare the values obtained. Which measure of central tendency would be appropriate for each distribution?
2. A chemical laboratory purchases 100 test tubes in January for an average cost of $0.90 per tube, 400 tubes in April for an average cost of $0.70 per tube, 300 tubes in July for an average cost of $0.80 per tube, and 200 tubes in October for an average cost of $0.85 per tube. What was the average cost per tube for all purchases?
3. The per unit production costs for each of 6 years is presented as shown. What would be a good estimate of per unit costs for 1986, assuming other factors remain constant?

Year	Cost
1980	45
1981	54
1982	60
1983	74
1984	82
1985	96

4. A production engineer wants to advertize the lower and upper limits where the middle 90% of samples would be with regard to percentage of carbon in a mixed powder. Therefore, 200 random samples of powder were taken and analyzed for percentage of carbon, with the results as presented in Table 3.2. What values would the engineer report?

Chapter 5

Measures of Variation

5.1 The Concept of Variability

The central tendency or central value provides no information with regard to the variability of the values, that is, how much they differ from the central value. Are the values clustered together, or are they spread apart? *Variability* refers to differences among the values: Low variability means that all values tend to be similar in magnitude and cluster together, resulting in what is defined as a *homogeneous distribution*; high variability means that the values vary considerably from low to high, resulting in what is defined as a *heterogeneous distribution*. Consider the two frequency distribution polygons in Figure 5.1. Both have the same mean, 50, but distribution A is homogeneous whereas distribution B is heterogeneous. The values in distribution A vary from a low of 40 to a high of 60, whereas the values in distribution B vary from a low of 20 to a high of 80.

The amount of variability is of interest not only as a descriptive property, but also is important in quality control and inferential statistics. In fact, statistics have been characterized as the study of variation (*1*). Kerlinger and Pedhazur (*2*) stated that the only thing a researcher has to work with is variance. In later chapters dealing with prediction and inferential statistics, the importance of variance will be shown because these statistics are based upon the variance and covariance of the variables being studied.

In experimental work and comparative studies, variation in values occur because of different conditions, circumstances, or sources, and the researcher is interested in determining and understanding what causes that variation. A statistical technique developed by Fisher (*1*), the analysis of variance, does exactly what the name implies. It

1453–0/88/0069$06.00/1

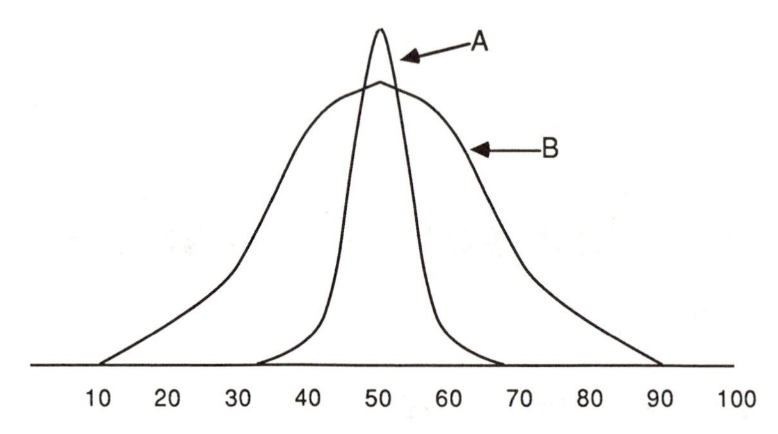

Figure 5.1. Polygons having the same mean but different variability.

analyzes the variation in different ways to determine sources of that variation so that potential causal circumstances can be identified. The variability within and between sets of data may be more important and may reveal more information about the data than the measures of central tendency. Besides, measures of central tendency become more meaningful and less susceptible to misinterpretation when accompanied by measures of variability. The complete description of a set of data requires that measures of both central location and the variation of the values are provided. The amount of variability of a set of data can be described quantitatively by various measures, the most common of which are the total range, various short ranges, variance, and the standard deviation. Variance and standard deviation are the most common and useful for both descriptive and inferential statistics. For these measures of variability, measurements are assumed to have been made at the level of an ordinal, interval, or ratio scale.

For nominal data, the only approach to describing variability is to show the number of different categories into which the observations fall. The concept of variability implies a set of categories with a fixed order of magnitude and meaningful relations to each other. Such is usually not the case with nominal data. Because numbers in a nominal scale may be assigned completely arbitrarily, statistics such as the range and standard deviation would have no meaning.

5.2 The Total Range

The simplest index of the variability of a set of data is called the total range, or simply, the range. The *range* is the number of values between the lowest and highest values. The general equation for the range is

$$\text{total range} = X_H - X_L + 1 \quad (5.1)$$

where X_H and X_L are the highest and lowest values, respectively. For example, given the values 9, 9, 10, 12, 12, 13, 13, and 14, the range would be $14 - 9 + 1 = 6$. Six inclusive values occur from low to high. Another way of conceptualizing the range is to think in terms of the real limits of each value and the total distance covered by these values. Thus, the real limits are from 8.5 to 14.5, which is a total of six values. The total range could then be given by the equation

$$\text{total range} = X_{\text{url}} - X_{\text{lrl}} \quad (5.2)$$

where X_{url} is the upper limit of the highest value, and X_{lrl} is the lower limit of the lowest value.

The total range is easy to compute and to understand. However, it is not very stable because only the two end values determine its magnitude. Any fluctuation in these two end values could cause large differences in the range. The total range is also said to be a biased statistic because the range tends to vary with sample size. Large samples tend to include more extreme scores than smaller samples; thus, the range would tend to be higher for large samples. Consequently, ranges calculated on samples of different numbers of cases are not directly comparable. The range is also limited with regard to further analyses. A few nonparametric statistics use the range in making inferences from samples to populations, but these are quite limited.

5.3 Short Ranges

To reduce the shortcomings of the total range with regard to being a function of sample size and lack of stability, several short ranges have been advocated. Essentially, the short range cuts off the bottom and top extreme values and then is based on the distance between the cut-off limits. The most common short range is the interquartile range, which cuts off the bottom 25% and the top 25% of the values. The *interquartile range* is the distance or number of values that includes the middle 50% of the values. When the interquartile range is divided by two, the quotient is known as the *semi-interquartile range*, or the *quartile deviation*, and is commonly abbreviated *Q*.

The semi-interquartile range is interpreted in a similar way as the standard deviation is interpreted (i.e., given the median, 50% of the values fall between ±1 semi-interquartile range from the median). Other short ranges could be set up to cut off the bottom and top 5% or the bottom and top 10% of the values. Respectively, the equations for the interquartile range, the short range cutting off the top and bottom 5% of the values, and the short range cutting of the top and bottom 10% of the values are

$$\text{IQ range} = X_{P_{75}} - X_{P_{25}} \tag{5.3}$$

$$\text{short range} = X_{P_{95}} - X_{P_{05}} \tag{5.4}$$

$$\text{short range} = X_{P_{90}} - X_{P_{10}} \tag{5.5}$$

From Section 4.3, percentile points for P_{10}, P_{25}, P_{75}, and P_{90} were calculated to be 29.83, 32.70, 37.17, and 39.50, respectively.

Thus, by subtraction, the interquartile range is 4.47 values, and the semi-interquartile range is 2.24. The short range cutting off the top 10% and bottom 10% of the cases is 9.67, between which 80% of the cases fall. Calculating short ranges is often useful in quality control studies to determine if product standards fall within designated limits.

5.4 Variance and Standard Deviation

The most important and most widely used measures of variation or dispersion are the variance and standard deviation. Each is based on every value in a distribution and both are based on the average squared deviations of the values from the mean. Both of these measures are widely used in advanced descriptive and inferential statistics.

By definition of the mean, the sum of the deviations about the mean is equal to zero. To overcome this difficulty, the deviations are squared to result in a set of positive values. The sum of those squared deviations is called the *sum of squares*, SS, and because it is used in nearly all analyses, the sum of squares is one of the most important concepts to learn. Two basic equations can be used to compute the sum of squares. The *deviation method* is computed by subtracting the mean from each value, squaring this difference, and then summing the result over all n cases. Computer programs often use this procedure, although this is tedious by hand. The second approach is the

raw score approach and is more convenient to use when calculating by hand or desk calculator. The equations for the raw score approach are

$$SS = \sum x^2 = \sum (X - \bar{X})^2 \quad (5.6)$$

$$SS = \sum X^2 - \frac{(\sum X)^2}{n} \quad (5.7)$$

$$SS = \sum X^2 - n\bar{X}^2 \quad (5.8)$$

where $x = X - M$, X is each observed value, n is the number of cases in a sample, and $\bar{X}$ is the mean of the observed values.

For the data in Table 5.1, the sum of squares is 76, which is identical for all of the methods. In this example, with a small n and a mean equal to a whole value, either calculation is satisfactory. However, if n was large and the mean was carried out to two decimal places such as 5.73, the deviation approach by hand becomes quite tedious. If the mean is known, then equation 5.8 could be used. However, usually most calculators are programmed to provide n, ΣX, and ΣX^2 automatically, which are the ingredients for both the

Table 5.1 Hypothetical Data for Calculating Sum of Squares

n	X	X^2	x	x^2
1	9	81	4	16
2	8	64	3	9
3	7	49	2	4
4	7	49	2	4
5	6	36	1	1
6	5	25	0	0
7	4	16	−1	1
8	2	4	−3	9
9	1	1	−4	16
10	1	1	−4	16
10	50	326	0	76

$n = 10$ $\quad M = 50/10 = 5$ $\quad S = 2.91$
$\Sigma X = 50$ $\quad SS = 326 - (50^2/10) = 76$ $\quad var = 8.44$
$\Sigma X^2 = 326$ $\quad SS = 326 - (10 \times 5^2) = 76$
$\Sigma x = 0$
$\Sigma x^2 = 76$

mean and sum of squares. Therefore, computing the mean first and then using equation 5.8 isn't necessary. Although the sum of squares enters into many statistical calculations, it is not used as a descriptive index because its size varies with n. The variance and standard deviation derived from the SS are the most important measures of variability for both descriptive and inferential statistics.

Variance is the mean of the squared deviations from the mean, and two general equations are

$$\text{variance} = \sigma^2 = \text{SS}/N \tag{5.9}$$

$$\text{variance} = S^2 = \text{SS}/(n - 1) \tag{5.10}$$

Equation 5.10 is used for populations, which in statistical terminology use the capital letter N to denote the number of cases and σ to denote the standard deviation. Equation 5.10 is referred to as an unbiased estimate of the population variance and is to be used with sample data. The reason for using $n - 1$ is that for small-size samples, extreme cases would not be included and thus the result would underestimate the actual population variance. To compensate for this, $n - 1$ is used. The quantity $n - 1$ is also known as *degrees of freedom*, and is another important concept in statistics. The number of values that are free to vary is what degrees of freedom refers to. For example, take the five values 4, 8, 12, 14, and 17 for which the mean is equal to $55 \div 5 = 11$. For the sum of these values to always total 55, four of the values are free to vary and could be any number. However, once these four are set, the last value is not free to vary. For example, suppose one sets the first four values to be 8, 9, 10, and 12. The last value has to be 16 for the sum to equal 55. This last value is not free to vary. Thus, four numbers can vary, and for this example, the degrees of freedom would be equal to $n - 1$ or 4.

Equation 5.9 should be used for all samples of data to provide an unbiased estimate of the population variance, and this equation is used for all inferential statistics.

The variance is described in terms of squared units and is used in this form for most statistical analyses. However, because it is based on the squared values, variance is not generally useful in describing the variability in a set of scores. To bring this measure back to the original units, the square root can be taken to yield the *standard deviation*. Two commonly used symbols for standard deviation are S and SD. Both symbols will be used throughout this book.

$$S = \sqrt{\text{variance}} \tag{5.11}$$

Because two equations are used for computing the variance, two standard deviations could be attained from the same set of data. If a population is considered, then equation 5.9 would be used, and the resulting standard deviation would accurately represent the population and would be labeled as small sigma, σ. If samples are dealt with, then equation 5.10 would result in an unbiased estimate of the population standard deviation. Because most studies deal with samples of data, equation 5.10 is preferred. As mentioned previously, Greek letters are usually used to denote population values or parameters, and English letters to denote sample values or statistics.

For the data in Table 5.1, the sum of squares is 76; thus, the variance obtained by using equation 5.9 would be 7.60. Whereas, by using the equation for samples, or equation 5.10, the variance would be 8.44. Thus, using the unbiased estimate of the population variance results in a larger variance because using degrees of freedom in the denominator compensates for the fact that extreme values are not likely to be included in small-size samples. The respective standard deviations obtained by using these two variances would be 2.76 and 2.91.

For grouped frequency distributions, the equation for SS is

$$SS + \sum fX^2 - \frac{(\sum fX)^2}{n} \qquad (5.12)$$

where X is the midpoint value of each interval, f is the frequency within each interval, and n is the total number of cases in the sample. Applied to the data in Table 3.2, the sum of squares would be

$$SS = 3513.488 - \frac{(836.70)^2}{200} = 13.15$$

Thus, by using equation 5.10, the variance would be 0.0661 and the standard deviation 0.257.

The standard deviation is a distance along the scale values that is usually interpreted by assuming that the distribution is approximately normal. If this assumption holds, then the distance from one standard deviation below the mean to one standard deviation above the mean will include approximately two-thirds of the cases included in the sample. The standard deviation is interpreted in terms of the normal curve table, which will be described in Chapter 7.

5.5 Combining Standard Deviations

Occasionally, statistics such as means and standard deviations are obtained for a number of subgroups, and these statistics are desired for the total combined group. In Chapter 4, a procedure of weighting means was presented to determine what the total mean would be. The same weighting procedure applies to combining standard deviations. The mean standard deviation (SD_w) can be computed as

$$SD_w = \sqrt{\frac{n_a(S_a^2 + M_a^2) + n_b(S_b^2 + M_b^2) + \ldots n_k(S_k^2 + M_k^2)}{n_a + n_b + \ldots + n_k} - M_T^2} \qquad (5.13)$$

where S is the standard deviation for each sample, M is the mean for each group, M_T is the weighted mean for all groups, and n is the number of cases in each group. To illustrate the use of this equation, consider the following production records from four plants, where n is the number of days of productivity per year, and the yield of a product is given in terms of the mean and standard deviation.

Plant	n	Mean	S.D.
A	180	48	4.2
B	200	43	5.5
C	210	50	6.4
D	300	46	4.0

For this data, the weighted mean calculated by using equation 4.4 would be

$$\bar{X}_w = \frac{180(48) + 200(43) + 210(50) + 300(46)}{180 + 200 + 210 + 300} = 46.67$$

The weighted average standard deviation would be

$$S_w = \sqrt{\frac{180(4.2^2 + 48^2) + 200(5.5^2 + 43^2) + 210(6.4^2 + 50^2) + 300(4^2 + 46^2)}{180 + 200 + 210 + 300} - 46.67^2} = 5.65$$

5.6 Relative Standard Deviation

When several series of values are expressed in the same unit of measurement and have approximately the same average value, the measures of variability, such as S, or range can be used to compare the variability among that series. However, if the series are measured in different units or have quite different average values, then the measures of variability expressed in the original values may not be appropriate. Even if the unit of measurement is the same, in any series the mean and standard deviation tend to change together. That is, series having large values and thus a high average value tend to have greater deviations from the mean than a series of small values. This situation results in differences in size of the standard deviations. To compare the variability of series of values that are expressed in different units or that have different average values, a measure of variability is needed that is independent of the unit of measurement and that takes into account the averages of each series. Several abstract measures of variability can be used for comparative purposes. The most common of these is the *relative standard deviation* (RSD), which is also referred to as the *coefficient of variation*, or CV. The CV is defined as the standard deviation divided by the mean and is usually expressed as a percentage. The general equation is

$$\mathrm{CV} = S(100)/M \qquad (5.14)$$

Because the standard deviation is expressed as a fraction of the mean, the CV provides a measure of variability relative to the average value and is independent of the unit of measurement.

As an example of where the relative standard deviation would be used, consider the question as to whether there is the same or more variability in the heights of 4-year-old children or 16-year-old children. If in a large group of 4-year-old children, the mean height is 32 in. with a standard deviation of 1.4 in. compared to the mean of 66 in. and a standard deviation of 2.8 in. for 16-year-old children, the older children would appear to be more variable with regard to height. However, applying the relative standard deviations yields

$$\mathrm{RSD} = 1.4(100)/32 = 4.38$$

for the 4-year-old children, and

$$\mathrm{RSD} = 2.8(100)/66 = 4.24$$

for the 16-year-old children. Thus, relative to their mean heights, the two groups of children show about the same variability, whereas consideration of only the standard deviations results in the appearance that the older children were more variable with regard to height.

The CV can also be used for comparing the variability of a series of data measured in different units of measurement, such as variation in height measured in inches and variation in weight measured in pounds, or the variation of the yield of one chemical measured in pounds compared to the variation in the yield of another chemical measured in liters.

Although the coefficient of variation is susceptible to misinterpretation because of its reliance on the mean value, it does provide valuable information for evaluating the results of studies (i.e., to compare results with what is typically obtained). For example, in fertilizer studies, although the yield of agricultural crops varies with regard to mean yield, and standard deviation of yield varies with locations and conditions, the CV is usually between 5% and 15%. If values were obtained outside of these limits, then the results might be suspect. Thus, the values of CV serve as a check on the results of a study.

5.7 Measures of Skewness and Kurtosis

Measures of skewness and kurtosis provide quantitative descriptions of nonnormal variation and are helpful in determining whether a distribution departs too much from normality to be analyzed by standard parametric methods.

Skewness and kurtosis depend upon how the values in a distribution vary from the mean. The different types of distributions were presented in Figure 3.7. Several methods are used to determine whether a distribution deviates from normal. Skewness usually can be detected by inspection of the frequency polygon. However, peakedness is more difficult to determine from inspection because this characteristic may result from the choice of dimensions for the vertical and horizontal scales. Thus, visual inspection of a frequency distribution would be the first step in determining whether nonnormality is suspect, but often this step is not very precise.

As will be shown in Chapter 7, another way of examining nonnormality is to compare the proportion of cases included within standard deviation intervals for the sample distribution compared to the theoretical distribution. However, this method serves as a preliminary analysis and does not result in a quantifiable measure of either skewness or kurtosis.

The most commonly used measures of skewness and kurtosis are based upon moments about the mean. The term "moment" originates in mechanics and is used to denote a measure of the tendency of a force to cause rotation of an object about a center point. Because this strength depends upon the amount of the force and the distance from the center point, the moment is the product of force times distance. When the sum of the moments tending to cause rotation in one direction is equal to the opposite force, then the object is said to be in balance. Applied to frequency distributions, the mean is the center point and the negative and positive deviations are analogous to the two different rotational forces.

Applied to frequency distributions, the first four moments about the arithmetic mean (m_1, m_2, m_3, and m_4) are

$$m_1 = \frac{\sum (X - M)}{n} = 0 \tag{5.15}$$

$$m_2 = \frac{\sum (X - M)^2}{n} \tag{5.16}$$

$$m_3 = \frac{\sum (X - M)^3}{n} \tag{5.17}$$

$$m_4 = \frac{\sum (X - M)^4}{n} \tag{5.18}$$

where X is each observed value, M is the mean of those values, and n is the number of cases included in the sample.

To compute m_3, each x value in Table 5.1 would be cubed. For the first value of $X = 9$, $(X - M)^3 = 64$, and for the last value of $X = 1$, $(X - M)^3 = -64$. The sum of these cubed deviations equals -48. Then, $m_3 = -48/10 = -4.8$. The same procedure would be used to compute m_4, but now each X value would be taken to the fourth power. For the first X value, $(X - M)^4 = 256$. These resulting

10 values taken to the fourth power sum to 964, and then m_4 = 964/10 = 96.4.

The measure of skewness makes use of the third moment and is defined as

$$\text{skewness} = \frac{m_3}{m_2\sqrt{m_2}} \tag{5.19}$$

where m_2 and m_3 are, respectively, the second and third moments about the mean. For a symmetrical distribution, the sum of the deviations above the mean when raised to the third power equals the sum of the deviations below the mean when these are also raised to the third power. Thus, both m_3 and the measure of skewness, sometimes symbolized as g_1, equal zero. If the distribution is not symmetrical, then these sums will not balance. For a positively skewed distribution, g_1 is positive, and g_1 is negative for a negatively skewed distribution. This measure of skewness is comparable for distributions that differ in variability and is independent of the scale of measurement used because g_1 uses the quantity $m_2(m_2)^{1/2}$, which is analogous to converting raw values to standard scale values. Thus, g_1 values computed from different sets of measurement and for distributions that have different variances are comparable.

The measure commonly used for kurtosis (g_2) uses the fourth moment, and is defined as

$$\text{kurtosis} = g_2 = \frac{m_4}{m_2^2} - 3 \tag{5.20}$$

where m_2 and m_4 are the second and fourth moments about the mean, respectively, and 3 is a constant. For a normal distribution, $g_2 = 0$. For a leptokurtic distribution, $g_2 > 0$ (i.e., a positive value), and for a platykurtic distribution, g_2 is a negative value.

For the data in Table 5.1, $m_2 = 76/10 = 7.6$, $m_3 = -48/10 = -4.8$, and $m_4 = 964/10 = 96.4$. Skewness would then be

$$\text{skewness} = \frac{-4.8}{7.6\sqrt{7.6}} = -0.23$$

which indicates a slight negative skew. Kurtosis would be

$$\text{kurtosis} = \frac{96.4}{7.6^2} - 3 = -1.33$$

which indicates a slightly platykurtic or flat distribution.

5.8 Summary

In this and the previous chapter, summary descriptive measures were presented for describing, comparing, and communicating the results of a study by using only one number per measure. The four measures—mean, standard deviation, skewness, and kurtosis—can be used to fully describe a set of data by four numbers. However, as was pointed out, these measures can be used in a manner that is representative of the data or can be used to spuriously inflate or conceal features of the data. Thus, users of statistics should be aware of the appropriate and inappropriate applications of these measures, as is true for any statistic.

The mean and standard deviation are by far the two most commonly used descriptive statistics. Given distributions that are relatively normal, these two measures are reliable and will accurately represent the sets of data. However, for nonnormal distributions or for data sets that are not based on at least a scale that simulates an interval scale, the results should be interpreted with caution.

References

1. Fisher, R. A. *Statistical Methods for Research Workers;* Hafner: New York, 1973.
2. Kerlinger, F. N.; Pedhazur, E. J. *Multiple Regression in Behavioral Research;* Holt, Rinehart & Winston: New York, 1973.

Problems

1. The following is a distribution of the number of filaments produced by 10 different grades of molten polymer. Describe this distribution in terms of central tendency, variability, skewness, kurtosis, and interquartile range.

Grade of Polymer	Strength of Filaments
10	9
9	9
8	8
7	8
6	8
5	7
4	5
3	3
2	2
1	1

2. Daily samples of the strength of different production runs of a dyestuff were taken in four plants. Strength was measured by the extent of coloring power in each sample. The means and standard deviations were available for each plant. The desired result is to obtain a composite mean and standard deviation for all four groups combined. What would the composite mean and standard deviation be?

Plant	n	Mean	SD
1	20	80	8
2	30	75	10
3	25	65	7
4	40	90	12

3. Compute the mean and standard deviation for the data in Table 3.2.

Chapter 6

Correlation and Regression

6.1 Relationship between Two or More Variables

In chemistry and in nearly all other areas, for any reaction or phenomenon, many different variables are usually involved. In some instances, two properties may be theoretically functionally related, and the problem is to determine the actual extent of this relationship. In other instances, the interest is in determining whether two properties that appear to be unrelated are actually related, a fact that could then be used to advantage. For example, two procedures may be used for producing a synthetic chemical. If one is less expensive than the other, and a high degree of correlation of the results for both procedures is found, then the less expensive procedure could be used because the resulting yield would be the same for both procedures.

The study of how two or more variables are related involves two closely related statistical analyses: correlation and regression. *Correlation* deals with describing the degree or magnitude of the relationship between two or more variables, and the statistic that describes the extent of this relationship is the *correlation coefficient*. The correlation coefficient reflects how closely values of the different variables vary together. *Regression* deals with estimating or predicting one variable from knowledge of its relationship with another variable. For example, how is the yield of a manufacturing process affected by various concentrations of a catalyst? Pairs of yields and concentration of the catalyst are gathered over a period of time to determine if a correlation exists between these two variables, and if it does, production yields can be estimated for other concentrations of that catalyst.

1453–0/88/0083$10.50/1

Many examples of correlation and regression analyses can be found in chemical studies. Some recent examples include studies of the repeatability or reliability of measurement procedures, which is one of the more common uses of correlation. Klompmakers and Hendriks (*1*) found that the correlation (r) between glycosaminoglycans (GAG) present in intestinal specimens at two levels of measurement (535 and 633 nm) was $r = 0.974$. Gavino et al. (*2*) found a similar correlation, $r = 0.96$, between two methods of determining the activity of acetone in standard solutions for reduced and unreduced samples. Robbat et al. (*3*) used the stepwise multiple linear regression procedure to generate molecular connectivity equations for predicting the retention characteristics of mononitrated polycyclic aromatic hydrocarbons. Multiple regression was also used by Barrett (*4*) to determine an equation for estimating hot corrosion attack for a series of nickel-cast turbine alloys. Nonlinear regression procedures were used by Arena and Rusling (*5*) for analyzing individual electrochemical response curves obtained at spherical electrodes.

6.2 The Correlation Coefficient

A number of different measures of relationship have been developed, but the most widely used is the *Pearson product-moment correlation coefficient* (r). Many of the other measures of relationship are special cases of the Pearson r, or simply r, and are interpreted in a similar way. In this chapter, the Pearson r for two variables will be presented first, followed by other measures of relationship.

Some basic concepts have to be understood when dealing with the correlation coefficient. First, r is an index number that can vary only from -1.00 through 0 to $+1.00$. The value of r, either negative or positive, indicates the strength of the relationship between the two variables. The sign of r indicates the direction of the linear relationship. A positive r means that low values on one variable correspond to low values on the other variable, and high values correspond to high values on the other variable. A negative r means just the opposite. As values increase on one variable, they decrease on the other (i.e., an inverse relationship).

Possibly the best way to describe what the various correlation coefficients communicate is to illustrate by means of bivariate frequency scatterplots such as in Figure 6.1. Conventionally, the inde-

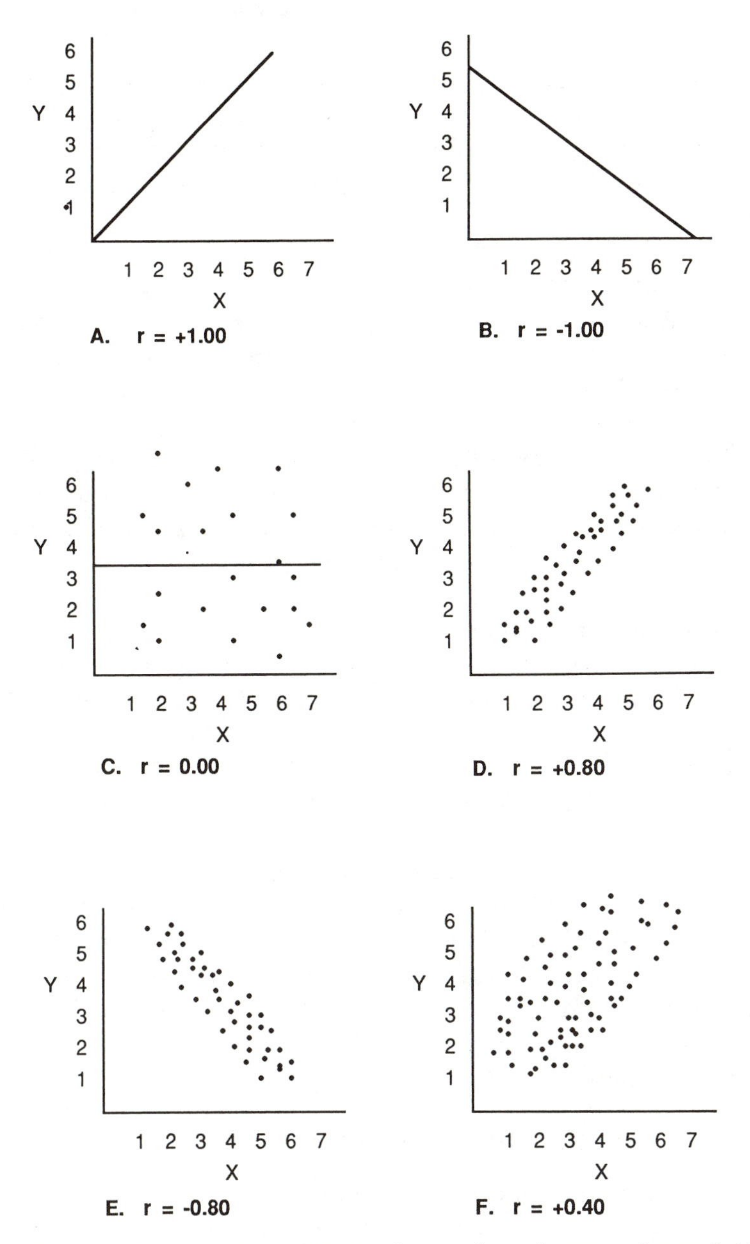

Figure 6.1. Scatterplots illustrating various degrees of correlation.

pendent variable, X, is plotted on the horizontal scale, and the dependent variable, Y, is plotted on the vertical scale. For the example mentioned in Section 6.1 dealing with the effect of various concentrations of a catalyst, Figure 6.1 shows that the various concentrations of the catalyst would be the X variable, and the production yield would be the Y variable. For some correlation problems (e.g., relating two manufacturing procedures), researchers are interested only in the relationship between the two variables and thus either could be X or Y.

For any correlation study, there is no substitute for a bivariate scatterplot of the data as the best way to get some idea of the type of relationship (i.e., linear or curvilinear), the shape of the results, and the type and degree of correlation. Some of the possible results are presented in Figure 6.1. Figure 6.1A depicts $r = +1.00$, whereas Figure 6.1B depicts $r = -1.00$. Both are defined as perfect relationships because all of the values would fall in a straight line, and one value can be predicted perfectly from knowledge of the other value. The only difference is that A is a direct relationship, whereas B is an inverse relationship. Figure 6.1C shows the distribution for $r = 0$. Here, high values of X are not consistently associated with low or high values of Y. Thus, high values of X do not correspond with high values of Y, and low values of X do not correspond with low values of Y. No trend can be found for those variables to vary together. In fact, as will be shown in Section 6.5, when $r = 0$, the best estimate of what the Y value would be for any X value is the mean of the Y values. Thus, the horizontal regression line is at the mean value of Y.

Figures 6.1D and 6.1E depict substantial correlations between X and Y (i.e., $r = 0.80$ and -0.80, respectively). Figure 6.1F depicts a low degree of relationship ($r_{xy} = 0.40$).[1] In each figure, the relationship predicted is linear, which is one in which a straight line would best fit the data. More will be said about this when regression is discussed. The subscripts indicate that r is the correlation between the two variables X and Y.

6.2.1 Computation of Pearson Correlation

The *Pearson product-moment correlation coefficient* is the average of the products of the moments of the two variables if each were

[1]Lower case letters are usually used as subscripts for r, because the basic formula given by Pearson was expressed with deviation scores.

measured in standard score units. In standard score form, r is the sum of the products divided by $n - 1$, or

$$r_{xy} = \frac{\sum z_x z_y}{n - 1} \qquad (6.1)$$

where z_x and z_y are the standard scores on each variable; n is the number of pairs of values, or the sample size; and r_{xy} denotes the Pearson product-moment correlation coefficient between two variables, X and Y.

In practice, this equation is laborious to use because the user must compute the mean of each distribution, determine the deviation of each value from the mean for each distribution (z_x and z_y), compute the products of these pairs of deviation values, sum these products, and then divide by $n - 1$. More convenient formulas can be used to compute the correlation coefficient. Because $z_x = (X - \overline{X})/S_x$, and $z_y = (Y - \overline{Y})/S_y$, where S_x and S_y are the standard deviations of X and Y, respectively, equation 6.2 can be obtained from equation 6.1 by substitution involving simple algebra.

$$r_{xy} = \frac{\sum (X - \overline{X})(Y - \overline{Y})}{(n - 1)S_x S_y} = \frac{\sum (X - \overline{X})(Y - \overline{Y})}{\sqrt{\sum (X - \overline{X})^2 \sum (Y - \overline{Y})^2}}$$

$$= \frac{\sum xy}{\sqrt{\sum x^2 y^2}} \qquad (6.2)$$

where x and y are deviations from the respective means $\overline{X}$ and $\overline{Y}$.

The raw score formula for Σxy is

$$\sum xy = \sum XY - \frac{(\sum X)(\sum Y)}{n} \qquad (6.3)$$

where X and Y are the raw values for the two variables. The maximum value of Σxy occurs when the deviation score values of X and Y are in the same order and equal to each other in each pair. Graphically, this result would happen as depicted in Figure 6.1A and would result in a perfect positive correlation, or $r = 1.00$.

Equation 6.2 requires the conversion of all values to standard score form, which will be discussed in Chapter 7. This equation is sometimes convenient to use for computer programs but is laborious by desk or hand calculator because it requires the conversion of all raw values to standard score form. The following equation for calculating r is best to use for hand calculators:

$$r_{xy} = \frac{n \sum XY - (\sum X)(\sum Y)}{\sqrt{[n \sum X^2 - (\sum X)^2][n \sum Y^2 - (\sum Y)^2]}} \qquad (6.4)$$

where n is the number of pairs of values, ΣXY is the sum of the cross products of each pair of values, ΣX is the sum of the X values, ΣY is the sum of the Y values, ΣX^2 is the sum of the squared X values, and ΣY^2 is the sum of the squared Y values.

Consider the data in Table 6.1 from a study to determine whether two colorimetric methods of determining a chemical constituent are closely related. The data are from 10 pairs of samples, each of which was measured by the two methods, X and Y.

For purposes of showing all computations, Table 6.1 presents all of the values needed for computing r by using equation 6.4. Writing

Table 6.1. Correlation of Ungrouped Data for Two Colorimetric Methods

Sample	X	Y	X^2	Y^2	XY
1	12	13	144	169	156
2	10	12	100	144	120
3	4	5	16	25	20
4	4	6	16	36	24
5	5	6	25	36	30
6	10	13	100	169	130
7	7	9	49	81	63
8	4	6	16	36	24
9	8	8	64	64	64
10	11	11	121	121	121
	75	89	651	881	752
Mean	7.50	8.90	$r = 0.953$		
S.D.	3.14	3.14			
Var.	9.83	9.88			
SS	88.5	88.9			

down all of the squared values and sums of products for each pair of values is not necessary because many hand and desk calculators sum these values automatically or compute r automatically. Usually for such studies, the means and standard deviations for each variable are presented to fully describe the variables. All of the sums of values required for these additional statistics are provided. To aid such computations, recording these sums as presented in Table 6.1 might be advisable.

For the data in Table 6.1, the computed r is equal to

$$r = \frac{10(752) - (75)(89)}{\sqrt{[10(651) - (75)^2][10(881) - (89)^2]}} = 0.953$$

6.2.2 Interpretation of Correlation Coefficients

Several methods can be used to interpret a correlation coefficient, and several factors must be taken into consideration in that interpretation. First, a bivariate scatterplot of the values will indicate whether the relationship between the two variables is linear and will display the relationship for a visual check on the computed correlation. Figure 6.2 shows the scatterplot for the data in Table 6.1. Visually, the scatterplot resembles Figure 6.1D. The two measures appear to be linearly related and have a high degree of relationship, and the relationship is direct and thus positive.

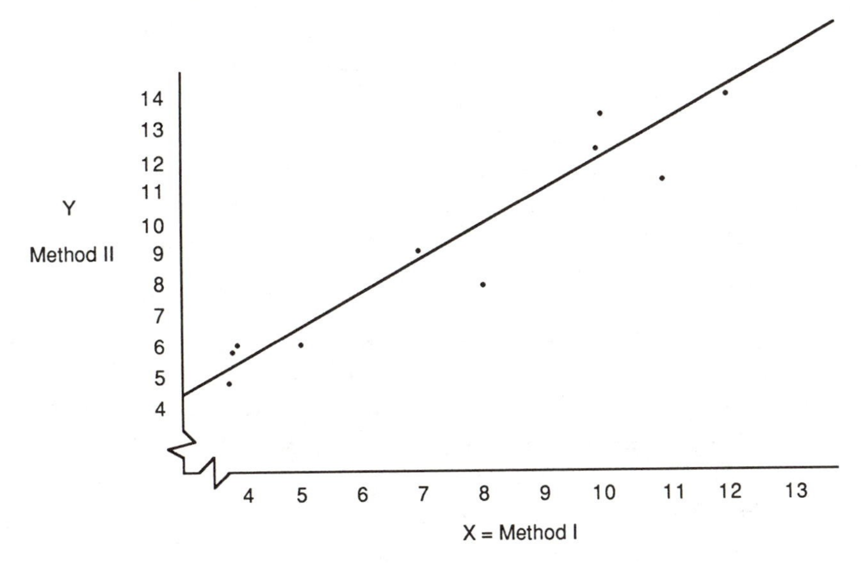

Figure 6.2. Scatterplot data of Table 6.1.

The second way of interpreting r is to compare the results obtained with what is expected or with the results of other studies. In our example, if both colorimetric methods have been shown to be reliable and valid for other studies, then a high consistency would be expected between these two methods. The high r indicates consistency in these two measures, and therefore either method could be used. Thus, the results are as expected.

The correlation coefficient is an index number, not a proportion. For example, a coefficient of 0.70 does not represent 70%, nor does this coefficient represent twice the relationship as a coefficient of 0.35. Similarly, the difference between coefficients of 0.60 and 0.70 does not represent the same change in degree of relationship as does the difference between coefficients of 0.30 and 0.40. However, an informative way to interpret the correlation coefficient is in terms of the proportion of variance associated with, or in common between, two variables. If the relationship between the two variables is linear, the square of the correlation coefficient, r_{xy}^2, indicates the proportion of variance in X associated with, or in common with, the variance in Y, and vice versa. If $r_{xy} = 0.70$, then $r_{xy}^2 = 0.49$. Therefore, 49% of the variance of X (0.49×100) overlaps, or is common with, the variance of Y.

The amount of common or shared variance can be shown by *Venn diagrams*. If circles represent the total variance of each variable X and Y, these circles can be drawn to show the degree of overlap. Figure 6.3 shows the overlap of variance between variables X and Y for correlation coefficients ranging from 0.00 to 1.00. When $r_{xy} = 0.00$, the two variables are completely independent of each other (i.e., no correlation between these two variables is found); hence, no commonality or overlap occurs (Figure 6.3A). Figures 6.3B–D show that as the correlation coefficient increases, the amount of overlap or commonality increases. For a perfect correlation ($r_{xy} = 1.00$), complete overlap occurs because all of the variance in X is accounted for by the variance in Y (Figure 6.3E). For the data in Table 6.1, $r_{xy} = 0.953$, which results in $r_{xy}^2 = 0.91 \times 100 = 91$, which in turn indicates that 91% of the variance in X overlaps with the variance in Y. Another way of interpreting this result is that 91% of whatever X is measuring is also being measured by Y, as shown in Figure 6.3F. In statistical terminology, r_{xy}^2 is also known as the *coefficient of determination*. Because r_{xy}^2 gives the proportion of variance that is common between the two variables X and Y, the uncommon or unique variance is the remainder, and this is known

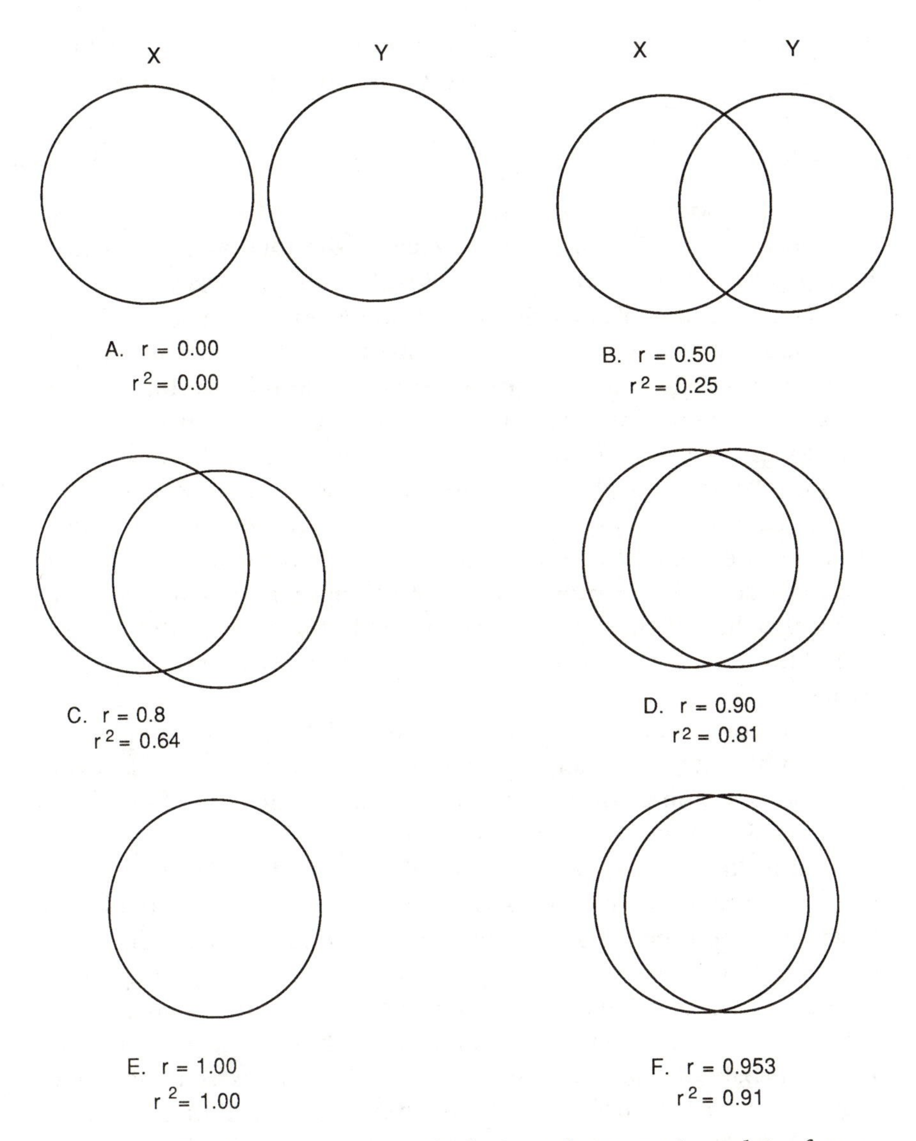

Figure 6.3. Venn diagrams, which show the percent overlap of variance between two variables.

as the *coefficient of nondetermination* and is usually symbolized as $k^2 = 1 - r^2$. For example, if $r_{xy} = 0.80$, then 64% common variance exists between these two variables and each has 36% of its variance that does not overlap, or is not common with, the other variable.

Some statisticians refer to the *coefficient of alienation*, k, which indicates the degree of lack of relationship. The formula for k is

$$k = \sqrt{1 - r_{xy}^2} \tag{6.5}$$

If both sides of this equation are squared ($k^2 = 1 - r_{xy}{}^2$) and transposed, then $r^2 + k^2 = 1.00$. That is, the total variance would be equal to the common variance plus the unique variance.

The correlation coefficient is an index number and does not have additive properties as regular numbers do. For example, we cannot say that $r = 0.60$ represents twice the relationship as $r = 0.30$. Nor is an increase from $r = 0.30$ to $r = 0.50$ equivalent to an increase from $r = 0.70$ to $r = 0.90$. In other words, the values of r do not represent equal distance units along a linear scale. However, as will be pointed out later, r values can be converted to a linear scale (Section 9.6). As an index number, the correlation coefficient does not represent a proportion. For example, $r = 0.50$ does not mean that half of the time the two variables will be related or that half of the variation of one variable can be accounted for by the other variable.

The correlation coefficient is always relative to the conditions under which it was obtained and should be interpreted with those conditions in mind. For example, r depends on the type of measurement used and may vary depending on what measures have been used, how these were used, and by the variability of the obtained values. If the range of scores on one or both of the variables is restricted, then the magnitude of r may be reduced. For example, if placement test scores are being correlated with job performance where people are hired only if they score higher than some cutoff value, then the range of placement scores is being restricted; hence, r will probably be low. On the other hand, if the extreme low and high pairs of values are selected, then the size of r will be spuriously inflated. If the data tend to be curvilinearly related, then r would underestimate the true relationship between those two variables. For curvilinearly related data, eta (η), or the correlation ratio, should be used (*see* Section 6.8).

The correlation coefficient merely indicates the degree of linear relationship between two variables. It does not tell us anything about the causal relationship between the two variables (i.e., why the relationship exists). Although the correlation coefficient is very useful in causal relationship studies, certain conditions must be present before one can say that X caused Y. First, the researchers must be sure that X preceded Y in time. Often in correlational studies, measures on both variables are obtained at the same time; thus it is difficult to determine which came first. For example, in personnel studies, a high r might be found between self-concept and job performance. Rather than jumping to the conclusion that only people with high self-concept should be employed, researchers should consider the possibility that people with high job performance gained a high self-concept while working on the job. Thus, which caused which? However, for chemical studies in which variation of X is controlled to see how this variation correlates with changes in Y, a cause-effect relationship can be inferred if other factors are controlled. This is the second factor in determining cause-effect relationships: the control of other factors that might cause or influence Y. As is true in any area of study, many factors could influence an event, and all of these factors would have to be ruled out in a tightly controlled study before definite cause-effect relationships could be concluded.

6.3 Regression

The correlation coefficient is very useful in exploratory research in which it is desired to determine which variables are related so that these relationships can be further researched. However, if a relationship is found between the variables, then that relationship can be used as a basis for generating an equation that predicts one variable from the other. The presence of a correlation between two variables X and Y implies that from knowledge of X, we know something about Y because these variables would vary together. The accuracy of that knowledge is directly related to the degree of correlation between the two variables. If $r = 1.00$, then perfect prediction of one variable is obtained from the other, and as r approaches 0, knowledge of one variable contributes no information about the other variable. In general, as the correlation increases, the accuracy of prediction also increases, and the error of prediction becomes smaller.

Correlation and regression are closely related, but it is not necessary to compute r to run a regression analysis. However, the calculation of both is similar and is based upon the same sums of squares, so computing both is often useful because the correlation coefficient could support the interpretation of regression. Whereas r merely describes the degree of relationship, regression deals with questions such as the following:

- How much does one variable change when the predictor variable changes a given amount?
- How accurately can the variable be predicted?
- If several predictor variables are found, what is the relative contribution of each?

In regression analysis, the variable that needs to be predicted is usually designated the Y variable, and the predictor variable is designated as X. Thus, Y can be predicted from X. Also, the predicted variable is usually designated $\hat{Y}$, called "tilda Y" or "hat Y". Usually, statisticians are mainly interested in predicting Y from X. However, two regression equations, one for predicting Y from X and the other for predicting X from Y, can be computed. The predicted variable is sometimes called the *dependent variable* and the predictor variable as the *independent variable*. For example, in a toxicological study, the percent of insects killed could be thought of as the Y variable (dependent) and the concentration of a drug as the X variable (independent) because the concentration of the drug is varied by the researcher. For the example of two colorimetric methods in Section 6.2.1, either method could be designated the Y variable. Thus, these terms would not be applicable for that study.

Linear regression involves dealing with a straight line relationship. In mathematics, the term "Y is a function of X" is used, but in statistics the term *regression* is used. The general definition of a straight line is

$$\hat{Y} = bX + a \tag{6.6}$$

where $\hat{Y}$ is the predicted value, X is the predictor variable, and b and a are computed constants. The regression equation is stated in words as follows: The predicted Y value is equal to b times the value of X plus a. The b term is the slope of the line and can be expressed as the ratio of the distance along the vertical axis to the distance along the horizontal axis (i.e., the ratio of how much Y increases for each increase in one unit of X). The a value positions the regression line on the chart and its value is the point at which the regression line intersects the Y axis. Examples of straight line equations and their plotted regression lines are presented in Figure 6.4.

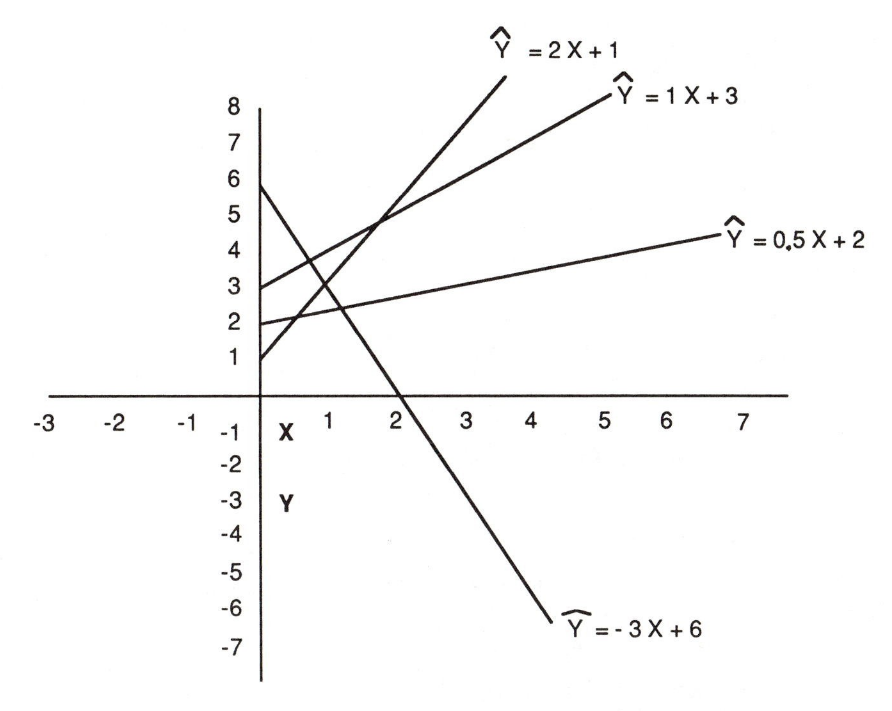

Figure 6.4. Bivariate scatterplots showing various functions of $\overline{Y} = bX + a$.

When using the regression equation for any study, the b and a values have to be computed and are known as regression coefficients. The derivation of the regression coefficients involves a good deal of elegant theory, and Guilford (*6*), McNemar (*7*), and Snedecor and Cochran (*8*) gave detailed expositions of the derivation of these formulas. For our purposes, regression coefficients are computed so that the sum of squares of the deviation of the actual values from the straight regression line is the minimum possible value. The regression equations take into account the fact that X and Y are often expressed in different scale units. The b coefficients are different because each b is equal to r_{xy} times the ratio of the standard deviation of the predicted variable to the standard deviation of the predictor variable. This relationship can be noted from equations 6.7 and 6.9. The slope for the regression line for predicting Y from X (b_{yx}) is

$$b_{yx} = r_{xy}(S_y/S_x) \tag{6.7}$$

where r_{xy} is the correlation coefficient for X and Y, S_y is the standard deviation of Y values, and S_x is the standard deviation of X values. If r_{xy} and the standard deviations have not been computed, then

$$b_{yx} = \frac{n \sum XY - (\sum X)(\sum Y)}{n \sum X^2 - (\sum X)^2} \tag{6.8}$$

The slope for the regression coefficient for predicting X from Y (b_{xy}) is

$$b_{xy} = r_{xy}(S_x/S_y) \tag{6.9}$$

where the terms are identical to those in equation 6.7. From raw scores, the equation is

$$b_{xy} = \frac{n \sum XY - (\sum X)(\sum Y)}{n \sum Y^2 - (\sum Y)^2} \tag{6.10}$$

The term b_{yx} is read as "the b coefficient for predicting Y from X", whereas the term b_{xy} is read as "the b coefficient for predicting X from Y". Equations 6.7 and 6.9 show that when the standard deviations are different for the X values compared to the Y values, the b regression coefficients will also be different. If both b coefficients are computed, $b_{yx}b_{xy} = r_{xy}^2$. By taking into consideration rounding errors, this expression provides a computational check for the b coefficients. However, usually only one b coefficient is needed because in the majority of studies, the goal is predicting Y from X.

The a value in the regression equation is a constant and must be included so that the mean of the predicted values will equal the mean of the observed values. As with the b coefficients, two equations can be used to compute for a: one for predicting Y from X, and the other for predicting X from Y. Equation 6.11 is used to calculate the a value for predicting Y from X, and equation 6.12 is used to calculate the a value for predicting X from Y.

$$a_{yx} = M_y - b_{yx}(M_x) \tag{6.11}$$

$$a_{xy} = M_x - b_{xy}(M_y) \tag{6.12}$$

where M_y is the mean of the Y variables, M_x is the mean of the X variables, b_{yx} is the b value for predicting Y from X, and b_{xy} is the b value for predicting X from Y.

Suppose a study was done to determine the linear regression of the percentage nickel content and the toughness of a certain type of alloy steel. For this study, the dependent variable Y is the toughness of the alloy, and the independent variable X is the percentage nickel. In this study, only the nickel content is varied. Table 6.2 presents the 11 pairs of values for X and Y, and the necessary sum of squared values and cross products. Using equation 6.4, $r = 0.965$. The regression coefficients presented in Table 6.2 have been computed both ways by using equations 6.7–6.10 and are equivalent considering rounding errors. By using the b and a values obtained from equations

Table 6.2. Data for Alloy Steel–Nickel Content Regression Equation

Sample	Nickel Content (X)	Steel Hardness (Y)	X	Y	XY
1	4.0	20	16.00	400	80.0
2	3.8	22	14.44	484	83.6
3	3.6	20	12.96	400	72.0
4	3.4	18	11.56	324	61.2
5	3.2	18	10.24	324	57.6
6	3.0	17	9.0	289	51.0
7	2.8	14	7.84	196	39.2
8	2.6	15	6.76	225	39.0
9	2.4	12	5.76	144	28.8
10	2.2	10	4.84	100	22.0
11	2.0	10	4.00	100	20.0
	33.0	176	103.40	2986	554.4
Mean	3.0	16			
S.D.	0.66	4.12			

$$b_{xy} = 0.965\left(\frac{0.66}{4.12}\right) = 0.155 \qquad b_{yx} = 0.965\left(\frac{4.12}{0.66}\right) = 6.024$$

$$b_{xy} = \frac{[11(554.4)] - [(33)(176)]}{[11(2986)] - (176)^2} = 0.155 \qquad b_{yx} = \frac{[11(554.4)] - [(33)(176)]}{[11(103.4)] - (33)^2} = 6.00$$

$$a_{xy} = 3.0 - 0.155(16) = 0.52 \qquad a_{yx} = 16 - 6.000(3.0) = -2.00$$

6.8 and 6.10 and computed and presented in Table 6.2, the complete regression equation for predicting Y from X is

$$\hat{Y} = 6.000X + (-2.000)$$

For predicting X from Y, the regression equation would be

$$\hat{X} = 0.155Y + 0.52$$

For $X = 4$, $\hat{Y}$ would be predicted to be

$$\hat{Y} = 6.000(4) - 2.000 = 22.00$$

To plot the regression line, high and low values of X would be taken, the $\hat{Y}$ values would be computed for each, and these values would be plotted on the bivariate scatterplot as shown in Figure 6.5. The same procedure could be used for predicting X from Y. However, because the main purpose is to predict Y, only this regression line will be plotted. The ***regression line*** is the best fitting straight line for the regression of Y on X and is the line from which the sum of the squared deviations of where the actual values fall and where they are predicted to fall is at a minimum. As can be noted from Figure 6.5, because there is a high degree of relationship ($r = 0.965$), there is a high degree of accuracy of prediction. The actual error of prediction can be estimated and will be presented in Section 6.4. For the purposes of this section, the regression line intersects the Y axis at the value of about 8, not -2.000, which is the computed a coefficient. The reason for this discrepancy is that neither the X nor the Y scale starts at a zero point. If this was the case, then the regression line would have intersected the Y axis at $Y = -2.000$. The regression equation shows that $\hat{Y}$ must be 6 times the X value -2.000 to position the line at the appropriate height on the chart.

The regression line can be extended to predict Y values for other values of X that were not included. For example, for $X = 5$, the predicted Y value would be 28.00. However, researchers have to be cautious in overextending such predictions. The limits within which the factors operate must be considered. For example, such predictions are possible only with the assumption that the relationship between X and Y will remain linear for higher and lower values of X. A safer estimate would be for predicting Y values for other X values between the limits of those studied (e.g., 2.5 or 3.5).

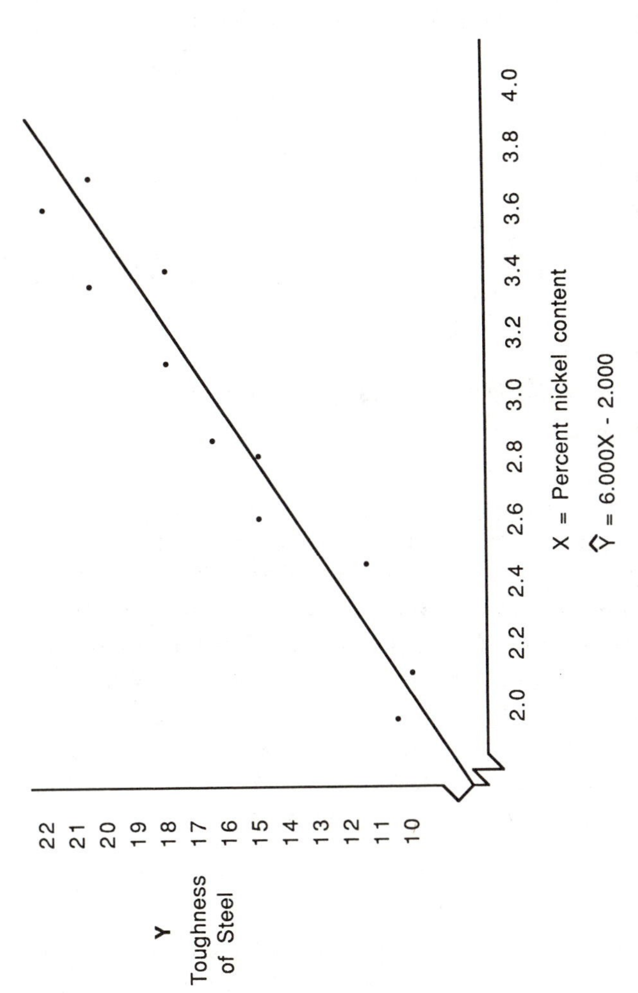

Figure 6.5. Scatterplot for alloy–nickel regression analysis.

One more use of the regression equation is fairly common and is based on the fact that the equation can be solved for either of its unknown values (e.g., Y or X). Suppose for the regression equation shown in Figure 6.5, rather than predicting Y for a given X, the question is what X value would result in $Y = 15$. This question could be answered by entering the Y value in the equation and solving for X.

$$15 = 6.000(X) - 2.000$$

$$6.000(X) = 15 - 2.000$$

$$X = 2.17$$

Thus, if a toughness of 15 is desired, the nickel content would have to be 2.17%.

6.4 Standard Error of Estimate and Measurement

If $r_{xy} = 1.00$, a perfect relationship exists between the X and Y variables, and no errors would occur in predicting one variable from the other because of the perfect covariation between the two variables. As r_{xy} approaches 0, the estimation of one variable from the other will vary from the actual observed values of the estimated variable, and this difference is known as the *standard error of estimate.* When Y is predicted from X, the variance of the actual values of Y from the predicted values of Y, or $(\hat{Y} - Y)$, is denoted as ${S_{yx}}^2$. The square root of this variance is known as the standard error of estimation (SE_{yx}) and may be computed by

$$SE_{yx} = S_y\sqrt{1 - r_{xy}^2} \qquad (6.13)$$

for predicting Y from X. For predicting X from Y, the formula is

$$SE_{xy} = S_x\sqrt{1 - r_{xy}^2} \qquad (6.14)$$

where SE_{yx} and SE_{xy} are the standard errors of estimation, respectively, for predicting Y from X and for predicting X from Y; S_y and S_x are the respective standard deviations for the Y and X values; and r_{xy} is the correlation between these two variables. Different symbols are used for the standard error of estimate in different statistic textbooks (e.g., S_{yx}, SE_{yx}, and SE_{yx}). Throughout this book, the letters SE for standard error and S for standard deviation will be used to avoid confusion and to distinguish between these two statistics.

For the data in Table 6.2, the standard error for predicting Y from X would be

$$S_{yx} = 4.12\sqrt{1 - 0.965^2} = 1.08$$

The standard error is interpreted in a similar way as the standard deviation discussed in Chapter 5. If no obvious inequalities of variances within columns or rows occur, which is referred to as homoscedasticity, then about two-thirds of the observed values are expected to be within the limits of ± 1.08 values from $\hat{Y}$.

When $r_{xy} = 0$, the best estimate of Y for any given X would be the mean of the Y values, and the standard error of estimate would be equal to the standard deviation of the Y values. Thus, no reduction occurs in the variability of the predicted Y values from the actual variability of the Y values, and with $r = 0$, no efficiency is found in trying to estimate Y for any given X.

For sample sizes less than $n = 50$, a correction should be used for possible sampling bias due to the small sample size. This correction can be done by multiplying the standard error by $\sqrt{n/(n - 2)}$, or to incorporate this correction into equation 6.13 as

$$S_{yx} = S_y \sqrt{(1 - r_{xy}^2)\frac{n}{n - 2}} \qquad (6.15)$$

For the example,

$$S_{yx} = 4.12 \sqrt{(1 - 0.965^2)\left(\frac{11}{9}\right)} = 1.19$$

Thus, the standard error of estimation adjusted for possible bias due to a small sample size is slightly larger than the standard error without the adjustment (1.19 compared to 1.08).

An index of *forecasting efficiency* indicates the percentage reduction in errors of estimation due to the correlation between the two variables. This forecasting efficiency (E) is defined as

$$E = 100(1 - \sqrt{1 - r^2}) \qquad (6.16)$$

For the example in Table 6.2,

$$E = 100(1 - \sqrt{1 - 0.965^2}) = 73.7$$

Thus, the margin or error in predicting Y from X with $r_{xy} = 0.965$ is only 73.7% as large as the margin of error in predicting Y with no knowledge of X.

As a quick reference in showing the effect of r_{xy} for the various indexes, Table 6.3 shows the value of the coefficient of determination (D) and the index of forecasting efficiency (E) for various values of r from 0.0 to 1.00. As can be noted from Table 6.3, the correlation coefficient has to be relatively high in order for both of these indices to be of practical value. For example, for $r = 0.30$, only 9% overlap occurs, and the reduction of the error of forecasting is only 4.6%.

Table 6.3. Coefficients of Determination and Forecasting Efficiency for Various Correlation Coefficients

r	*D*	*E*
0.00	0.00	0.0
0.05	0.00	0.1
0.10	1.00	0.5
0.15	2.25	1.1
0.20	4.00	2.0
0.25	6.25	3.2
0.30	9.00	4.6
0.35	12.25	6.3
0.40	16.00	8.3
0.45	20.25	10.7
0.50	25.00	13.4
0.55	30.25	16.5
0.60	36.00	20.0
0.65	42.25	24.0
0.70	49.00	28.6
0.75	56.25	33.9
0.80	64.00	40.0
0.85	72.25	47.3
0.90	81.00	56.4
0.95	90.25	68.8
0.98	96.00	80.1
0.99	98.00	85.9
0.995	99.00	90.0
0.999	99.80	95.5

For $r = 0.60$, only 36% overlap or shared variance and a 20% reduction in error of estimating occur. As will be discussed in Chapter 8, often a statistic may have theoretical consequences but have little practical use (e.g., forecasting efficiency).

6.5 Multiple Linear Regression

The concepts of correlation and regression between two variables can be extended to situations in which more than one single independent or predictor variable occurs. The traditional view of studying the relationship between one independent variable and one dependent variable is limited, especially when studying phenomena that are complex. By "complex" is meant that a phenomenon has many sources of variation and possible causes of variation. Multiple linear regression is a very useful analysis to explain or account for the variance in a dependent variable by estimating the contributions to that variance from two or more independent variables. The basic concepts are the same as for the correlation and regression between two variables, and the results are interpreted in a similar way. Multiple regression and correlation analysis has four general purposes:

- to derive an equation that provides the best estimate of the dependent variable from combinations of several independent variables,
- to determine the error of estimate involved in using this multiple regression equation,
- to determine which independent variables in a subset give the best linear prediction equation, and
- to determine the proportion of variance in the dependent variable that is accounted for or explained by each of the independent variables and for all of them combined.

For example, the output for a chemical product may involve a number of factors such as several reactants; different inorganic materials; and different concentrations in the reaction mixture, temperature, pressure, and reaction time.

Two basic principles apply to multiple regression. The multiple correlation increases as the magnitude of the correlations between the dependent and each independent variable increases, and the multiple correlation increases as the magnitude of the intercorrelations among the variables decreases. These principles can be shown by Venn diagrams as in Figure 6.6. In Figure 6.6A, zero correlation occurs between the two predictor variables. Thus, there is no overlap between the two independent variables, and in this special case, the total overlap or variance accounted for in the predicted variable is equal to 0.32, or $R^2 = r_{12}{}^2 + r_{13}{}^2 = 0.40^2 + 0.40^2 = 0.32$. As shown, the total variance of Y accounted for by the two independent predictor variables is 32%. When the two predictor variables are correlated, then they also overlap each other as well as with the predicted variable. Thus, the total variance accounted for in the predicted variable is less than the sum of the two variances accounted for added together. Actually, the total predictor variance accounted for would be around 23% for Figure 6.6B.

Because the calculations become complex and extensive, only the basic concepts will be presented because multiple regression is usually run with one of the computer statistical packages such as SPSS, BMDP, or SAS. Additional information on multiple regression can be found in the references listed for Chapter 16. Chapter 16 also lists computer packages that would include programs for computing multiple correlation and regression. The multiple regression for predicting Y from a series of X variables is given by the prediction equation

$$Y = b_1X_1 + b_2X_2 + \ldots + b_kX_k + a \tag{6.17}$$

where b is the regression coefficient computed for each X variable, X is each predictor variable, k is the total number of variables in the series, and a is computed by

$$a = \overline{X}_y - b_1\overline{X}_1 - b_2\overline{X}_2 - \ldots - b_k\overline{X}_k \tag{6.18}$$

Thus, the basic concept of estimation is the same as for just two variables. Computer programs usually provide values for the following variables: the coefficient of multiple correlation (R); the proportion of Y variance predicted from the regression equation (R^2, also known as the coefficient of multiple determination; the intercept on the Y axis (a); the partial regression coefficients for raw values ($b_1 \ldots b_k$); the standard partial regression coefficients ($\beta_1 \ldots \beta_k$); the standard error of the estimate ($S_{1.234}$); the F ratio (F) for determining if a variable is adding a significant change in R^2; and the t ratio (t) for determining the significance of each standardized regular

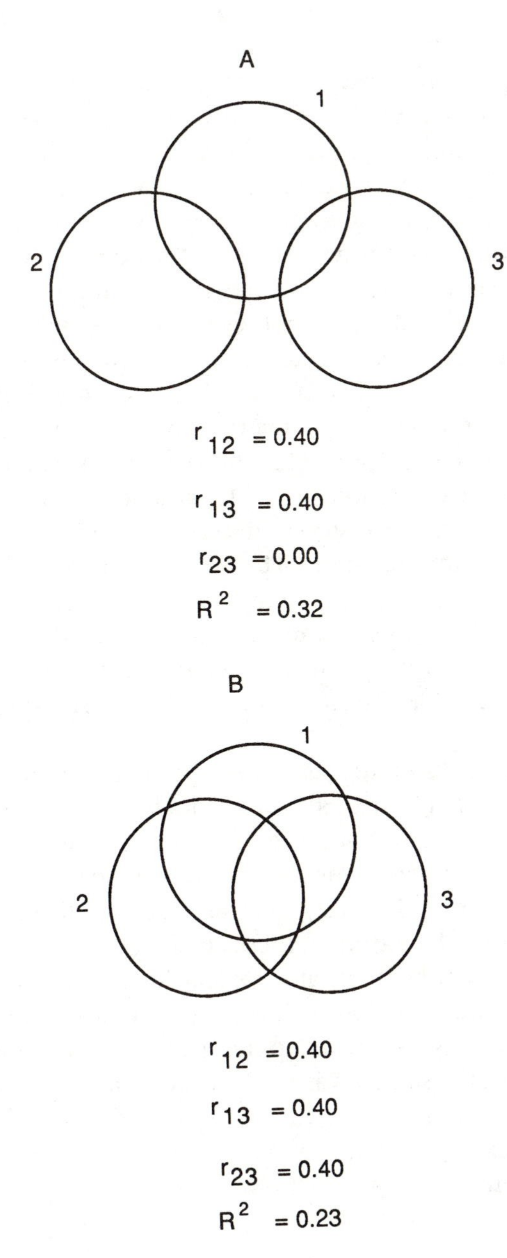

Figure 6.6. Overlap of predictor variables in multiple regression.

coefficient. Also provided are various descriptive statistics such as the mean, standard deviation, standard error of the beta coefficients, scatterplots, and the correlation matrix. These statistics show all of the intercorrelations between each pair of variables.

Most statistical packages of computer programs have several options for running multiple regression. If the number of variables is small or if a preliminary analysis is needed to get an idea as to the contribution of each independent variable, a multiple regression equation using all of the variables could be run. Possibly the most common procedure is to run a stepwise multiple linear regression, which enters and removes variables from the equation in a stepwise manner on the basis of selected criteria. Two methods are used to do this. The first method is *forward stepping*, which begins with no independent variables, then adds the one independent variable that has the highest correlation with the dependent variable, then adds the next highest correlated variable, and so on. At each step, each independent variable is tested to determine if it is adding a significant contribution to the equation. The second method, *backward stepping*, begins with all independent variables and then removes those that are not making a significant contribution to the multiple regression equation. Most programs also have a provision to enter desired independent variables regardless of their contribution to the equation.

Although multiple regression is a powerful analysis, precautions must be followed in its use as with any analysis. The ease of computer-run analyses may tempt some researchers to run such analyses indiscriminately and search for something significant. Possibly such use is acceptable for preliminary analyses, but application of multiple regression should be based on theoretical considerations, knowledge of the areas of application, and logical judgment. Practical consideration should be considered; thus, researchers should search for the best combination of as few independent variables as possible. Finally, statisticians must also be aware of how combinations of independent variables might be influenced by these factors interacting at various levels. Thus, extrapolation from researched levels to other levels has to be done with caution.

6.6 Partial Correlation

Closely related to multiple linear regression and correlation is partial correlation. In fact, multiple correlation uses partial correlation to enter and remove variables in the stepwise procedure. *Partial correlation* is used to determine the correlation between two variables Y and X while eliminating the effects of a third or a number of other variables upon both the variables being correlated.

When only one variable is held constant, the first-order partial coefficient is used. The general equation is

$$r_{12.3} = \frac{r_{12} - (r_{13})(r_{23})}{\sqrt{(1 - r_{13}^2)(1 - r_{23}^2)}} \tag{6.19}$$

where $r_{12.3}$ is the correlation between variables 1 and 2 with the effect of variable 3 removed, and r_{12}, r_{13}, and r_{23} are the correlation coefficients between pairs of the three variables.

As an example, the purity of a manufactured chemical may depend on a number of factors such as the proportion of the active ingredient and the time of reaction. The correlation between purity, X_1, and proportion of active ingredient, X_2, would also include the effect of the time of reaction, X_3, if this was varied for different batches. The partial correlation coefficient enables statisticians to partial out the effect of time so that the correlation between X_1 and X_2 would not be influenced by time of reaction. If the correlation between purity and amount of active ingredient is $r_{12} = 0.60$, the correlation between purity and time is $r_{13} = 0.70$, and the correlation between amount of active ingredient and time is $r_{23} = 0.40$, then

$$r_{12.3} = \frac{0.60 - (0.70)(0.40)}{\sqrt{(1 - 0.70^2)(1 - 0.40^2)}} = 0.49$$

Thus, if reaction time is taken into account, the correlation is considerably less than 0.60, which is the original r_{12} between purity and proportion of active ingredient. The partial correlation coefficient for purity and time could also be determined by holding constant the proportion of active ingredient.

$$r_{13.2} = \frac{0.70 - (0.60)(0.40)}{\sqrt{(1 - 0.60^2)(1 - 0.40^2)}} = 0.63$$

Thus, for this example, purity is more closely related to reaction time ($r_{13.2} = 0.63$) than to the proportion of active ingredient ($r_{12.3} = 0.49$).

The partial correlation can be extended to control for two or more factors. The equation for controlling the effects of two additional independent variables is

$$r_{12.34} = \frac{r_{12.3} - (r_{14.3})(r_{24.3})}{\sqrt{(1 - r_{14.3}^2)(1 - r_{24.3}^2)}} \tag{6.20}$$

where $r_{12.34}$ is the second-order partial correlation coefficient and is based on the three first-order partial correlation coefficients.

6.7 *The Phi Coefficient*

When the two variables to be correlated represent genuine dichotomies (e.g., yes/no responses), the phi coefficient (ϕ) can be used to determine the degree of relationship between the two variables. For example, ϕ can be used whenever data such as pass/fail, male/female, living/dead, or heavy/light designations need to be correlated. Phi is also known as the *fourfold point correlation*, because this setup is used to calculate it. As an example, suppose sex and pass/fail rates are to be correlated. The number of units passing and failing an inspection criterion would be tabulated in Figure 6.7 for both men and women. Hypothetical data are shown in the fourfold table.

The equation for ϕ from frequencies tabulated into the fourfold table is

$$\phi = \frac{(B)(C) - (A)(D)}{\sqrt{(A + B)(C + D)(A + C)(B + D)}} \tag{6.21}$$

For the data in the fourfold table,

$$\phi = \frac{(14)(43) - (36)(7)}{\sqrt{(36 + 14)(43 + 7)(36 + 43)(14 + 7)}} = 0.17$$

	Pass	Fail	
Men	A 36	B 14	50
Women	C 43	D 7	50
	79	21	

Figure 6.7. Hypothetical data for ϕ.

The value of ϕ is interpreted in a similar way as the Pearson correlation coefficient. In fact, ϕ is the Pearson r between two dichotomized variables. However, ϕ nearly always underestimates the Pearson r, and this discrepancy tends to increase as the frequencies depart from a 50:50 tally for the margin frequencies. Also, if any cell is equal to 0, ϕ is indeterminate.

If there are more than two categories for either or both variables, then another measure of correlation, the *contingency coefficient* (C) could be used. However, use of this coefficient would be limited in chemical research. The formula for C is based on the chi-square statistic, which will be covered in Chapter 10.

$$C = \sqrt{\frac{\chi^2}{\chi^2 + n}} \qquad (6.22)$$

where χ^2 is the chi-square for the data, and n is the total number of cases included. The value of C is also interpreted like a Pearson r.

6.8 Spearman Rank-Order Correlation

Often, one or both of the variables is measured by an ordinal scale, for example, in product ratings, a ranking of taste and appearance. can be used to determine if these two variables are related. The rankings of both taste and appearance may be obtained for a group of people who are trying a new product. If they are related, then one variable, for example, appearance, influences the other variable, taste. If a high degree of relationship is found, then both factors have to be taken into consideration for successful marketing. The equation for the Spearman rho (ρ) is

$$\rho = 1 - \frac{6(\sum D^2)}{n(n^2 - 1)} \qquad (6.23)$$

where 1 and 6 are constants, ΣD^2 is the sum of the squared differences in rank order, and n is the number of cases.

Rho can be used whenever one or both of the variables is measured by an ordinal scale, or in situations in which one or both of the variables can be changed from interval to rank order. Often when this happens, two or more cases will have the same value and result in tied rankings. The average rank is then assigned to the involved cases. For example, if there were three cases of $X = 30$, and these cases were the third, fourth, and fifth cases in order from high to low, then all three would be given the rank of 4.

For our hypothetical taste–appearance example, suppose the rankings for 15 consumers are presented as in Table 6.4. Each scale has been changed from a rating value to a rank order of the values shown in columns for taste rank and appearance rank.

$$\rho = 1 - \frac{6(53.5)}{15(15^2 - 1)} = 0.904$$

The computed $\rho = 0.904$, indicates a high degree of relationship between these two variables. Possibly, appearance influences taste, or taste influences appearance. The result would indicate that for consumer acceptance, both taste and appearance should be considered.

Table 6.4. Spearman Rho for Rankings of Taste and Appearance Data

Consumer	Taste Rating	Appearance Rating	Taste Rank	Appearance Rank	D	D^2
1	47	7	10	8.5	1.5	2.25
2	52	8	7.5	6	1.5	2.25
3	35	4	15	12	3	9
4	60	9	3	3	0	0
5	49	6	9	10	1	1
6	52	8	7.5	6	1.5	2.25
7	63	10	1	1	0	0
8	46	5	11	11	0	0
9	54	9	5	3	2	4
10	39	3	14	13	1	1
11	58	9	4	3	1	1
12	44	2	12	14.5	6.25	6.25
13	53	7	6	8.5	6.25	6.25
14	61	8	2	6	16	16
15	42	2	13	14.5	2.25	2.25
						53.50

$$\rho = 1 - \frac{6(53.5)}{15(15^2 - 1)} = 0.904$$

6.9 The Correlation Ratio

All of the coefficients of relationship presented so far have been based on the assumption of a linear or straight line relationship. However, often in real situations, two variables are related in a nonlinear way. For example, temperature and reaction time may be linearly related up to a certain point, but then the trend may not continue. For nonlinear relationships, the correlation ratio, eta (η), is the index of correlation. The equation is

$$\eta = \sqrt{1 - \frac{\sum (Y - \overline{Y}_c)^2}{\sum (Y - \overline{Y}_T)^2}} \qquad (6.24)$$

where Y is the value of the predicted or dependent variable for each case, $\overline{Y}_c$ is the mean of the Y values for cases within each category, and $\overline{Y}_T$ is the mean of all Y values. To calculate η, categories must first be set up for the predictor or X values. This setup is usually done by inspection and depends upon the range of these values. Usually, at least 7 but no more than 12 to 14 categories are formed. If there are fewer than seven categories, η may not be sensitive to the data (6, 7).

The computation procedure is more complex than for other correlation coefficients, and the procedure is outlined here for the data in Table 6.5. For this example, Y is the yield of a chemical and is the predicted value, and X is the time factor.

1. Categorize the values on the predictor variable (X) into equal intervals in the same way as a grouped frequency distribution. For this example, $i = 3$.
2. Within each category of X, compute the mean for the Y values falling in that category ($\overline{Y}_c$).
3. Compute the mean for all Y values ($\overline{Y}_T$).
4. For each subject or case within each category, obtain the difference between the mean of the Y values for that category and each Y value ($Y - \overline{Y}_c$).
5. Square each of these differences $(Y - \overline{Y}_c)^2$.
6. Obtain the difference between the total mean of the Y values and each Y value, ($Y - \overline{Y}_T$).

7. Square each of these differences, $(Y - \bar{Y}_T)^2$.
8. Sum the values obtained in steps 5 and 7.
9. Compute η.

For the data in Table 6.5, step 5 gives $\Sigma(Y - \bar{Y}_c)^2 = 38$, and for step 7, $\Sigma(Y - \bar{Y}_T)^2 = 1724$. Then,

$$\eta = \sqrt{1 - \frac{38}{1724}} = 0.989$$

The value of η is interpreted exactly the same as the Pearson r.

Table 6.5. Data for Calculating Eta

Subject	X	Y	$\bar{Y}_c$	$(Y - \bar{Y}_c)$	$(Y - \bar{Y}_c)^2$	$(Y - \bar{Y}_T)$	$(Y - \bar{Y}_T)^2$
1	5	4	5	−1	1	−14	196
2	5	5	5	0	0	−13	169
3	7	6	5	1	1	−12	144
4	8	6	7	−1	1	−12	144
5	10	8	7	1	1	−10	100
6	11	10	11	−1	1	−8	64
7	12	11	11	0	0	−7	49
8	13	12	11	1	1	−6	36
9	14	16	19	−3	9	−2	4
10	14	18	19	−1	1	0	0
11	15	20	19	1	1	2	4
12	16	22	19	3	9	4	16
13	18	26	27	−1	1	8	64
14	19	28	27	1	1	10	100
15	20	28	30	−2	4	10	100
16	21	32	30	2	4	14	196
17	21	30	30	0	0	12	144
18	23	27	26	1	1	9	81
19	24	26	26	0	0	8	64
20	25	25	26	−1	1	7	49
				0	38	0	1724

NOTE: The spaces between rows define the different categories of the predictor variable (X).

However, η is always positive because it is computed from squared values. The trend of the relationship has to be perceived from inspection of the scatterplot because the sign of η would not be meaningful even if it was available. The value of η indicates to researchers what they most want to know about the relationship, which is the closeness of the relationship or the goodness of fit of the data to a given trend. As with r^2, η^2 gives the proportion of variance in the Y variable that overlaps, or is common with, the X variable.

When correlating two variables by using η, two η values can be computed, depending on which variable is categorized. With r, only one value can occur. The value of η will almost always be larger than r. The reason for this is that in setting up the categories of X, η will be more sensitive to the complex up-and-down patterns of Y, whereas r assumes that one regression line will be the best fit of the data and thus is not as sensitive to these particular fluctuations. Finally, the size of η depends somewhat on the number of categories set up for the X variable. If the number of categories is too small, then η might underestimate the true relationship because the up-and-down patterns of Y would not be reflected by wide categories. However, if too many categories occur, then the mean value of Y in each class becomes less stable. As the mean values fluctuate, η tends to be inflated. If there are as many categories used as there are Y values so that only one Y value occurs per category, then the sum of the variances within each category would be equal to the total variance, and η would then be equal to 1.00.

The standard error of η (SE_η) is given by

$$SE_\eta = \frac{1 - \eta^2}{\sqrt{n - 1}} \qquad (6.25)$$

For the data in Table 6.5, the standard error would be

$$SE_\eta = \frac{1 - 0.989^2}{\sqrt{20 - 1}} = 0.005$$

Thus, the amount of relationship indicated by η is probably very close to the actual relationship between these two variables in the population because the standard error is very small.

The standard error of estimate for η can be computed by using the concept of the variance of the actual Y values from the mean of the Y values within each category of X. In Table 6.5, this variance is given in the column $(Y - \bar{Y}_c)^2$, which gives the sums of the squared deviations that total 38. The variance would then be $38 \div (n - 1) = 2.00$. The square root of this would be 1.41, which is the standard error of estimate for predicting Y from X.

Often, statisticians want to determine if a departure from a linear relationship occurs between two variables. Several tests of this departure are available, but one of the most frequently used is the F ratio (F):

$$F = \frac{(\eta^2 - r^2)(n - k)}{(1 - \eta^2)(k - 2)} \tag{6.26}$$

where η and r are, respectively, eta and the Pearson r computed on the same data; n is the number of cases included in the sample; and k is the number of categories of X. For the example in Table 6.5, $r = 0.92$, $\eta = 0.989$, and $k = 7$.

$$F = \frac{(0.989^2 - 0.92^2)(20 - 7)}{(1 - 0.9892)(7 - 2)} = 15.6$$

Interpretation of this F ratio will be discussed in Chapter 11. Briefly, the computed F ratio is compared to the expected F ratio obtained for η calculated by using random values of X and Y for $k - 2$ and $n - k$ degrees of freedom. By looking up the expected F value with 5 and 13 degrees of freedom, the table value separating chance from significant F ratios is 3.02 at the 0.05 level of significance. Thus, $F = 15.6$ indicates a significant nonlinear trend in the data, and η should be used to show the degree of relationship between these two variables.

6.10 Reliability and Prediction

The concept of reliability or consistency of measures used in research studies was mentioned briefly in Chapter 2, and several references were given in Section 2.4 that cover the theoretical issues involved. Generally in the physical sciences, the measures used are accurate and highly reliable. However, personnel and production studies often have to use measures that are less reliable than physical measurements (e.g., employee satisfaction, consumer preference, or personal aptitude). For such situations when less reliable measures have to be used, statistical analyses including the correlation coefficient are affected. Lack of reliability of the measures used for either one or both variables being correlated tend to lower or attenuate the true, or theoretical, correlation observed between the two variables compared to the correlation that might be obtained if perfectly reliable measures were available. A correction can be applied to estimate what the true correlation might be between these two variables if perfect measures were available (i.e., perfect in the sense that they are free from errors of measurement). This correction of an observed r_{xy} is the correction for attenuation and can be computed by using the equation

$$r_{Txy} = \frac{r_{xy}}{\sqrt{r_{xx} r_{yy}}} \quad (6.27)$$

where r_{Txy} is the hypothetical true correlation between variables X and Y if perfect measures were available for both X and Y; r_{xy} is the obtained correlation between X and Y for a given sample; and r_{xx} and r_{yy} are the reliability coefficients for the measures X and Y, respectively. Another way to conceptualize this equation is to consider that the theoretical maximum correlation between any two imperfect measures is the square root of the product of their reliabilities.

As an example, suppose the observed correlation between job aptitude and job performance obtained for a group of workers was computed to be $r_{xy} = 0.45$. If the reliability of the job aptitude test was $r_{xx} = 0.80$, and the reliability of the measure of job performance was $r_{yy} = 0.60$, then by applying equation 6.27,

$$r_{Txy} = \frac{0.45}{\sqrt{(0.80)(0.60)}} = 0.649$$

The correlation between perfect measures of X and Y would be expected to be 0.65 rather than the obtained correlation of 0.45. The obtained $r_{xy} = 0.45$ may be considered as being attenuated from $r_{Txy} = 0.65$ because of errors of measurement. Another way to consider this attenuated effect is to square these two coefficients to determine the loss in predictive capacity of the attenuated relationship. The squared coefficients for this example are, respectively, 0.2025 and 0.4212. Thus, a loss of about 22% in predictive capacity due to lack of perfect measurements of X and Y occurs.

If highly reliable measures of X and Y were dealt with, such as in the physical sciences where reliability of measures are very high, so that $r_{xx} = 0.95$ and $r_{yy} = 0.98$, then if the obtained correlation coefficient was $r_{xy} = 0.45$, the corrected correlation would be $r_{Txy} = 0.466$. This example indicates little attenuation in the obtained r_{xy} when highly reliable measures are used.

In real situations, statisticians often have to deal with measures, such as aptitude tests, that must be accepted as the best available. Thus, determining the attenuating influence of this measure may have little practical value. However, measures of the criterion variable such as job performance might lack reliability and could be improved. In such situations, statisticians might be interested in determining what the true correlation might be between these two variables when there is correction for attenuation for only the criterion measure. The one-sided correction for attenuation would then be

$$r_{Txy} = \frac{r_{xy}}{\sqrt{r_{yy}}} \tag{6.28}$$

where the terms are defined as in equation 6.27. For the same problem, where $r_{xy} = 0.45$ and the reliability of the criterion measure was $r_{yy} = 0.60$,

$$r_{Txy} = \frac{0.45}{\sqrt{0.60}} = 0.581$$

If the measure of job performance did not have measurement errors, then the true correlation between job aptitude and job performance would theoretically be $r_{Txy} = 0.581$, which was obtained by using the aptitude test available and not correcting for its measurement error.

The correction for attenuation is made on the basis that accurate estimates of the reliability of each variable are obtained. If either or both reliabilities are underestimated, then the correction for attenuation will be too large. If either or both estimates of reliability of the measures are overestimated, the correction for attenuation will be too small. Therefore, the correction for attenuation has to be applied with these considerations in mind. The references listed in Section 2.4 give situations where reliability indices might be too high or too low.

Another effect on the correlation coefficient between two variables is due to lack of reliability of the measures. This effect is limiting in that the maximum possible correlation between two variables X and Y is estimated by

$$r_{\max xy} = \sqrt{r_{xx} r_{yy}} \qquad (6.29)$$

where r_{xx} and r_{yy} are the reliability coefficients of the two measures being correlated. For the job performance problem, the maximum correlation that could be obtained between job aptitude and job performance given the reliabilities of the two measures as $r_{yy} = 0.80$ and $r_{yy} = 0.60$ would be

$$r_{\max xy} = \sqrt{(0.80)(0.60)} = 0.6928$$

This result would be an estimate of the maximum correlation that could be obtained between these two variables by using the measures with the given reliabilities.

For any single measure, the square root of the reliability coefficient is often used as the theoretical upper limit of correlation that measure could have with any other variable. Thus,

$$r_{\max xy} = \sqrt{r_{xx}} \qquad (6.30)$$

gives an indication of the upper limit of the correlation that the measure of X could have with any other variable. For the measure of job aptitude, the maximum correlation that could be obtained between that measure of job aptitude and any other variable would be

$$r_{\max xy} = \sqrt{0.80} = 0.894$$

For the measure of job performance,

$$r_{\max xy} = \sqrt{0.60} = 0.775$$

Thus, the theoretical maximum correlation that could be obtained between job performance and any other variable would be 0.775.

Lack of reliability of a measure also has an influence on the precision of an individual value obtained for any item or person. If a measure had perfect reliability (i.e., free from errors of measurement), then all measures obtained for each item or person would be identical. If the measure is not perfect, then repeated measures will vary from one measure to another. The *standard error of measure-*

ment is defined as the standard deviation of values an individual might be expected to obtain on a large number of measurements taken by the same instrument, assuming that the individual remains unchanged and is not affected by the measures. The standard error of measurement (SE_{meas}) can be calculated by

$$SE_{meas} = S_x\sqrt{1 - r_{xx}} \quad (6.31)$$

where SE_{meas} is the standard error of measurement, S_x is the standard deviation of the X measures, and r_{xx} is the reliability of the measure of X. For the job aptitude test example, if the standard deviation of the distribution of scores for this test is 10 and the reliability is $r_{xx} = 0.80$, then the distribution of scores for a person who had an aptitude score of 60 would have a standard deviation of about 4.5 points on repeated measures on this aptitude test. To be exact, the standard error would be

$$SE_{meas} = 10\sqrt{1 - 0.80} = 4.47$$

This standard error of measurement is interpreted in the same way as a standard deviation using the normal curve distribution. Thus, if the errors are assumed to be randomly and normally distributed, 68% of the repeated measures of this individual are probably going to fall in the interval from 55.5 to 64.5. The standard error of measurement is useful in determining the limits where repeated measures would likely fall for any given individual on that particular measure or test. However, errors of measurement tend to be higher near the middle range of values for that particular instrument and smaller for values that depart from the middle range. Thus, this tendency has to be taken into consideration when interpreting the standard error of measurement. A complete discussion of the standard error of measurement and its interpretation is contained in the references listed in Section 2.4.

References

1. Klompmakers, A. A.; Hendriks, T. *Anal. Biochem.* **1986,** *153,* 80–84.
2. Gavino, V. C. et al. *Anal. Biochem.* **1986,** *152,* 256–261.
3. Robbat, A. Jr. et al. *Anal. Chem.* **1986,** *58,* 2072–2077.
4. Barrett, C. A. *NASA Tech. Memo* **1985,** NASA–TM–87020, E–2491, NASA.15:87020.

5. Arena, J. V.; Rusling, J. F. *Anal. Chem.* **1986,** *58,* 1481–1488.
6. Guilford, J. P. *Fundamental Statistics in Psychology and Education;* McGraw–Hill: New York, 1965.
7. McNemar, Q. *Psychological Statistics;* Wiley: New York, 1969.
8. Snedecor, G. W.; Cochran, W. G. *Statistical Methods;* Iowa State University Press: Ames, IA, 1967.

Problems

1. A company wants to determine the repeatability of measures of lead in an aqueous solution by a testing laboratory. Samples of the aqueous solution were divided into two parts: vial A and vial B. The vials were then sent to the laboratory for analysis in a random order, and the results were then paired as they were measured. The following are the results of the measures of the 15 pairs of samples:

Vial A	Vial B
36	32
37	33
24	27
16	13
12	14
32	31
16	18
27	24
29	27
42	47
11	9
25	22
27	30
21	24
42	45

   How consistently does this laboratory measure the lead content in the aqueous solution, and what is the standard error of measurement?

2. A chemist wants to determine if a relationship exists between the amount of tin extracted from a product and the refluxing time

that the product is boiled with hydrochloric acid. The results of the amount of tin found in terms of milligrams per kilogram for the various refluxing times are as follows:

Refluxing Time (min)	Tin Extracted (mg/kg)
30	54
35	59
40	56
45	59
50	57
55	60
60	56
65	58
70	59
75	60

Is there a relationship between refluxing time and the amount of tin extracted? What is the standard error for estimating the amount of tin extracted?

3. Suppose that a research report shows that the correlation between amount of ammonium chloride and yield of an organic chemical is 0.80. The average amount of ammonium chloride per batch was 14.5% with a standard deviation of 1.5%, and the mean yield of the organic chemical was 160 pounds per batch with a standard deviation of 8 lb. Set up the regression equation for predicting yield of the chemical from amount of ammonium chloride, and then calculate the standard error of estimate for predicting yield. Use this information to predict what the yield would be if the amount of ammonium chloride per batch was set at 16%.

Chapter 7

Principles of Probability

Statistical inference is concerned with two broad areas of statistics: hypothesis testing and estimation. The basic purpose of statistical inferences is to make generalizations about a characteristic of a population from the study of a sample taken from that population. Only generalizations can be made when the entire population is not examined. The uncertainty in the generalization is mainly due to variation from sample to sample.

Statistical inferences are based on the notion of probability. Probability is important because all hypothesis testing and statistical estimation are derived from either theoretical or empirical distributions of how events would happen solely on the basis of random trials. In inferential statistics, statisticians always compare observed results to those that would happen by chance alone. For example, in testing a hypothesis, the conclusion is always made on the basis of what proportion of the time the event could have happened by chance. Estimates should be given in terms of probabilities. As Parzen (*1*) pointed out, probability theory provides the basis for hypothesis testing and estimation in such diverse fields as telephone communications, predictions of genetics, production control, thermal noise in electronic circuits, and the Brownian motion of particles immersed in a gas or liquid.

For statistical applications, the two most important and useful concepts are probability distributions and sampling distributions. These distributions give the probabilities that certain events may happen (e.g., what values may be expected to occur under repeated sampling to establish the frequency of those random events). *Probability distributions* usually refer to distributions of events, whereas sampling distributions refer to the sample-to-sample variation in sta-

1453–0/88/0121$06.00/1

tistics expected from an infinite number of random samples. *Sampling distributions* are basic to statistical inference and can be used, for example, to determine the probability that population values will lie within specified limits. An understanding of probability is essential for both of these conceptual uses. In this chapter, some of the basic principles of probability distributions will be discussed as applied to descriptive statistics.

7.1 Basic Probability Concepts

Probability can be approached from several ways, such as the subjective or personal approach, which nearly everyone has used. For example, we wonder whether or not it will rain tomorrow, if a particular person will have an automobile accident, if our team will win an important game, or if we will win a sweepstakes lottery. Usually, the probability of such events is expressed in general terms such as "it is likely to rain tomorrow", or in terms of percentages such as "there is a 30% chance of rain tomorrow." This same type of reasoning is applied in research studies, but for research, more precision must be obtained for estimation and hypothesis testing. Such precision is provided by using either theoretical or empirically determined probability distributions.

Theoretical determinations of probability are based on the knowledge of the number of possible outcomes and their relative likelihood of happening. For example, nearly all games of chance such as those involving dice, coins, or cards have a definite probability of outcome, and thus have a theoretical basis. For example, the probability (p) of getting a "4" in one throw of a fair die is 1/6, because there are six sides to the die marked "1" through "6." Thus, theoretically, $p(4) = 1/6$, or when a die is thrown, it will end with "4" on the top surface 0.1667 of the time.

Theoretical distributions are supported by empirical evidence or empirically determined distributions, which are based on observing the frequency of occurrence of the event relative to the total number of trials or opportunities for the event to happen. This approach to probability requires that a population or a very large number of events be defined so that the resulting distribution will accurately represent the population and be stable (i.e., provide a consistent probability referent of the population parameter). This concept is known as the *law of large numbers* and states that p, the probability, will approximate the true proportion of likely events in the population more closely as the number of trials increases. Thus, statisticians usually think in terms of probabilities that would occur over an infinite number of trials or what would happen if the trials were repeated a number of times rather than just once.

Probability is simply a measure of the likelihood that a chance event will happen. The classical definition of probability is that if an event can occur in N mutually exclusive and equally likely ways, and if n of those events possess a characteristic E, then the probability of E occurring is the ratio n/N, or the probability that E will occur is equal to the number of ways, n, that the event can occur divided by the total possible events, N. In equation form, this definition is usually written $P(E) = n/N$. For example, the probability of obtaining a "4" on a single toss of a fair die is equal to one out of six possibilities. The probability of getting a "head" when a coin is tossed is 1/2, or half of the time.

The basic definition has two important concepts that are fundamental to all probability:

1. "Mutually exclusive" means that one event has no effect whatsoever on any other event, and
2. "Equally likely" implies that the probability of any event happening is the same for all events.

Coin tosses are a good example of independent events, because the first toss is unlikely to influence the other tosses. "Equally likely" implies that the events happen on a random basis, and hence the importance of random sampling in inferential statistics. This basic definition of probability is used as the basis for many inferential statistics that will be discussed in the next chapter.

Two basic theorems are used extensively in inferential statistics. The first is the *addition theorem*, which states that the probability of occurrence of several mutually exclusive events is the sum of their separate probabilities. This addition theorem is used when the probability of either one event or another event is desired. For example, the probability that a tossed die will be either "1" or "2" is $1/6 + 1/6 = 2/6$, or one-third of the time. The probability that a tossed coin will be either a head or a tail is $1/2 + 1/2 = 1$, which is another point to remember about probability: The sum of the probabilities of all possible outcomes is equal to 1. The addition theorem is useful in research when decisions are of the either "this" or "that" type, and the key word here is the conjunction "or".

The other important theorem is the *multiplication theorem*, which uses the "and" conjunction. This theorem is useful when the probability of a specific sequence or combination of outcomes is to be determined, such as what is the probability of tossing a coin twice and getting a head followed by another head. This probability is given by the multiplication theorem, which states that the probability of several independent events occurring together is the product of their separate probabilities. Thus, the probability of getting two heads in a row would be equal to $1/2 \times 1/2 = 1/4$.

Although these two basic theorems provide the basis for most probability applications for inferential statistics, the frequency distribution of the probabilities associated with these three theorems is most important. In research, statisticians are usually not interested in random isolated events, but in the probability that particular values of a variable will fall within certain limits. As a basic example of how such probability distributions are developed and used, consider the situation in which a pair of dice are thrown, and it is desired to determine the probability that the two dice will sum to a particular value. First, the limits must be determined of the possible sums of the two numbers, which will be from $1 + 1 = 2$ to $6 + 6 = 12$. Then the total number of possible outcomes must be determined. In this case, this total is 36. Figure 7.1 shows all of the possibilities: the frequency of occurrence as to how each pair of dice could fall, the sum of the two dice, the frequency of each sum, and the probabilities associated with each sum. Thus, there is only one way out of the 36 possible ways that the two dice could land to sum to 2, only one way for the dice to land to sum to 12, but six ways that the sum could equal 7. Thus, a sum of 2 would be expected with a probability of 1/36, or in terms of a proportion, only about 0.028 or just less than 3% of the time. However, the sum of 7 would happen 6/36 or 1/6 of the time, or $p = 0.167$, which is just about 17% of the time.

The same type of probability frequency distribution could be developed for tossing several coins and recording the numbers of heads, or in drawing cards from a deck of 52 cards and tabulating the number of aces drawn. Because we are interested only in seeing how these probability distributions are useful for inferential statistics,

6						6/1					
5					1/5	1/6	6/2				
4				3/2	5/1	5/2	2/6	6/3			
3			3/1	2/3	2/4	2/5	5/3	3/6	6/4		
2		2/1	1/3	4/1	4/2	4/3	3/5	5/4	4/6	6/5	
1	1/1	1/2	2/2	1/4	3/3	3/4	4/4	4/5	5/5	5/6	6/6
Sums	2	3	4	5	6	7	8	9	10	11	12
f	1	2	3	4	5	6	5	4	3	2	1
p	$\frac{1}{36}$	$\frac{2}{36}$	$\frac{3}{36}$	$\frac{4}{36}$	$\frac{5}{36}$	$\frac{6}{36}$	$\frac{5}{36}$	$\frac{4}{36}$	$\frac{3}{36}$	$\frac{2}{36}$	$\frac{1}{36}$

Figure 7.1. Distribution of results of single tosses of two dice. The symbol f *denotes the frequency, and* p *denotes the probability.*

only those that are most applicable will be considered. One useful probability distribution and the one that is the basis for some descriptive and inferential statistics is the binomial equation, or ***binomial expansion***. As the name implies, this concept deals with probabilities associated with events that have only two possible outcomes, such as the toss of a coin. However, rather than individual coins being tossed, a number of coins can be independently tossed. For example, if three coins were tossed, what is the probability that all would fall heads up? To determine this probability, the binomial expansion can be used. The general binominal expansion is

$$(p + q)^n \tag{7.1}$$

where p is the probability for the occurrence of the event, q is the probability for nonoccurrence of the event, and n is, in this case, the number of coins being tossed. Applying the equation would yield

$$\begin{aligned}(1/2 + 1/2)^3 &= (1/2)^3 + 3(1/2)^2(1/2) + 3(1/2)(1/2)^2 + (1/2)^3 \\ &= (1/8) + (3/8) + (3/8) + (1/8)\end{aligned}$$

This equation follows how we would expect the coins to fall. The expected distribution, where H is heads and T is tails, can be displayed as follows:

H	H	H	1/8
H	H	T	
T	H	H	3/8
H	T	H	
T	H	T	
H	T	T	3/8
T	T	H	
T	T	T	1/8

For demonstrating the underlying concept of probability that is most important for inferential statistics, *Pascal's triangle*, which shows the result of the binomial expansion, is useful. The frequencies for all combinations of heads and tails for tosses of a single coin up to the toss of 10 coins at the same time are as in Table 7.1.

At the top of the table, with $n = 1$, or just one coin being tossed, there are two possibilities: a head or a tail. If two coins are tossed, then four possible outcomes (HH, HT, TH, or TT) may occur. If three coins are tossed, eight possible outcomes result, and the probabilities associated with each of these outcomes is as shown. Each entry in the table is equal to the sum of the two frequencies on either side of it in the row directly above that entry, except for the 1's at the extremes in each row.

The important concept to be gained from the binomial expansion is that the probability of any specific event or combination of events can be theoretically obtained, and these probabilities have been documented by empirical studies. For example, the probability for obtaining all heads in tossing 10 coins is only 1/1024, which means that this kind of an event would be expected to happen only once out of every 1024 tosses of 10 fair coins. Getting nine heads for such a toss of 10 coins would happen 10 times out of every 1024 tosses, and getting eight heads would happen 45 times out of every 1024 tosses. This concept of probability is the basis of all hypothesis testing and estimation in inferential statistics. Observed results are always compared to what would be expected to occur if only chance

Table 7.1 Pascal's triangle for cases from $n = 1$ to $n = 10$

n																						*Sum*
1										1		1										2
2									1		2		1									4
3								1		3		3		1								8
4							1		4		6		4		1							16
5						1		5		10		10		5		1						32
6					1		6		15		20		15		6		1					64
7				1		7		21		35		35		21		7		1				128
8			1		8		28		56		70		56		28		8		1			256
9		1		9		36		84		126		126		84		36		9		1		512
10	1		10		45		120		210		252		210		120		45		10		1	1024

was operating. Thus, all probability statements are based on various theoretical or empirical random distributions.

As with any distribution, the binomial distribution has a mean, which is equal to np. That is, the probability of the event E times the number of possible events. For example, in tossing 10 coins, what will be the average number of heads expected? The answer is $10 \times 0.5 = 5$. In tossing a six-sided die 10 times, what will be the mean probability of getting a "1" or a "2" value? The probability of getting either a "1" or a "2" is 2/6, or 1/3. Thus, the mean would be $10 \times 1/3 = 10/3 = 3.333$.

The standard deviation (S) of the binomial distribution is $\sqrt{npq}$, where n is the number of replications (in this case, number of tosses or number of coins) of the event; p is the probability of the favored event E; and $q = 1 - p$, or the probability of the event E not occurring. For the tossing of 10 coins,

$$S = \sqrt{10(1/2)(1/2)} = \sqrt{2.5} = 1.58$$

For the tossing a die 10 times and getting either a "1" or a "2",

$$S = \sqrt{10(1/3)(2/3)} = \sqrt{2.22} = 1.49$$

Although more convenient statistics are available for practical applications, an example of how the binomial distribution could be used can serve as an introduction to the more frequently used normal curve distribution. Suppose a market researcher wants to determine if there is a preference by the general public for brand A or brand B of a specific product. The market researcher observes 100 consumers who are buying this product, and 62 are observed to have purchased brand A. Does this result represent a preference for brand A or a chance deviation from no preferences for either brand, and hence a 50/50 selection? The mean and standard deviation expected for no preference or for the chance binomial distribution could be calculated, and then the obtained observation could be compared with what would be expected to occur by chance alone. For this problem, the mean chance frequency would be $100 \times 1/2 = 50$, and the standard deviation would be $\sqrt{100(1/2)(1/2)} = 5$. Equation 7.3 from the next section could then be used to determine if the deviation of the observed frequency of selection of brand A (62) is significantly different from what would be expected by chance alone (50).

$$z = \frac{62 - 50}{5} = 2.4$$

As will be shown in the next section, when $z = 2.4$, the selection of brand A by 62 people is very unlikely to represent a chance difference from the expected frequency of 50. Thus, there appears to be a preference for brand A by these 100 people.

7.2 The Normal Distribution

The most commonly used and probably the most important theoretical distribution is the normal, or Gaussian, distribution. Although Abraham de Moivre derived the function rule for the normal distribution in 1733, Karl Friedrich Gauss was one of the first to apply the "rediscovered normal curve" to problems in astronomy in the 19th century. Thus, the Gaussian distribution was named. This probability distribution is the distribution used for interpreting most descriptive statistics but is also used for many inferential statistics. As mentioned in Chapter 5, the mean and standard deviation are probably the most frequently used statistics to describe data, and the interpretation and implications of these will be described in this chapter. The main reason why this distribution is so frequently used is that many of the distributions found in observational and experimental work in the physical, biological, and behavioral sciences approximate the normal distribution. Also, as will be shown, many distributions of statistics also are distributed this way.

Abraham de Moivre derived the function rule for the normal distribution as the limit of the binomial distribution when $p = 0.5$ and n approaches infinity. This assumption has been demonstrated empirically and proven mathematically, but the proof will not be given here. A random variable, X, is normally distributed if its probability distribution is given by the function rule:

$$Y = \frac{1}{\sigma\sqrt{2\pi}} e^{-(X-\mu)^2/2\sigma^2} \qquad (7.2)$$

where Y is the frequency or height of curve for each value of X; $\pi = 3.1416$; $e = 2.7183$, which is the base of Napierian logarithms; μ is the mean of the distribution; and σ is the standard deviation of the distribution.

This equation is seldom used in practice because statisticians are usually interested in the areas under the normal distribution curve, and these areas are given in the table of the normal curve (*see* Table B in the Appendix). However, the equation points out some important features of the normal curve. Given a situation in which the mean and standard deviation are fixed, all elements in the equation are constant except X. Because statisticians are interested in the squared deviation of X from the mean, both negative and positive values of X will yield the same Y, and as X increases, the exponent of e decreases, and because the exponent has a negative sign, Y will decrease. Thus, Y is at its maximum when $X = \mu$. The resulting curve is symmetrical, unimodal, always has the same shape, and is asymptotic at the extremities (i.e., the curve approaches but never touches the horizontal axis and thus is said to extend from negative to positive infinity). However, the area under the curve is considered to be finite (i.e., the total proportion under the area is equal to 1).

The most frequent use of the normal curve in statistics is concerned with the proportion of the area under the normal curve in relation to specific distances along the baseline or horizontal axis. Three common examples are to

1. find the proportion of the area under the curve between the ordinate at the mean and the ordinate at any specified point on the baseline, either above or below the mean;
2. find the proportion falling under the curve between any two ordinates; and
3. find the proportion of the area falling above or below the ordinate at any point along the baseline.

The normal curve is usually defined as having $N = 1$, $\mu = 0$, and $\sigma = 1$. The table of the normal curve is given in these units. With $N = 1$, the total area under the curve is taken as unity, and all segments are given in terms of proportions. Because the normal curve is defined by $\mu = 0$ and $\sigma = 1$, all observed values must be transformed into these standard measures, or z values, before the normal distribution or its table can be used. For sample data, this transformation can be made by using the equation

$$z = \frac{X - M}{S} \quad (7.3)$$

where z is the standard measure, X is the observed value, M is the mean of the observed values, and S is the standard deviation of the observed values.

The normal curve, showing the proportions of the total area under the curve within units along the baseline, is given in Figure 7.2. For this curve, 34.13% of the cases always fall between the mean ordinate and $+z$ or $-z$, and the area falling under the curve between z and $2z$ is always 13.59%. These areas are always constant and provide the reference percentages for all areas under the curve for any distance along the baseline.

Table B in the Appendix gives the proportion of cases falling under the curve for various reference points along the baseline given in terms of z values (*see* column 1). Column 2 gives the area under

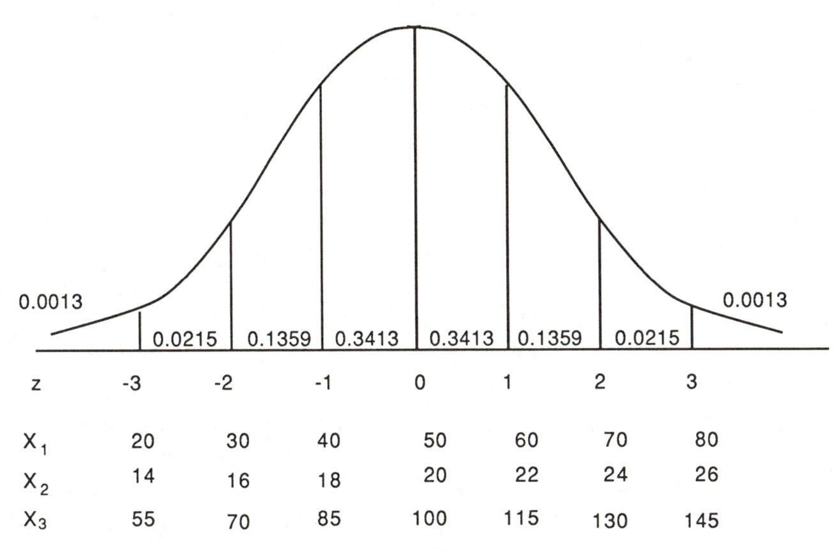

Figure 7.2. Proportions of total area under the normal curve within z units along the baseline. Three possible distributions of values are also shown.

the curve between the ordinate at the mean and the ordinate at each of these z values. For example, 15.54% of the distribution falls under the curve between the mean ordinate and an ordinate drawn from $z = 0.40$, and 40.15% of the area falls under the curve between the ordinate at the mean and the ordinate at 1.29σ away from the mean, or $z = 1.29$. Approximately two-thirds of the cases, 68.26% to be exact, fall in the area between $-z$ and $+z$. In psychometric theory, this area is known as the average range of the values for a given distribution.

Column 3 in Table B shows the proportion of cases that fall under the curve above and below specific points along the baseline. For example, for $z = +0.30$, 61.79% of the distribution would fall below this point and 38.21% above. For $z = -0.40$, 34.46% of the cases would fall below and 65.54% above that point.

The following examples can be considered by using the scale values shown in Figure 7.2. The first, X_1, is for a distribution of standardized values known as Z or T values, for which the mean is always 50 and the standard deviation is 10. By using equation 7.3, the z equation, a z value for all possible Z or T values can be computed. For example, for $Z = 60$, $z = +1.00$; for $Z = 35$, $z = (35 - 50)/10 = -1.5$. Once these z values are obtained, then Table B can be used to find the exact proportion of cases falling above or below that point, or the proportion of cases falling under the curve between that point and the mean. For the second set of scores, X_2, the mean is equal to 20 and the standard deviation is equal to 2. For this distribution $z = X - (20/2)$. The last distribution is for Weschler intelligence quotients, which use a mean of 100 and a standard deviation of 15.

For control charts and for inferential statistics, Table B can be very useful in determining points in a distribution that either includes or excludes a given proportion of the cases. For example, suppose the top 10% and the bottom 10% of any production run are considered too deviate with regard to set tolerance limits. The problem would be to find the exact values of the characteristic being measured at the lower and upper cutoff points. From column 3 in Table B, the area beyond z, when $p = 0.1003$, is $z = 1.28$. Thus, the cutoff point values would be $\pm 1.28\sigma$ from the mean. If the mean for this characteristic was 80 and the standard deviation was 11, then $1.28 \times 11 = 14.08$. The cutoff points would then be 80 ± 14.08, or 65.92 and 94.08.

7.3 Applications of the Normal Curve

The following applications assume that the characteristic being considered is approximately normally distributed, that the sample size is fairly large ($n < 100$), and that the events are independent of each other, mutually exclusive, and represent events as they would occur in a random or chance order. That is, no bias occurs in selection; thus, the sample does not contain only highly favored or biased events.

Suppose a corporation has 200 outlets and wants to set a minimum sales limit of 2000 items per year for each outlet. Over the past 5 years, the average number of sales per outlet was computed to be 3500 with a standard deviation of 1200 items. If this minimum sales limit is put into effect, what proportion and how many outlets would be terminated? To find this answer, the z value must be computed to determine the distance from the mean of 3500 that the 2000 value is in terms of the standard deviation of the values. To do this,

$$z = \frac{2000 - 3500}{1200} = -1.25$$

Referring to the table of the normal curve with $z = -1.25$, we find that the area beyond z is 0.1056. Thus, about 10.56% of the outlets would be terminated, or, for the 2000 outlets, a total of $2000 \times 0.1056 = 211$ outlets would be terminated.

For the same example, suppose the corporation decides to cut back the number of outlets by 20%. If this were done on the basis of sales, which outlets would be terminated? For this problem, the proportion desired to be cut back is given, and the minimum sales figure that will serve as the cutoff point is needed. Looking at Table B with 0.2000 in the area beyond z, we find the z value closest to 0.2000 is 0.84. Thus, -0.84σ below the mean will cut off the bottom 20% of the cases, or for this example, the lowest 20% of the outlets. Because the standard deviation is 1200, the cutoff value will be equal to $-0.84 \times 1200 = 1008$ points below the mean of 3500, giving a cutoff point of 2492 sales per year.

Another example using the same data might be to find the lower and upper sales limit between which 50% of the outlets would fall. Or to put the question another way, what are the sales records of the middle 50% of the outlets? Answering this question requires finding the points below and above the mean that will cut off the middle 50% of the distribution. Table B, column 2, gives 0.2500 as the proportion under the curve between the mean and that z value, which turns out to be approximately 0.67. Thus, $z = \pm 0.67$ will cut off the middle 50% of the cases. Finding the sales values can be done by multiplying the standard deviation of 1200 by 0.67 to get 804. Thus, the middle 50% of the outlets sold from 2696 to 4304 items each.

Reference

1. Parzen, E. *Modern Probability Theory and Its Applications;* Wiley: New York, 1960.

Chapter 8

Hypothesis Testing

8.1 Hypotheses: Definition and Description

Researchers, in attempting to solve theoretical and practical problems through empirical research, must be able to extend the findings of the research, usually done on a relatively small sample, to some specific population that the sample represented. However, a degree of uncertainty is always involved when making such generalizations. The techniques of inferential statistics make it possible in terms of mathematical probability to determine the degree of error involved in making those generalizations. The formulation and testing of hypotheses provide the conceptual basis for applying inferential statistics to theoretical and practical problems and also provide the basis for scientific reasoning.

Empirical research is a process that involves a number of steps, the first of which is usually to state the questions to be answered or to list the purposes of the research. For example, researchers might ask "What are the effects of a catalytic agent on the yield of adiponitrile?" or "What are the factors affecting strength of synthetic fibers?" Such questions help to identify and clarify problems. However, they need to be refined so that they can be researched. This refinement is usually done by restating the questions in the form of hypotheses.

A *hypothesis* is a tentative statement about an expected relationship between two or more variables. It is merely a supposition of what the relationship might be, and serves as a guide in formulating the research. Hypotheses can be developed from both deductive and inductive reasoning. From a given theory or sets of principles or laws, a researcher might deduce that certain relationships will occur in a

1453–0/88/0135$08.00/1

given situation or under certain circumstances. This deduction would lead to the formulation of a hypothesis that could then be tested empirically through observed reactions of phenomena. These observed reactions or phenomena may then stimulate the development of further hypotheses inductively. In most scientific research, a continuous combination of deductive and inductive reasoning occurs that involves the continuous testing of hypotheses. Thus, an intricate pattern of theory development and hypothesis testing is woven. In general, the term theory relates to a general explanation of phenomena, whereas hypotheses deal with specific aspects of the theory under specified circumstances. For any theory, there will usually be many hypotheses, which, when tested, will either support the theory or suggest alterations. Many hypotheses may be tested for any given theory or a single hypothesis, and any specific hypothesis may be either simple or complex. There are many sets of propositions or theories relating to phenomenon such as the yield or quality of a product, the resistance of materials to chemicals, or the power or fuel consumption of a process. Each of these theories could lead to specific sets of hypotheses, each of which are empirically tested because the extent to which they are correct is not known. For example, there are theoretical considerations regarding the yield of a product (e.g., penicillin) such as the conditions of agitation, aeration, temperature, and quality of ingrediencts. These considerations could lead to a study designed to test the hypothesis that there would be no difference in yield of penicillin when four concentrations of various sugars are used to produce the necessary mycelium.

The hypotheses described to this point are generally referred to as research or scientific hypotheses. Such hypotheses can be stated in a number of different ways, from a very general supposition or statement about a relationship to those that are very specific. Scientific hypotheses serve to focus the direction of a study, and generally have three characteristics:

- They state the expected relationship or difference between two or more variables, which is usually based on intelligent, informed guesses about the phenomenon of concern; the gueses are often based on theory.
- They are stated in such a way that the relationship or difference is clearly and unambiguously delineated. These statements are usually of the "if this, then this" form.

- They are testable (i.e., their truth or falsity can be determined by observation or experimentation).

If such hypotheses refer to phenomena that can be observed, then evidence can be collected to test each hypothesis. However, many scientific hypotheses refer to all members of a specified population, and testing the hypothesis *directly* by observing all members of the population is often impossible. If the hypothesis cannot be tested directly, then it can usually be tested by formulating a statistical hypothesis that can be tested by statistical inference. Statistical inference enables a researcher to make a reasonable decision regarding the probability that a statistical hypothesis is either true or false.

A statistical hypothesis is a statement about one or more parameters of a population distribution that requires verification, always refers to the population rather than to the sample, and is called a hypothesis because it refers to a situation that might be true. Statistical hypotheses usually are formulated from or are implied by the scientific hypothesis, but these two hypotheses are seldom equivalent. Scientific hypotheses generally refer to the phenomenon being investigated (i.e., an understanding of why the relationships or differences exist) and are usually stated to express that understanding. Statistical hypotheses are usually expressed as two mutually exclusive and exhaustive statements concerning some population parameter. The first statement is the statistical hypothesis that is tested and is called the *null hypothesis*. The second statement is called the *alternative hypothesis* and is the hypothesis that remains tenable if the first hypothesis is rejected. Usually the alternative hypothesis corresponds to the scientific hypothesis in that it is consistent with the relationship or difference expected.

An example may help to clarify the step-by-step sequence in formulating hypotheses. Suppose that the effect of the quality of ammonium chloride on the yield of an organic chemical is investigated, and the main interest is to compare the use of finely ground ammonium chloride to the usual coarsely ground ammonium chloride. There is reason to believe that the finely ground ammonium chloride will increase the yield of the organic chemical. The scientific hypothesis might be " An increase in yield of an organic chemical will occur if finely ground ammonium chloride is used compared to the usual coarsely ground ammonium chloride." Because this study is conducted so that the results will generalize to all instances where the finely ground ammonium chloride could be used under the same production conditions, all instances are unlikely to be observed. Thus, an indirect test of this hypothesis has to be made by statistical inference. Therefore, statistical hypotheses need to be formulated. For this example, the present yield obtained by using the coarsely ground ammonium chloride is 80 lb/batch, which is taken as the population mean for the present situation. A random sample of production runs could be taken for the finely ground ammonium chloride, and the mean of this sample can be used to estimate what the population mean might be when this type of ammonium chloride is used.

The null hypothesis (H_0), which is usually stated contrary to what is expected, could be "The population mean for the finely ground ammonium chloride is equal to or less than 80." In statistical notation, H_0: $\mu \leq 80$. This hypothesis is the one that is tested. The alternative hypothesis (H_1) would be "The population mean for the finely ground ammonium chloride is higher than 80", or H_1: $\mu > 80$. The alternative hypothesis is an expression of what would be expected for finely ground ammonium chloride and is thus consistent with the scientific hypothesis. Now, evidence has to be collected to make a decision regarding these two statistical hypotheses. This decision is made following an established set of decision rules that are based upon the concepts from mathematics known as decision theory.

8.2 Testing Hypotheses

The procedure traditionally followed in testing a statistical hypothesis can be summarized by the following steps:

1. Formulate the statistical hypotheses (i.e., the null hypothesis and an alternative hypothesis).
2. Select the appropriate sample statistic.
3. Specify the test statistic.
4. State the level of significance, α, to be used.
5. Specify the sample size, n, to be obtained.
6. Obtain a random sample of n cases, compute the sample statistic, and then compute the test statistic.
7. Make a decision about the hypothesis.

8.2.1 Formulating Hypotheses

The null hypothesis is the hypothesis that is always tested. If this hypothesis is rejected, the alternative hypothesis remains tenable. These hypotheses can be stated as either exact or inexact hypotheses. In the example dealing with ammonium chloride, if the null hypothesis stated that the mean for the population is equal to 80, this would be an exact hypothesis and would be designed as H_0: $\mu = 80$. An inexact null hypothesis would be H_1: $\mu \leq 80$ because it specifies a range of values, those equal to or less than 80, rather than a single value for μ. The alternative hypothesis, H_1, to the exact hypothesis can be either exact or inexact (e.g., H_1: $\mu = 85$ is an exact alternative hypothesis, whereas H_1: $\mu \neq 80$ is an inexact alternative hypothesis). If the null hypothesis is inexact, (e.g., H_0: $\mu \leq 80$), then the alternative hypothesis must also be inexact (e.g., H_1: $\mu > 80$). Exact hypotheses specify a specific value for the population parameter, whereas inexact hypotheses specify a range of possible values. Inexact alternative hypotheses are generally specified because a precise prediction regarding the outcome of a study usually cannot be made by specifying exact values. The procedures that are followed to test both of these hypotheses are the same.

The hypotheses that have been considered so far have dealt with population means. However, hypotheses can deal with any statistic (e.g., correlation coefficients, frequencies or proportions, variances, or differences between various statistics). Hypotheses should specify the parameters being estimated from computed sample statistics.

8.2.2 Selecting a Sample Statistic

The two general types of hypotheses are nondirectional and directional. The critical areas of the sampling distribution for testing these two types of hypotheses are shown in Figure 8.1 for the 0.05 level of significance. For the nondirectional hypothesis, the critical area is divided into two equal parts, whereas for the directional hypothesis, all of the area of rejection is in the direction specified by the alternative hypothesis.

A directional hypothesis is usually used when there is considerable evidence that the scientific hypothesis will actually be supported (i.e., that the result will be as anticipated or predicted). This evidence might be based on previous research, experience, general observations, or theory. The scientific hypothesis might be phrased

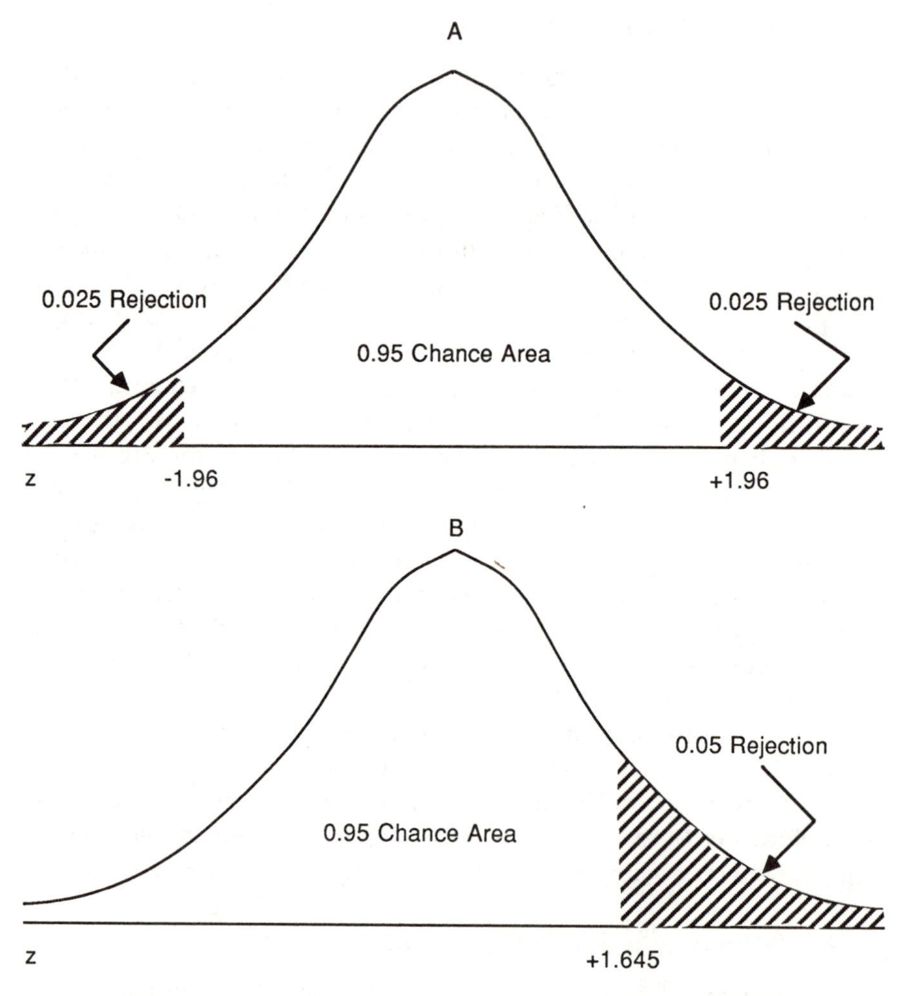

Figure 8.1. Areas of rejection (shaded areas) for the (A) nondirectional or two-tailed test, and (B) the directional or one-tailed test of the null hypothesis at the 0.05 level of significance.

by words such as "greater than", "less than", "increased", or "improved". Or possibly the researcher is interested only in a difference that is in one direction. For example, if one catalyst has been shown to be effective, then another catalyst would probably not be considered if it was less effective if all other factors were equivalent. The directional null hypothesis would be specified as H_0: $\mu_1 \leq \mu_0$, where μ_1 refers to the estimate of the population mean based upon the sample, and μ_0 refers to the actual or hypothetical population mean.

The alternative hypothesis would be H_1: $\mu_1 > \mu_0$. The directional hypothesis requires a one-tailed test as shown in Figure 8.1B.

If the researcher is not interested in specifying the direction of the difference, then a nondirectional hypothesis would be used. The statistical hypotheses would then be H_0: $\mu_1 = \mu_0$ and H_1: $\mu_1 \neq \mu_0$. A two-tailed test would be called for to test the null hypothesis.

A number of test statistics can be used to test hypotheses. The most common are the normal deviate z ratio, Student's t statistic, the chi-square statistic (χ^2), and the F statistic. Each of these refers to a specific sampling distribution that has the same name (e.g., a test statistic is called a t test if its sampling distribution is a t distribution).

8.2.3 Specifying the Test Statistic

The selection of the appropriate test statistic is based upon (1) the hypothesis to be tested, (2) what information is known about the population, and (3) the assumptions that have to be met in order to use each test statistic.

8.2.4 Level of Significance

The *level of significance* is the probability that a true null hypothesis will be rejected (i.e., the risk the researcher is willing to take that the null hypothesis will erroneously be rejected when in fact H_0 is true). The hypothesis testing procedure divides the sampling distribution of the test statistic into two mutually exclusive parts: the area of rejection and the area where researchers fail to reject the null hypothesis. If the null hypothesis is true, the test statistic will be distributed as specified by that particular distribution, and the probability of various values for that test statistic can be determined. Usually, the region of rejecting the null hypothesis is set so that the probability of getting specified values of the test statistic when H_0 is true is very small (i.e., usually at the 0.05 level or lower). The probability criterion set for rejecting the null hypothesis is the level of significance and is designated by the Greek letter alpha, α. The region set for rejecting H_0 is the *critical region*, and the values of the test statistic that set off this area in the sample distribution are the *critical values*. Setting the level of significance specifies a decision rule for determining whether the null hypothesis should or should not be rejected. For example, in the z distribution, the decision rule would be to reject the null hypothesis if $z = \pm 1.95$ at the 0.05 level of significance for a nondirectional test, and ≥1.645 for a directional test. A nondirectional or *two-tailed test* provides for the possibility of rejecting the null hypothesis for differences that are either larger or smaller than a hypothesized value. A directional or *one-tailed test* provides for the possibility of rejecting the null hypothesis only for differences in one direction (e.g., larger than a hypothesized value). These areas are shown in Figure 8.1.

As can be noted from Figure 8.1, the region of rejection for a two-tailed test is divided into two areas, whereas for the one-tailed test all of the critical region is in one tail of the distribution. The respective z ratios required for significance at the 0.05 level are 1.96 for the two-tailed test and 1.645 for the one-tailed test. Thus, the absolute value of the z ratio has to be larger to reach the critical region for the two-tailed test than for the one-tailed test. Therefore, a researcher is less likely to reject a false null hypothesis with a two-tailed test. As will be discussed in Section 8.5, the power of a statistical test refers to the probability of rejecting a null hypothesis when it is actually false and should be rejected. Because rejecting such a hypothesis is less likely with a two-tailed test compared to a one-tailed test, the one-tailed test is more powerful than a two-tailed test if the hypothesis about the direction of the true difference is correct. If the result is in the other direction, the area of rejection would not be in the direction of the outcome, and little if any chance would occur to reject the null hypothesis even when it is false and should be rejected. There is always some possibility that H_0 will still be rejected, even though the investigator would be wrong about the direction of the true difference. This error is sometimes referred to as a Type III statistical error, which is correctly rejecting a false H_0 but incorrectly interpreting the direction of the difference. The general suggestion with regard to choosing between a two-tailed and a one-tailed test is that if the researcher really does not know what to expect, a two-tailed test should be used. If sufficient evidence indicates that the result is going to be in one direction, or if the researcher is only interested in the result if it comes in that specified direction, then a one-tailed test would be appropriate.

The level of significance selected also specifies the probability of making a *Type I statistical error*, which is the error of rejecting a true null hypothesis. If $\alpha = 0.05$, the probability of making a Type I statistical error is 0.05. Statistical errors will be discussed in Section 8.3.

8.2.5 Specifying the Sample Size

Several factors have to be considered when specifying the sample size. As will be explained in Section 8.5, the power of a test statistic, which is the probability that it will reject the null hypothesis when that hypothesis is false and should be rejected, is related to several factors including sample size. Several procedures can be used for determining the sample size that would be necessary for the level of power desired for the test statistic. These procedures will be discussed in Section 8.6. In general, the size of the sample should be so large that the study will be sensitive enough to detect differences in parameters that are also large enough to be considered of practical importance. A consideration also has to be made regarding what test statistic is used. If the z distribution is used, then the sampling distribution of means will be approximately normally distributed if the population distribution of that variable is normal or if the sample size is fairly large (i.e., $n > 100$). Thus, a large sample size is generally required when using the z distribution as the test statistic.

8.2.6 Conducting the Study

After the sample size has been specified and all other conditions for conducting a study have been made, the study is begun. The study should be conducted so that all relevant factors are controlled so that the only difference between or among conditions are those that are specified in the hypothesis. For the yield of penicillin example, all relevant factors, such as the conditions of agitation, aeration, and temperature, should remain constant, and only the types and concentrations of the various sugars should be varied. Data are collected with regard to the yield of penicillin for the types and concentrations of sugar specified. If 20 samples are to be taken from each batch of penicillin produced by each combination of type and concentration level of sugar, then data from these samples would be collected and used for the calculation of the test statistic.

8.2.7 Making a Decision about the Hypothesis

The last step in hypothesis testing is to make a decision regarding the null hypothesis on the basis of the test statistic computed in step 6. This decision would be made following the decision rule specified in step 4. If the test statistic falls in the critical region, then H_0 would be rejected and the alternative hypothesis would be considered as tenable. Any decision always has to be made with the consideration that a Type I or a Type II statistical error is possible. That is, every decision rule contains a certain probability of being wrong, and this probability always has to be considered. Statistical errors will be discussed in Section 8.3.

Generally, the interpretation of the decision made can be described in terms of chance. When the null hypothesis is actually true, then the test statistic will be distributed as a random variable following the specified distribution. The critical region can then be thought of as the area in which the test statistic will fall for repeated samples a very small proportion of the time. This area is equal to the level set for that decision. Thus, the probability that the test statistic would fall into this area by chance alone is very small. The critical area for rejecting the null hypothesis is sometimes referred to as the nonchance area.

If the test statistic falls in the critical region, or the nonchance area, then the probability of obtaining this statistic on the basis of purely random events would be equal to or fewer than 5 times out of every 100 possible trials at the 0.05 level of significance, or equal to or fewer than 1 time out of every 100 trials at the 0.01 level. The conclusion would be that the null hypothesis is rejected at the selected level of significance, and that the finding is statistically significant. Rejection of the null hypothesis implies the acceptance of the alternate hypothesis.

If the test statistic falls in the nonrejection or chance area, the finding is not significant; the null hypothesis is not rejected, and the conclusion is that the event could have happened by chance alone, which means that the null hypothesis is probably true. That is, the null hypothesis is not rejected, but it is not accepted on the basis of the evidence used to test the hypothesis. In other words, statisticians do not prove either the null or the alternative hypothesis, but merely decide whether either is tenable.

8.2.8 Summary

For all hypothesis testing, the important concept to keep in mind is that the null hypothesis is the hypothesis that is always tested. That is, statisticians always assume that the null hypothesis is true until evidence indicates otherwise, the evidence being that the probability is very low that the result would have happened by chance alone.

The second important concept is that in testing hypotheses, all that is known is whether the outcome could be considered as a random chance event or what statisticians consider as beyond chance because its probability of occurrence by chance alone is very small. The testing of hypotheses is based solely on probability estimates; nothing is provided with regard to why the result happened. The reasons for the result have to be based on other considerations such as the research design used, the control of extraneous factors, and how the data were collected.

Finally, in testing hypotheses, statisticians always refer to population parameters, not to sample statistics. This terminology permits generalizing the results to other samples and similar situations.

8.3 Statistical Errors

One of the most important concepts in inferential statistics is that every decision is based on probability, and for most uses, probability is based on what would happen if only random, chance variation was operating. Conclusions about population parameters based on sample information will always be subject to sampling errors because each sample drawn from a given population will vary within some specified limit if these samples are drawn at random. Even though research designs and statistical analysis procedures can be used to minimize these errors, the researcher is always confronted with the possibility that the outcome of the research is due to sampling error rather than any real relationship or difference among the variables in the study. In making decisions on the outcomes of statistical analyses based upon sample data, statisticians always have the risk of making an error in their estimates of population parameters or in testing hypotheses.

Two kinds of statistical errors are possible every time the null hypothesis is tested by using sample data. If statisticians reject the null hypothesis when in fact it is actually true, then a *Type I error* will result. If statisticians fail to reject a null hypothesis when in fact it is false and should be rejected, then a *Type II error* will result. In testing hypotheses, statisticians never know whether they have made an error. However, knowing the probability of making either type of error may be possible.

If the hypothesis is rejected, there are two possibilities:

1. the null hypothesis is actually false and should be rejected because a real relationship or difference exists between the variables being studied, or

2. the null hypothesis is actually true and should not have been rejected. In case 2, the researcher concludes that a relationship or difference exists when in fact none actually exists; thus, the researcher has made a Type I statistical error.

The probability of making a Type I error is set by the researcher. If the α level is set at 0.05, then the probability of making a Type I error will be 5% of the time for repeated samples under the same conditions. The probability that statisticians correctly do not reject a true null hypothesis is also determined by the level of significance set because this probability is equal to $1 - \alpha$.

If the researcher fails to reject the null hypothesis, the two possibilities are as follows:

1. The null hypothesis is really true and should not have been rejected, or
2. The null hypothesis is false and should have been rejected.

If the first situation exists, then a correct decision has been made. If the second situation is true, then the researcher has failed to reject a false null hypothesis and has made a Type II statistical error. The probability of a Type II error is usually represented by the lower case Greek letter beta, β. The possible decision outcomes that are possible in testing any hypothesis and their respective probabilities are shown in Figure 8.2. If the true or actual situation is such that the null hypothesis is really true, then if H_0 is rejected, a Type I error has been made because statisticians would be rejecting a true

	True or Real Situation	
	H_0 is True	H_0 is False
Test Decision Reject H_0	Type I error $P = \alpha$	Correct decision $P = 1 - \beta$
Fail to Reject H_0	Correct decision $P = 1 - \alpha$	Type II error $P = \beta$

Figure 8.2. Decision outcomes for tests of hypothesis.

null hypothesis. If the true situation is that H_1 is true, then H_0 should be rejected, and if it is not, then a Type II error will be made. The probability of a Type I error is determined by the α level set by the researcher. The probability for a Type II error is determined by

1. The size of the sample,
2. The size of the population standard deviation,
3. The alpha level set by the researcher,
4. The magnitude of the difference between the statistics being compared (e.g., in this case, the means for the null and alternative hypotheses), and
5. Whether a directional or nondirectional hypothesis is being tested.

The basis for both of these errors is that under the condition that H_0 is true, the sampling distribution of the test statistic is distributed so that all values of that statistic are possible and can be considered as being randomly distributed. Then for hypothesis testing, statisticians divide this distribution into two mutually exclusive areas: the critical area or the area of rejection where H_0 is rejected, and the area where H_0 is not rejected. These areas are sometimes referred to as, respectively, "the area beyond chance" and "the area that could be expected on the basis of chance". Thus, a test statistic that falls into the area set aside as the critical area could reflect a real difference or could be one of the results expected on the basis of random sampling. At the 0.05 level of significance, 5 out of every 100 random results are designated as being significant. Nothing is indicated with regard to which of those 5 out of every 100 results will fall in the critical "beyond chance" area. It could be the first result, the twenty-first, the thirty-second, or any other individual result that could fall in this area. However, logically, statisticians would probably not expect the first result obtained with repeated random sampling to fall into the "beyond chance" area, although this is always a possibility. If researchers get into this area with their first result, then this value is usually considered as representing a result that is beyond chance. However, any result that falls into this area could actually be one of the chance results, and if this has happened, then researchers would be making a Type I statistical error. A Type I statistical error is always a possibility whenever the null hypothesis is rejected.

8.4 Choosing the Significance Level

The significance level is chosen by the researcher on the basis of the consequences of making the two kinds of statistical errors. As indicated in Section 8.3, the probability of making a Type I error is set by the level of significance selected by the researcher. However, the probability of making a Type II error is often unknown because the exact value of the alternative hypothesis is seldom known. If all other factors are considered equal, including the sample size and the difference between the mean for H_0 and the mean for H_1, the probability of a Type II error increases as the alpha level is set at lower levels (i.e., from 0.05 to 0.01 to 0.001). Thus, the level of significance is set to balance the consequences of each type of error. The two most frequently used levels are the 0.05 and 0.01 levels of significance. However, other levels are often used based on the consequences of making one of the two types of statistical errors.

In general, researchers view making a Type I error as more serious than making a Type II error. This view is due to significant results tending to be accepted as genuine; decisions based on these results are often made without further investigations. However, the nature and type of problem under study have to be considered when deciding which type of error should be minimized. For example, in exploratory studies, a higher alpha level is often set so that potential effects can be identified for further research. For such studies, a Type II statistical error might cause the researcher to eliminate potential effects rather than retaining them for further study.

If costly changes are to be made on the basis of the results of research, then the consequences of making a Type I error are large because expensive changes would be made that in reality are not justified. Thus, a very small probability level should be selected or replications of the study should be done to determine if the results are consistent. The importance of replications for any particular type of study becomes apparent when the multiplication theory is applied to the individual probability levels. If a study is replicated and the same results are found, then the researcher could ask the question "What is the probability of that result happening two times in a row?" If each study was significant at the 0.05 level, then the probability that these results will follow each other sequentially is $0.05 \times 0.05 = 0.0025$. If the study was repeated for a third replication with the same result, then the probability that the result would come out three times in a row by chance alone would be $0.05 \times 0.05 \times 0.05$

= 0.000125. If the results were significant at the 0.01 level and three replications were made with the same result, then the probability of getting three replications with the same result by chance alone would be $0.01 \times 0.01 \times 0.01 = 0.000001$. Thus, replications are one way to guard against the possibility that the significant result of just one study might just be a Type I statistical error.

The consequences of a Type II error may be the loss of an important idea or concept based upon an erroneous nonsignificant finding because often no follow up is done on the concept being tested if the null hypothesis is not rejected. However, the consequences of a Type II error may be very serious in some situations. For example, if a new medication or therapy is erroneously thought to not be effective in the treatment of a serious or fatal disease, a Type II error might result in withholding the medication or therapy from patients who desperately need it. Thus, the loss functions associated with both types of errors have to be known before making a decision as to which error should be minimized.

A recent trend, especially when using computer statistical packages, is to just report the actual probability level for the computed test statistic without comparing this obtained probability with a predetermined level of significance. This method has the advantage of letting the reader decide the level of significance that should be used and thus whether the hypothesis should be rejected. However, because the criterion for rejecting a null hypothesis is not specified, then hypotheses may or may not be rejected on the basis of what a researcher wants rather than on the basis of a prespecified objective criteria.

8.5 Power of a Statistical Test

The power of a statistical test is the probability that the null hypothesis will be rejected when in reality it is false and should be rejected. Power is defined as $1 - \beta$ (i.e., as 1 minus the probability of making a Type II statistical error). The power of a statistical test depends on the level of significance that is used, the sample size, the true value of the population parameter being estimated or tested, and whether a one-tailed or a two-tailed test is used. In practice, usually the population value is not known. Thus, determining the exact power of a test statistic is often difficult.

To illustrate the concept of power and the probabilities that are associated with this concept, consider the following example. First, review the probabilities α, $1 - \alpha$, β, and $1 - \beta$ that are associated with each decision outcome for tests of hypotheses as shown in Figure 8.2.

Suppose that a directional hypothesis is being tested so that H_0: $\mu \leq 50$, and H_1: $\mu > 50$. Assume that the basic population distribution of measurements is normal with $\sigma = 20$, and $n = 100$ elements have been randomly selected from this population. The decision rule for testing the null hypothesis would be set up as depicted by the left curve (1) in Figure 8.3 when H_0 is actually true and when $\alpha = 0.05$. If the z distribution is used to test H_0, then H_0 would be rejected if the sample mean is equal to or higher than 53.29 because in the z distribution, the critical value at the 0.05 level of significance is equal to 1.645 for a one-tailed test. Thus, the critical value in terms of the measurements used would be equal to 2, the standard error of the mean, times 1.645 added to the hypothesized mean (50), which would equal 53.29. If H_0 is actually true, then the probability of making a correct decision is equal to $1 - \alpha = 0.95$. If the sample mean happened to be greater than 53.29, and if H_0 is really true, then a Type I statistical error would be made. The probability that this might happen is equal to the alpha level set for the study, which in this case is equal to 0.05.

If the H_0 is really false, then the probabilities associated with the outcomes can be computed. For any alternative hypothesis that might be specified, β, which is the probability of a Type II error in testing the null hypothesis when the alternative hypothesis is true

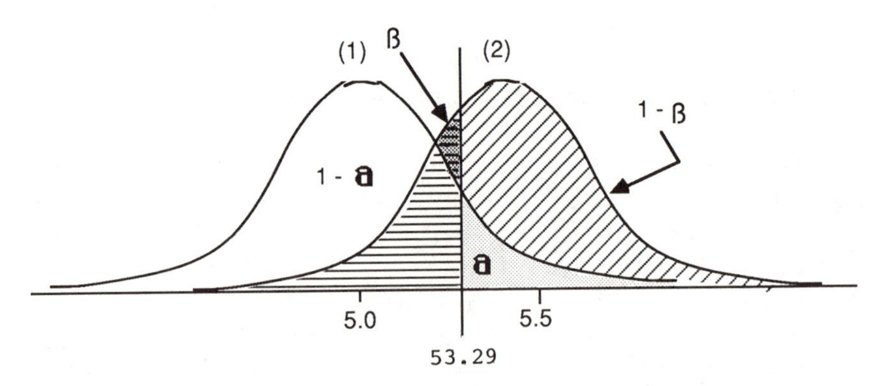

Figure 8.3. Regions corresponding to decision probabilities when (1) H_0 is true and when (2) H_1 is true, where $\mu_0 = 50$, $\mu_1 = 55$, and the critical value for the mean at the 0.05 level is 53.29.

(i.e., the probability of not rejecting H_0 when in fact it should be rejected), can be determined. Likewise, if the alternative hypothesis is true and the null hypothesis is correctly rejected, then $1 - \beta$ gives the probability that the decision to reject H_0 was correct.

If the alternative hypothesis is exact and specifies a specific value, say $\mu_1 = 55$, then the means from that population would be distributed as shown by the right curve (2) in Figure 8.3, given that the distribution for these measures is also normal with the same standard deviation of 20. The probabilities can be computed by determining how far the critical value needed for rejecting H_0 deviates from the specified alternative H_1 in terms of standard error units. For $H_1\colon \mu_1 = 55$, therefore,

$$z = \frac{53.29 - 55}{2} = -0.855$$

where 2 is the standard error of the mean based on $\sigma = 20$ and $n = 100$. Thus, the critical value, 53.29, needed to reject H_0 is 0.855 standard error units below the alternative hypothesized mean of 55. From Table B in the Appendix, the area under the curve from -0.86 to the mean is 0.3051. Therefore, $1 - \beta = 0.8051$, which is equal to $0.3051 + 0.5000$, which is above the mean of 55. Thus, the power of the test to reject H_0 is 0.8051, given the decision rule and the specific alternative hypothesis that $\mu_1 = 55$. Thus, in the long run for many samples of $n = 100$ that could be drawn from this population, approximately 80% of the time, rejection of $H_0\colon \mu \leq 50$ will be a correct decision if the real mean is 55.

Because the situation in which H_1 is really true is being considered, H_0 should be rejected. For this example, β, the probability of not rejecting a false H_0, is 0.1949, which is the area of the normal curve below $z = -0.86$. Thus, the probability of making a Type II statistical error is 0.1949 when H_1 is actually true.

The power function of a statistical test is often displayed as a power curve as shown in Figure 8.4. For the power curve in Figure 8.4, the level of significance is the 0.05 level, and a nondirectional null hypothesis is used. The base of the curve gives differences that might be found between the hypothesized population means in terms of standard deviation units, or the effect size expected for different outcomes; the vertical axis gives the power of the test. The five curves show the power function for sample sizes of $n = 4$, 10, 50, and 100. As can be read from this chart, if the differences between the hy-

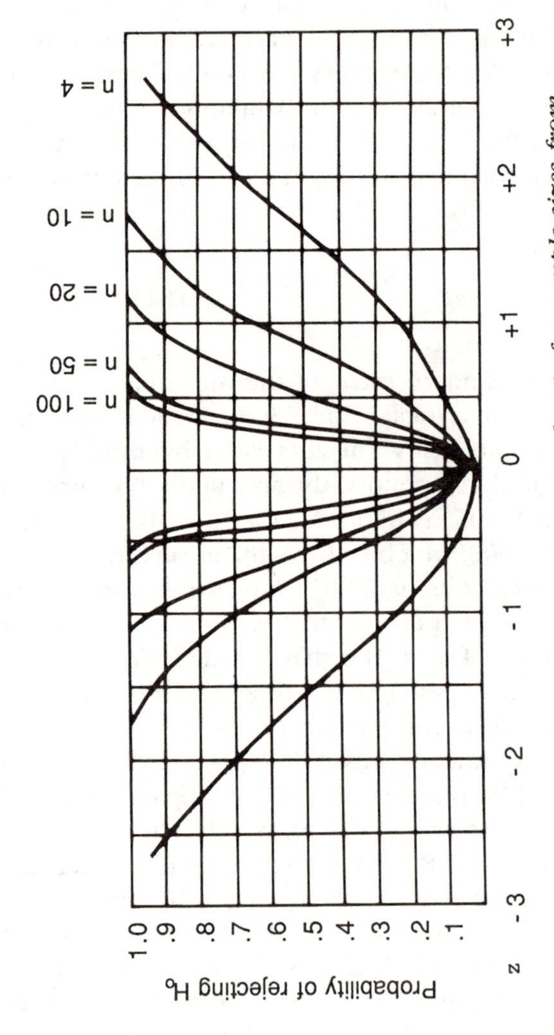

Figure 8.4. Power curves for statistical tests for sample sizes from n = *4 to* n = *100 for mean differences expressed as* z *values.*

pothesized population means is half of the population standard deviation, the power of the test is only 0.25 for a sample of $n = 10$, 0.50 for a sample of $n = 20$, and 0.90 for a sample of $n = 50$. If the difference between the hypothesized means was 1 standard deviation, then the power of the test would be 0.65 for $n = 10$, 0.94 for $n = 20$, and almost certain (i.e., 1) for $n = 50$ and $n = 100$.

Given the same conditions, a one-tailed test is more powerful than a two-tailed test. For example, had a one-tailed test been used in the previous example, the power function for $n = 20$ would have been around 0.70 from other tables for a difference between hypothesized population means that is equal to 0.5 a standard deviation, and 1.00 for population means that were different by 1 standard deviation.

As stated before, a significance level set at the 0.05 level will result in more power for the test statistic than a test set at the 0.01 level if all other factors are held constant. However, as the alpha level changes from 0.05 to 0.10 or 0.20, the probability of making a Type I error increases. Because the consequences of making Type I errors are usually greater than for Type II errors, α is seldom set at a level greater than 0.05. Other advanced statistics books provide (*1–5*) a more thorough discussion of the power function of statistical tests and the calculation procedures.

8.6 Estimating Sample Size

In the physical sciences, very homogeneous populations are often dealt with in which all elements of a population are considered to be similar to each other in most respects. If all aspects relating to the elements selected for a study are exactly the same, and if all factors important to a study are exactly identical (i.e., no variation whatsoever), then a sample of one case would be sufficient for a study, and the researcher could generalize to the total population of perfect cases. However, even though certain properties of a pure chemical or element are taken for granted, statisticians can never guarantee for certain that every sample is perfect or that all aspects of a study are invariant. Thus, the scientist always suspects a certain amount of variation or heterogeneity in the population being studied and in the procedures used in a study. Therefore, determining the extent of this variation and how it influences the outcome of a study is essential.

The extent of this variation has to be determined by obtaining a random sample of elements in order to generalize or make inferences to other elements that might be considered, or specifically to the population of elements. The question then is how large a sample is needed to estimate the variance in the population and, in turn, various population parameters. Determining the sample size needed must be done while planning a study and is based on a number of factors, the most important of which are the variance of the variables in the population, the direction of and the minimum treatment effect to be detected, the consequences of making statistical errors, the number of subgroups or treatments that are required in the study, and economic factors related to data collection. All of these factors have to be taken into consideration, and, as will be shown, the first three factors listed are interrelated.

The size of the sample has to be large enough to detect differences in the population parameters that are considered large enough to be of practical value, and, at the same time, retain adequate power to reject H_0 when in fact it is false and should be rejected. However, the size of the sample can also be too large and result in rejecting null hypotheses that have little practical use. One criticism of the procedure used to test hypotheses is that, if the sample size is large enough, nearly any null hypothesis can be rejected. That is, differences could be found that are significantly different from a statistical point of view but have no practical significance in that they are too small to be useful in the real world. Thus, a need arises to determine the size of the sample needed so that practical differences can be found, and, at the same time, to be efficient with regard to time and resources.

As pointed out in Section 8.5, the factors that affect the power of a test include the population standard deviation of the variable being considered, the alpha level set by the researcher, the probability of correctly rejecting a false H_0 $(1 - \beta)$, the difference between the means for H_0 and H_1, and the sample size. These factors are interrelated, and if four of the factors are known or can be specified, then the fifth factor can be computed. Of the four factors that are needed to solve for n, α and $1 - \beta$ are determined by the researcher, σ is sometimes known or can be estimated by the researcher, but often the difference between the means specified by H_0 and H_1 is not known. However, often the difference the researcher would like to detect can be specified. For the example in Section 8.5, suppose that a difference of 5 lb/batch is the smallest difference that would be considered to justify any production change. Because any change in production procedure would probably be costly, the researcher would want to set a small alpha level to guard against making a Type I statistical error. The desired level of power might be set at 0.90. Now that the four factors have been specified, n can be calculated by using the equation given by Kirk (*5*):

$$n = \frac{(z_\alpha - z_\beta)^2}{(\mu_1 - \mu_0)^2/\sigma^2} \quad (8.1)$$

where z_α is the value of z that cuts off the critical region of the sampling distribution; z_β is the value of z that cuts off the β region of the sampling distribution; $\mu_1 - \mu_0$ is the difference between hypothesized means, or the difference to be detected; and σ^2 is the population variance.

For the example in Section 8.5, for a one-tailed test, $z_\alpha = 1.645$, and z_β for a power of 0.90 would be -1.28. With $\sigma = 20$ and a five-point difference specified,

$$n = \frac{[1.645 - (-1.28)]^2}{(55 - 50)^2/20^2} = \frac{8.556}{0.0625} = 136.90$$

or $n = 137$, which would be the minimum sample size needed to detect a five-point difference. If a power of 0.80 was considered, then

$$n = \frac{[1.645 - (-0.84)]^2}{(55 - 50)^2/20^2} = \frac{6.175}{0.0625} = 98.8$$

or a sample size of 99 would be needed.

Often in exploratory research, no information is available on the population parameters μ and σ. Thus, these values could not be specified for equation 8.1. For such situations, an alternative procedure that uses a hypothesized effect size (HES) can be used. HES is a ratio similar to the z score discussed in Chapter 5, and is computed by dividing the minimum difference that the researcher wants to detect by the population standard deviation. If σ is not known, then it can be estimated as described in Chapter 5. Cohen (*2*) refers to HES as an effect size and used the symbol d to express the magnitude of the difference desired divided by σ. Usually an effect size of 0.3

or larger is considered as representing a practical difference. By using HES, equation 8.1 simplifies to

$$n = \frac{(z_\alpha - z_\beta)^2}{\text{HES}^2} \tag{8.2}$$

Applying this equation to the example, if an effect size of 0.6 is desired, then for a power of 0.80,

$$n = \frac{[1.645 - (-0.84)]^2}{(0.6)^2} = \frac{6.175}{0.36} = 17.15$$

For this example, a difference of five points was desired. With $\sigma = 20$, HES would be $5/20 = 0.25$. Then,

$$n = \frac{[1.645 - (-0.84)]^2}{(0.25)^2} = \frac{6.175}{0.0625} = 98.8$$

or 99, which is the same n required by equation 8.1.

A number of other procedures and tables can be used to estimate the size of the sample needed for a given study, depending on the type of study and the test statistic used. For example, tables showing the sample size required for comparing means and proportions are given in various resources (*6–8*) for various power levels. These references plus others (*2, 5, 9–13*) provide detailed discussions on procedures for estimating sample sizes.

Natrella (*7*) provided tables for determining sample sizes needed to detect differences for effect sizes from 0.01 to 3.0 for power levels from 0.50 to 0.99. Two of these tables have been reproduced as Tables L and M in the Appendix. Table L is for two-tailed tests, and Table M is for one-tailed tests. Thus, for a two-tailed test, to detect a difference between means that would reflect an HES equal to 0.4 and with power set at 0.90 and $\alpha = 0.05$, a sample size of 66 would be required. For the example, from Table M, a one-tailed test with $1 - \beta = 0.80$, HES = 0.6, and $\alpha = 0.05$ yields a sample size of 18.

8.7 Sampling Procedures

The importance of random sampling has been emphasized in Chapter 7 as well as in this chapter. To make valid inferences, the samples from a specified population must represent that population if they are to provide accurate estimates about the characteristics of that population. Samples should be similar to the population as much as possible in all characteristics in order to provide unbiased and consistent estimates of the population parameters. A sample would be unbiased if, for an infinitely large number of random samples, the mean of the sample statistic equals the value of the population parameter being estimated. The sample would be said to be consistent if the sample statistic tends to get closer to the population parameter as the size of the sample increases. This rule is the law of large numbers. If a sample is to fulfill these two requirements, then elements must be selected from that population in a random, unbiased way. Random sampling is based on the principle that such samples are typically more representative of the population than other types of samples, and that such sampling permits the researcher to estimate the accuracy of the sample by applying the principles of probability theory. Correct procedures in sampling help prevent incorrect inferences being made from the sample data.

The most widely used method of selecting elements from a population is simple random selection. Kish (*13*) operationally defined a simple random sample as any sample taken from a population so that every element has an equal and independent chance of being chosen. The completely randomized design involves randomization in three stages:

- random selection from a defined population,
- random assignment of the selected elements into subgroups, and then
- random assignment of subgroups to the different treatments.

These aspects are depicted in Figure 8.5. The assumption is that the sample as well as each subgroup represents the population and that any differences that might exist among the subgroups are merely chance differences within a given range for a given level of probability. Thus, at the beginning of the study, the groups are assumed to be equivalent, and the null hypothesis of no significant differences among groups is assumed to be true. At the end of the treatment period, the researcher hopes that differential effects have occurred due to the different treatments and that the null hypothesis can then be rejected. If the hypothesis is rejected, then strong evidence exists

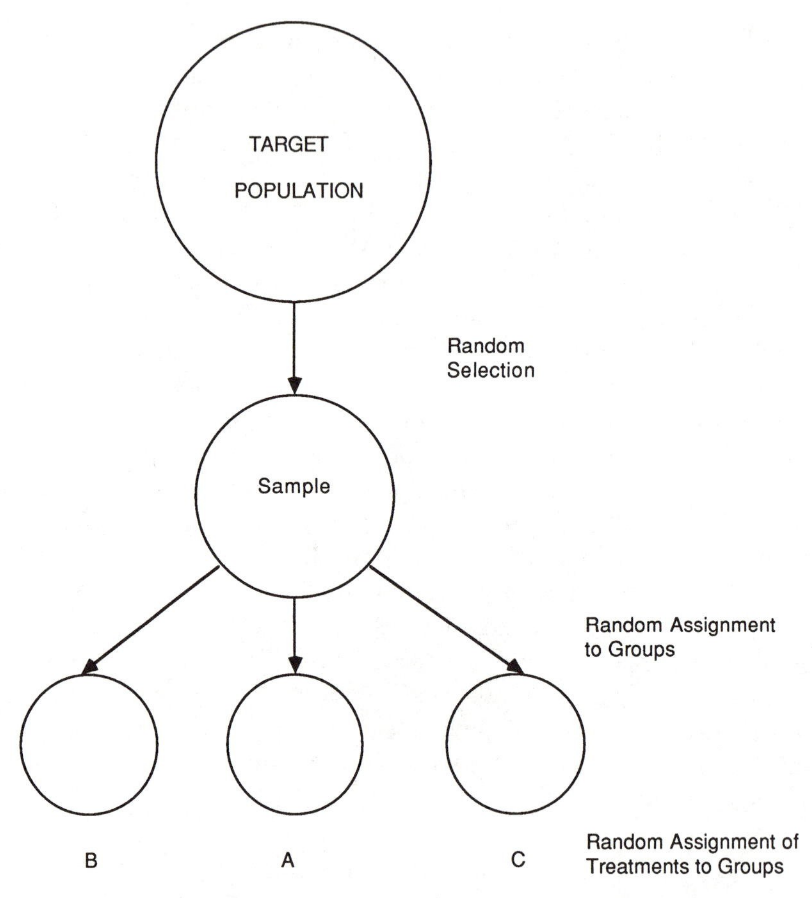

Figure 8.5. Complete randomized design.

for cause-effect relationships because factors important to the study were controlled through randomization.

In practice, the completely randomized design is often not possible to use. For example, when studying productivity ratios of males and females, or for workers at two or more different plants, random assignment to groups is not possible, as depicted in Figure 8.6. In this situation, elements can be randomly selected from the two defined populations (e.g., males and females or plants A and B), so that each group studied represents the respective population. However, if the hypothesis of no significant difference between or among groups is rejected, then the significant differences may be due to inherent differences between or among the populations involved as well as treatment effects. Thus, determining cause-effect relation-

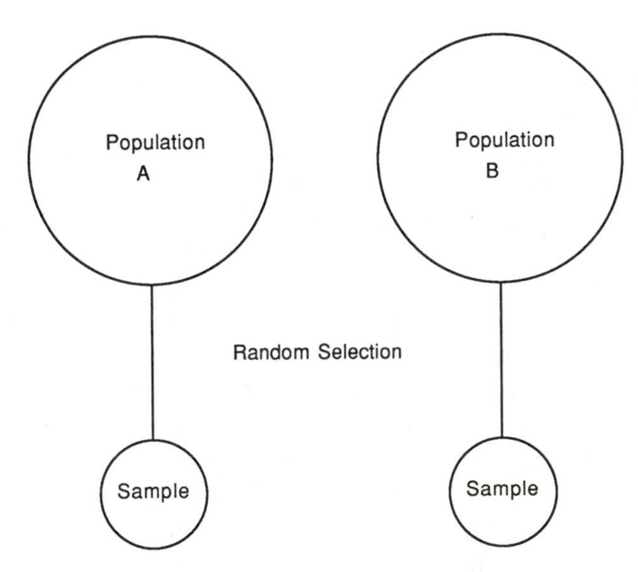

Figure 8.6. Example of random selection but no random assignment to groups.

ships is more difficult. For example, if productivity rates for two procedures at two plants are found to be significantly different, and a completely randomized design could not be used to compare these two procedures, then the differences might be due to the procedures, plant differences, or a combination of these two factors.

Many different sampling procedures can be used. In sampling theory, an *element* or case is defined as the unit about which information is collected and that provides the basis for analysis. Elements could be individuals, specimens taken from a production run, groups of workers, an entire plant taken as a unit, or an entire community taken as a unit. A *population* is the specified aggregation of all elements and often is designated as the *sampled population*, from which the elements are taken, and the *target population*, to which the results are to be generalized. Ideally, these populations should be the same and in practice are very close. For example, a sample from a target population of plastic produced by a specific production process may be only a sample from a few production lots, which is the sampled population rather than samples from all production lots. However, if a large number of production lots are sampled from, then the differences between sampled and target populations will be small and probably inconsequential.

The basic sampling method is *simple random sampling* (SRS).

This method is usually done by using a table of random numbers such as Table A in the Appendix or as found in statistics textbooks, or in published tables (*14–16*).

To obtain a SRS, the researcher first sequentially numbers each element in the population to be sampled 1 to N, which is the total number of elements in the population. Then after deciding n, the number of elements to be included in the sample, the researcher randomly selects a place to begin obtaining numbers from the table of random numbers. The general rule is that once a starting place is selected, the numbers are read in any direction from that starting place, and the numbers between 1 and N are recorded until n usable numbers are obtained with no repetitions. Because these tables are considered as sampling with replacement, the same number can be randomly placed at any point in the listing. If repetition of a number is found, then this number is discarded and the next usable number is included in the sample.

Simple random sampling can be used when dealing with a finite population, and all elements can be numbered. However, this situation is often not possible, especially when dealing with quality control or when dealing with infinite populations. For such situations, *systematic sampling* can be used. For this type of sampling, every kth element is chosen for inclusion in the sample (e.g., every 100th item, or 10 out of every 500 items). If done in an acceptable way, this type of sampling can be considered as identical to random sampling for all practical purposes. One caution must be exercised when using systematic sampling: the arrangements of elements in the population, known as *periodicity*. If the population listing is based on a cyclical pattern (e.g., listing of production runs in terms of yield so that the first listed is always the most productive), then by taking the sixth production run and then every sixth run, the highest five production runs would never be included. For such a situation, a simple random order that would be different for each production run could be taken. Systematic sampling is usually more convenient and often is just as accurate as simple random sampling. As with simple random sampling, statisticians must make decisions regarding the frequency of sampling and to ensure that sampling is done in a way to totally represent the target population in an unbiased way.

If the population is known to contain elements that can be divided into a number of subpopulations, or strata, then a random sample may be taken from each of these subgroups. This procedure is known as *stratified sampling*. Stratified sampling is based on the principle that the sample will be more representative of the population if it represents each segment of that population. This type of sampling also permits the researcher to study each subgroup as well as the total population. For example, if there are three plants that vary in size, then samples of production runs can be taken in proportion to the size of each plant.

Other types of sampling that can be done were described by Kish (*13*) and Cochran (*17*). For example, *cluster sampling* can be used in which the population is divided into a number of subgroupings usually based on geographical area, as for marketing research. Then, entire subgroups or clusters are taken at random to be the sample. For example, in testing the acceptance of a new product, several cities could randomly be selected rather than randomly selecting individuals from all cities. Often, this group sampling is done by a multistage sampling procedure in which a random sample of 10 states might be selected; then five cities within each state randomly selected; and finally, all people living in each of five randomly selected city blocks in each city are surveyed. The main advantage to cluster sampling is the reduced cost of collecting data. The same type of sampling could be used to obtain water samples, exposure to hazardous wastes, and other studies involving large geographic areas. Sometimes a double or split sample is taken that provides for a check of sampling equipment or procedures for accuracy.

All of these sampling procedures are based on probability sampling theory, which is based on random sampling. Other types of sampling might be used in special circumstances. However, these nonprobability sampling procedures limit the types of inferences that can legitimately be made. One type of nonprobability sampling is *purposive* or *judgment sampling*, in which elements are selected on the basis of the researcher's knowledge of the population. The risk here is that another person may not select the same type of elements, and selection bias results. Another type of sample is to rely on those available, which could result in volunteers or a sample that is convenient but not representative. Finally, some studies rely on *quota sampling*, which often happens in opinion or marketing research. Those collecting data are instructed to obtain a specified number of elements, and often those that can be obtained with the least effort are included in the sample. Because these nonprobability sampling procedures could result in nonrepresentative, biased samples, none are recommended for collecting data for statistical analyses from which inferences are to be made.

References

1. Belz, M. H. *Statistical Methods for the Process Industries;* Macmillan: London, 1973.
2. Cohen, J. *Statistical Power Analysis for the Behavioral Sciences;* Academic: New York, 1977.
3. Glass, G. V.; Stanley, J. C. *Statistical Methods in Education and Psychology;* Prentice Hall: Englewood Cliffs, NJ, 1970.
4. Hays, W. L.; Winkler, R. L. *Statistics;* Holt, Rinehart & Winston: New York, 1971.
5. Kirk, R. E. *Elementary Statistics;* Wadsworth: Belmont, CA, 1984.
6. Keppel, G. *Design and Analysis: A Researcher's Handbook;* Prentice Hall: Englewood Cliffs, NJ, 1973.
7. Natrella, M. G. *Experimental Statistics;* U. S. Government Printing Office: Washington, DC, 1966.
8. Owen, D. B. *Handbook of Statistical Tables;* Addison–Wesley: Reading, MA, 1962.
9. Davies, O. L.; Goldsmith, P. L. *Statistical Methods in Research and Production;* Longman Group: London, 1976.
10. Dixon, W. J.; Massey, F. J. *Introduction to Statistical Analyses;* McGraw–Hill: New York, 1969.
11. Guenther, W. C. *Am. Statistician* **1981,** *35*(*4*), 243–244.
12. Guilford, J. P.; Fruchter, B. *Fundamental Statistics in Education and Psychology;* McGraw–Hill: New York, 1978.
13. Kish, L. *Survey Sampling;* Wiley: New York, 1965.
14. Beyer, W. H. *Handbook of Tables for Probability and Statistics;* Chemical Rubber Company: Cleveland, OH, 1966.
15. Fisher, R. A.; Yates, F. *Statistical Tables;* Hafner: New York, 1953.
16. Rand Corporation. *A Million Random Digits;* Free Press: Glencoe, IL, 1955.
17. Cochran, W. G. *Sampling Statistics;* Wiley: New York, 1963.

Chapter 9

Inferences from One Data Set

9.1 Estimation of Population Parameters

Inferential statistics is frequently used in chemical studies to estimate population parameters from one sample of data. For example, the average sulfur content of a product might be determined by taking a sample from several production lots, or the purity of a chemical could be tested by analyzing a sample from several production runs. Estimating parameters from single samples provides the basis for quality control charts such as those developed by Shewhart (*1*).

The use of sample statistics as estimators of population parameters is based upon a number of principles and procedures, and each parameter can be estimated from its sample counterpart (e.g., X and S). Estimators of population parameters are random variables that vary from sample to sample and are distributed in a specific way. This distribution is referred to as the *sampling distribution* of that statistic. An estimator is computed by using specific rules, usually in the form of equations, that must be followed so that the estimates will be the *best unbiased estimates* of those parameters.

Three major characteristics or properties are desired for a sample statistic to be considered as a good estimator of a population parameter. The first property is that it should be unbiased. An estimator is unbiased if, for a large number of random samples, the mean of the estimates equals the mean of the population parameter it estimates. The second property is consistency, which deals with the accuracy of the estimator. A sample estimator should be closer to the population parameter as the sample size increases, and the prin-

1453–0/88/0163$07.00/1

ciple underlying this concept is the *law of large numbers*. As the size of n increases, and if other factors are equal, the standard error (SE) of the estimator will decrease, and thus the estimator becomes more accurate. The third property deals with relative efficiency. If two different sample statistics can be calculated from the same data, and if both are unbiased estimators of the same parameter, the estimator that has the smaller standard error will be more efficient, if other factors are held constant. This property is determined by comparing the standard errors of the two statistics by forming a ratio of the two standard errors. For example, the mean has been demonstrated to be a more efficient estimator of the population mean than the median, even though both have the same value in a unimodal normal distribution. This characteristic is also referred to as the minimum variance unbiased estimator.

Two types of estimation approaches are commonly used. The first approach provides an answer to the question, "What is the best guess as to what the population parameter might be. This approach uses a single value to estimate the population parameter and is known as *point estimation*. The value used is the sample statistic corresponding to the parameter being estimated. For example, if a researcher were interested in estimating the average sulfur content in a product, then he or she would use the sample mean, $\overline{X}$, to estimate the population mean, μ. Although the sample mean provides a point estimation of the population mean and is used in inferential statistics, especially with regard to hypothesis testing, the sample mean varies from one sample to another. The sample mean usually does not exactly represent the population mean, and this difference always has to be taken into consideration.

Because any sample statistic is subject to sampling fluctuation, the second approach to estimating population parameters shows the extent of that fluctuation. The second approach provides an answer to the question, "Can an interval be constructed that, with a specified degree of confidence, would bracket the population parameter?" This approach uses the point estimation as the best single estimate of the population parameter, and then determines the range of values from that point that could be expected on the basis of chance fluctuations. This determination is done by attaching a plus or minus value ($\pm$) to the point estimation to indicate the amount of variation or error involved. This range of values, called a *confidence interval* or the fiducial limits, provides the lower and upper limits to which the population parameter has a high probability of being included. The probability that the interval includes the parameter is the *confidence coefficient*. The confidence coefficient is set by the researcher, and this value determines the size of the interval. As with hypothesis testing, the two confidence coefficients usually used are the 95% and 99% levels corresponding to, respectively, the 0.05 and the 0.01 levels of significance. The confidence coefficient is usually expressed as $100(1 - \alpha)$.

These limits can be calculated for any statistic, such as a standard deviation or a proportion, and such limits provide a more useful and revealing guide to interpreting estimates of population values than point estimates. All of these interval estimates are based on the concept of random sampling and standard errors that can be computed for each statistic, and all use a known theoretical or empirical distribution as the basis for determining the confidence interval. For example, approximately 68% of the values in a normal distribution are expected to fall in the interval from one standard deviation unit below the mean to one standard deviation unit above the mean. Thus, this range could be considered as the confidence limit within which 68% of the values would fall, provided that the distribution is normally distributed. The confidence limits for other confidence coefficients are interpreted similarly by using the appropriate percentages.

All of the confidence intervals are interpreted similarily on the basis of the probability that the interval will cover or include the parameter being estimated. That is, the rule for computing the confidence limits is designed so that if the same equation is used for all possible random samples of the same size drawn from a population with a given parameter, the confidence interval will include the parameter in a specified percentage of the samples. A 95% confidence interval is interpreted as one of the possible intervals that could be computed, and 95% of such intervals would contain the population parameter. Or, one could say that the statistician is 95% confident that a parameter falls within the confidence interval. The parameter being estimated is fixed and the intervals vary from sample to sample. Thus, for the 95% confidence coefficient, if 100 confidence intervals were computed, 95 of them would include the parameter.

The confidence intervals are centered around a sample statistic, not the population parameter. For example, if the sample mean was computed to be 40 and the confidence limits were computed so that the lower limit was 38 and the upper limit was 42, it would be incorrect to say that the population mean falls within this interval at a given level of probability. Once the sample mean is included in the confidence statement, that specific confidence interval does or does not include the population mean. Thus, for a 95% confidence interval, it is not correct to say that the probability is 0.95 that the population mean falls within this interval. The correct statement would be that a 95% confidence interval for the population mean is 38–42, or that one is 95% confident that the population mean falls within this interval.

Confidence intervals provide more information than point estimation, which is used in testing hypotheses about a population parameter. A confidence interval can be thought of as an interval containing all values of a parameter for which the null hypothesis would not be rejected at the corresponding level of significance. Only values falling outside of the confidence limits would lead to rejecting the null hypothesis. Thus, a confidence limit leads to the same conclusion as testing a null hypothesis dealing with the same parameter, but provides more information. The test of a null hypothesis is a yes/no decision, whereas the confidence interval gives the range of values for the parameter. As with directional and nondirectional hypotheses, confidence intervals can be constructed for a one-sided or a two-sided interval estimate. Because confidence intervals provide more information than a test of a null hypothesis, confidence intervals are recommended to be computed and reported in research reports.

Freund (*2, 3*), Hays and Winkler (*4*), and Kirk (*5*) provided detailed information on estimating population parameters. Natrella (*6*) gave examples of sampling distributions for estimating the population mean for sample sizes of $n = 4$, 100, and 1000. These distributions graphically illustrate the concept of random intervals computed from random samples for each sample size.

9.2 The Population Mean

Estimating what a population mean might be is probably the most often used procedure in research studies. Usually, two types of problems are encountered. The first problem is when no information is available as to what the population mean might be, and the second problem is when the population mean is known and a researcher wants to know if a sample can be considered as a random sample from that population. The first problem involves estimating what the value of the population mean might be, and the second problem involves testing a hypothesis regarding the hypothesized mean. Both problems are based on establishing a confidence interval for the mean by using the standard error of the mean.

9.2.1 Estimating Population Mean

To make an inference about the population mean, a researcher needs to know the sampling distribution of the sample means, and to know this, the researcher has to specify the form of the population distribution. However, often information about the shape of the population is not known. This lack of knowledge is not a problem because the sampling distribution of sample means is given in the *central limit theorem*. This theorem states that if a population has a finite variance σ^2 and mean μ, then the distribution of sample means from random samples of n independent observations approaches a normal distribution with standard deviation $\sigma\sqrt{n}$ and mean equal to μ as the sample size increases (*4*, p. 292; *5*, p. 228). Thus, regardless of the shape of the population distribution, if the sample size is large enough, the means will be approximately normally distributed. The sampling distribution for many of the commonly used sample statistics will also be approximately normally distributed with increasing sample size. The size of n needed depends upon the shape of the population distribution. However, the sample distribution of the mean will be approximately normal even for moderate-size samples such as when $n < 30$ (*2*, p. 205).

The sampling distribution of the mean, as stated in the central limit theorem, has a standard deviation of $\sigma\sqrt{n}$. To avoid confusion of using the term standard deviation to refer to both the measure of dispersion of values in a frequency distribution and statistics in sampling distributions, the term *standard error* is usually used for sampling distributions. Thus, σ_m or SE_M will refer to the standard error of the mean, whereas other subscripts will be used for other statistics. When the population standard deviation is known, then

$$SE_m = \sigma/\sqrt{n} \qquad (9.1)$$

where σ_m is the standard error of the mean, σ is the population parameter standard deviation, and n is the number of cases in the sample.

However, in practice, the population standard deviation is rarely if ever known and has to be estimated. Equation 5.7 was recommended as providing an unbiased estimate of the population variance. For large sample sizes (i.e., $n > 100$), the standard error of the mean can be estimated by

$$\mathrm{SE}_m = S/\sqrt{n} \tag{9.2}$$

where S is the unbiased estimate of the population σ, and n is the sample size. The value S is computed by

$$S = \sqrt{\frac{\sum x^2}{n - 1}} \tag{9.3}$$

where Σx^2 is the sum of squared deviations about the mean. The estimate of the population standard error of the mean can be estimated directly from the sum of squares by

$$\mathrm{SE}_m = \sqrt{\frac{\sum x^2}{n(n - 1)}} \tag{9.4}$$

The standard error of the mean, σ_m or SE_M, is the standard deviation of the distribution of sample means obtained from repeated samples of n cases taken at random. In other words, the standard error shows where sample means would probably fall for random samples of size n taken from the same population. The SE_M is interpreted in the same way as S: by using the normal curve table. This interpretation can be relied upon because the central limit theorem states that the distributions of means can be approximated very closely with a normal curve having a mean μ and standard deviation $\sigma/\sqrt{n}$ (*2*, p. 205). For estimating the population mean, a confidence interval can be set up by using SE_M that will include the population mean in a given percentage of all possible random samples.

9.2.2 Confidence Intervals or Limits

Confidence intervals or limits are set up at a given level of confidence set by the researcher. The two most commonly used levels are the 95% and the 99% confidence limits. Two approaches are used to determine these confidence limits. When the population standard deviation is known, or for large samples, which most statisticians define as any sample size greater than $n = 100$, the confidence interval for the 95% confidence coefficient can be computed as

$$\text{CL} = \overline{X} \pm 1.96(\text{SE}_M) \tag{9.5}$$

and for the 99% confidence limits,

$$\text{CL} = \overline{X} \pm 2.58(\text{SE}_M) \tag{9.6}$$

where CL denotes the confidence limits. For other significance levels, the appropriate z value from the table of the normal curve would be used. For example, at the 50% confidence limits, $z = 0.6745$ would be used in place of 1.96 in equation 9.5.

When the population standard deviation is not known, or for small samples, (i.e., samples with $n \leq 100$) for which an estimate of the population has to be used because this population parameter is seldom known, the confidence limit is calculated by using the t statistic rather than z in equations 9.5 and 9.6. Thus, for small samples, the confidence limits are computed by using the general equation

$$\text{CL} = \overline{X} \pm t_p(\text{SE}_M) \tag{9.7}$$

In this equation, t_p is the t value taken from Table C in the Appendix for the level of confidence desired for df $= n - 1$, where df is the degrees of freedom. The reason for using the t value in place of the z value is that, in computing the confidence interval, two variables are used, the sample mean and the estimated population standard deviation. For small samples, these two variables may not be exactly normally distributed, and in fact, tend to be increasingly leptokurtic as the size of the sample becomes smaller. For large-size samples, the sample distributions are usually normally distributed, if the distribution is normal in the population, so the normal curve statistic will be applicable. For small-size samples, the t statistic applies even for very small samples. As can be noted from Table C, as the sample size increases, the t value approaches the normal curve value. For samples of $n = 100$, the t value would be around 1.98 compared to $z = 1.96$ at the 0.95 confidence level.

To estimate the population mean by using a large sample, consider the following example. A company wishes to determine the expected lifetime of tires made from a new synthetic material. A simple random sample of 100 tires is subjected to a forced life test.

The lifetime is recorded, and the mean and standard deviation are computed for each of the 100 lifetime values. The question to be answered is what 95% confidence limit or interval estimate can be made for the mean lifetime of all tires made from that material. The answer will be obtained by calculating the interval given by equation 9.5.

Suppose the mean for the sample of 100 tires was 32,500 mi, and the standard deviation was computed as 3000 mi by using equation 5.7. For this data, the standard error of the mean would be computed by using equation 9.2:

$$\mathrm{SE}_M = \frac{3000}{\sqrt{100}} = 300$$

Thus, 95% of the means from random samples of $n = 100$ each would fall in the interval given by equation 9.5, or, for this example,

$$\mathrm{CL} = 32{,}600 \pm 1.96(300)$$

$$\mathrm{CL} = 32{,}600 \pm 588 = 32{,}012 - 33{,}188$$

As depicted in Figure 9.1, the confidence limits for this sample are from 32,012 to 33,188 mi. Thus, μ would be included within these limits for 95% of the sample intervals that could be computed on the basis of samples of $n = 100$. If the researchers wanted to be more confident that the estimate would include the population mean, they could use the 0.99 confidence limits. These limits would then be

$$\mathrm{CL} = 32{,}600 \pm 2.58(300)$$

$$\mathrm{CL} = 32{,}600 \pm 774$$

or from 31,826 to 33,374 mi.

For this same example of estimating the mean tire lifetime for the population, suppose only 20 tires rather than 100 could be tested. Also, suppose that the same sample mean of 32,500 mi and standard deviation of 3000 were obtained. For this small sample,

$$\mathrm{SE}_M = \frac{3000}{\sqrt{20}} = 671.14$$

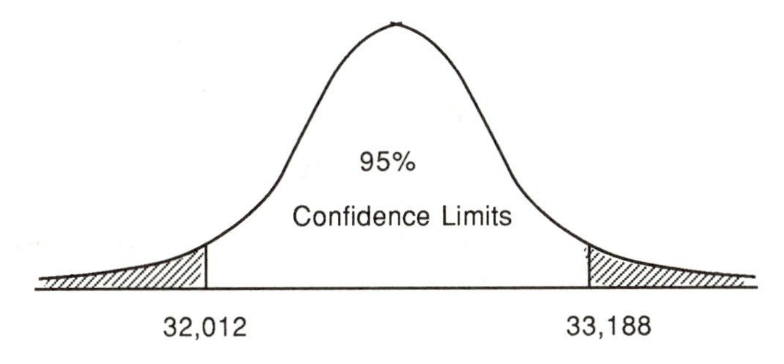

Figure 9.1. Confidence limits for sample means based on samples of n = *100 for lifetime of tires.*

Because this sample is considered a small sample, the t value should be used in place of z to determine the confidence limits. When using the t test, the degrees of freedom associated for each use of the t distribution must be computed. *Degrees of freedom* refers to the number of observations minus the number of restrictions placed upon those observations. Usually, these restrictions are statistics that must be calculated from the sample data necessary to calculate the statistic being used. The usual rule for computing the degrees of freedom is that one degree of freedom is lost for each statistic used to calculate another statistic. For this use of t, the mean has been used as the central point below and above which the confidence interval is set. Thus, for this application of t, there will be $n - 1$, or $20 - 1 = 19$ degrees of freedom. From Table C, the t value required at the 0.05 level or for the 95% confidence level is 2.093. By using equation 9.7, the confidence limits would then be

$$\text{CL} = 32{,}600 \pm 2.093(671)$$

or from 31,195.6 to 34004.4 mi. This estimate is wider for two reasons. First, other conditions remaining constant, the standard error (SE) is larger for small samples because it varies inversely with n. Second, using t, which is larger than z for small samples, results in a larger confidence interval, which is considered a more conservative estimate.

9.2.3 Probability Questions

Closely related to determining the confidence limits within which the population is likely to fall are questions dealing with the probability that the mean is likely to fall below or above certain values, or determining the probability that means might differ from the population mean by a specified amount. The procedures that can be used to answer these types of questions are similar to those regarding the distribution of values in a sample as described in Sections 7.3 and 7.4. The only difference is that the standard error of the mean is used rather than the standard deviation because these questions deal with the distribution of sample means rather than individual data points or values.

The first type of question that is frequently encountered regards the probability that sample means of a specified value or higher might occur. To answer this type of question, the mean and standard error of the mean must be computed. As an example, suppose that the mean lifetime of light bulbs with a new type of fusing process is to be determined. Specifically, an electrical engineer wants to know how often samples with a mean lifetime of 1200 h or longer might be obtained. A sample of 144 light bulbs is tested, and the mean lifetime is 1168 h with a standard deviation of 240. The standard error of the mean is needed, and for this problem would be $240/\sqrt{144}$, or $\sigma_m = 20$. A ratio could then be formed by taking the difference between the sample mean of 1168 and the expected mean of 1200, divided by the standard error of the mean. For these data,

$$z = \frac{1200 - 1168}{20} = 1.6$$

The engineer would then go to the table of the normal curve, which is Table B in the Appendix. For $z = 1.6$, 0.0548 of the area lies under the normal curve at and above that point. Thus, in random sampling of samples with $n = 144$, about 5.5% of the samples of 144 light bulbs would be expected to have a mean life of 1200 h or longer.

Closely related to this problem is one frequently encountered in quality control work that might state that fewer than 10% of sample averages or means could be below a specified standard, otherwise the total order might be rejected. For the same light bulb data, suppose the minimum acceptable mean lifetime would be 1150 h, and not more than 10% of the means for samples taken can be below this value. What is the probability that samples of this new product will be rejected by this requirement? To answer this type of question, the same data are required, and the z ratio is used. The only difference is that now the percentage of the means that will fall at or below a certain point is given, and the problem is to find that point. From Table C, the z value below which 10% of the cases fall must first be found. This $z = -1.28$, or 1.28 standard errors below the sample mean. The product -1.28×20, which is the standard error of the mean, yields a time of -25.6 h. This value is then subtracted from the sample mean of 1168 for a point value of 1142.4 h. Thus, 10% of the means are likely to fall at or below the value of 1142.4, which is below the acceptable standard. Thus, more than 10% of the samples of bulbs would have an expected mean lifetime of less than 1150 h, and the total order would probably be rejected.

Another type of problem sometimes encountered deals with bid specifications, in which the lower and upper limits must be specified within which a specified percentage of a product will fall. For the same light bulb problem, suppose the bid requirement is for the 80% confidence limits. The question then is between which two mean lifetime levels will 80% of the sample means fall. To solve this problem, the z values must be found that cut off the bottom and top 10%, leaving the middle 80%. From Table B this value is $z = 1.28$, which is the same as for the previous problem, but now the limits are calculated from points below and above the sample mean. For this problem, the limits would be $1168 \pm 1.28(20)$, or mean lifetime limits of 1142.4–1193.6 h.

9.3 Hypothesis Testing for a Single Mean

In many situations, a given or known standard serves as the desired level of performance or requirement for some process. In other situations, a hypothetical or theoretical value should be attained. For example, in quality control, standards may be set that have to be adhered to. In such situations, the question to be answered is whether a sample of some variable or a factor is within tolerable limits to be considered as a random sample from a population with the stated values.

To answer questions of this type, the question can be stated in the form of a statistical hypothesis. For example, for a situation involving a specification that the mean level of impurity in a chemical will not exceed 1.6%, the hypothesis could be that no significant difference occurs between the mean taken for a sampling of the chemical and the specified criteria that would be considered the population mean. In symbolic form, the null and alternative hypotheses (H_0 and H_1, respectively) would be

$$H_0: \leq 1.6 \text{ and } H_1: > 1.6$$

The null hypothesis states that the difference between the estimated population mean and the population mean of 1.6 will be equal to or less than zero. In order to test this hypothesis, a random sample of n cases would have to be taken from production runs of that chemical, and then the mean and standard error of the mean would have to be found for the sample of n cases. Then, dependent on the sample size, equation 9.8 could be used for samples with $n >$ 100 or equation 9.9 could be used for samples with $n <$ 100. Both of these equations form a ratio of the difference between the sample mean and the population mean to the differences expected in sample means due to chance alone or random variation, which is defined as the amount of error variance. The resulting ratio is then compared to the theoretical distribution for such ratios to determine the probability that this ratio could occur because of chance variation.

On the basis of chance, statisticians know that differences will exist, and if nothing but chance is operating, then the denominator and the numerator should be equivalent, and the ratio should then be equal to about $z = 1$ or $t = 1$. That is, if both the difference between the sample mean and the population mean and the difference that could be obtained purely on the basis of chance are both due to just random variations or sampling error, then these two estimates should be equivalent. If there is a real difference, then the ratio should be greater than 1, indicating that the obtained difference is larger than what would typically be obtained by chance alone. The two ratios for testing the hypothesis comparing sample with populations means are

$$z = \frac{\overline{X} - \mu}{SE_M} \tag{9.8}$$

$$t = \frac{\overline{X} - \mu}{SE_M} \tag{9.9}$$

where $\overline{X}$ is the mean of the sample, SE_M is the standard error of the mean, and μ is the population or hypothetical mean.

For the impurity example, if 144 samples of the chemical were taken, and the mean and standard deviation for this sample were, respectively, 1.68 and 0.6, then, because a large-size sample is used, equation 9.8 can be used. The standard error of the mean (SE_M) would be estimated as

$$SE_M = \frac{0.6}{\sqrt{144}} = \frac{0.6}{12} = 0.05$$

Then,

$$z = \frac{1.68 - 1.60}{0.05} = 1.60$$

For this example, researchers are only interested if the level of impurity exceeds the specified level set, a one-tailed test would be made. From Table B, at the 0.05 level of significance, a $z = 1.645$ is needed for a one-tailed test. The obtained z does not equal or exceed the z value required; thus, the null hypothesis is not rejected and on the basis of the sample data, the chemical is concluded to be within the required specifications for maximum allowable impurities. A 95% confidence interval ($CI_{0.05}$) for μ obtained on the basis of this one sample would be, for a one-tailed test,

$$CI_{0.05} = 1.68 - 1.645(0.05) = 1.598\text{–}1.68$$

Because the specification standard of 1.60 would be included in this confidence interval, the researchers would come to the same conclusion that, based on the data collected from 144 samples of this chemical, the chemical does not deviate from the specified level of impurities. The researchers can be 95% confident that their result will be correct.

As another example, suppose that a new batch of ore is to be tested for its nickel content to determine if it is consistent with the usual 3.25% that has been found for previous batches. Ten samples are taken with the following results:

Sample	1	2	3	4	5	6	7	8	9	10
% *Nickel*	3.27	3.23	3.21	3.24	3.26	3.20	3.22	3.21	3.25	3.23

For these data, $\overline{X}$ = 3.232, S = 0.0236, n = 10, and σ_m = 0.0075. By taking μ = 3.25%, the hypothesis that $\overline{X} - \mu = 0$ at the 0.01 level of significance can be tested. The mean and standard deviation are computed for this sample to describe the sample in terms of mean level of nickel content and the variation within the sample. The standard error of the mean can be computed by using equation 9.2, and then, because the sample is small, the t ratio can be computed by using equation 9.9. With a sample mean of 3.232, standard deviation of 0.0236, and n = 10, the standard error of the mean is equal to 0.0075. Then,

$$t = \frac{3.232 - 3.25}{0.0075} = -2.40$$

With n = 10, the degrees of freedom would be $n - 1$, or $10 - 1 = 9$. From Table C, for nine degrees of freedom at the 0.01 level of significance, $t = \pm 3.250$ is required to reject the null hypothesis. For the data, because $t = -2.40$, researchers would fail to reject the null hypothesis and conclude that the sample is not significantly different from the 3.25% nickel content expected for such ores. Had the null hypothesis been tested at the 0.05 level of significance, the researchers would have concluded that the sample ore is significantly different in nickel content because the obtained $t = -2.40$ is greater than the table value of $t = 2.262$ at the 0.05 level of significance.

As a check on the conclusion and to show where sample means are likely to fall for repeated samples, the confidence limits of the mean can be shown. At the 99% confidence limit, which corresponds to the 0.01 level of significance for testing the null hypothesis, and by using equation 9.7, the lower and upper limits would be 3.232 ± 3.25(0.0075), or 3.232 ± 0.024 = 3.208–3.256. The interval brackets the population mean of 3.25; thus, the null hypothesis would not be rejected. This conclusion is the same as the one reached with $t = -2.40$. At the 0.05 level of significance corresponding to the 95% confidence limits, the lower and upper limits would be 3.232 ± 2.262(0.0075), or 3.232 ± 0.0170 = 3.215–3.249. For this level of confidence, the interval does not bracket the population mean of 3.25, and thus the null hypothesis would be rejected as it was at this level of significance.

As noted in Section 9.1, hypothesis testing involves point estimation because the statistician is interested in a specific population value. For many decisions, this estimate is all that is required. However, interval estimation provides the same type of information plus information about many possible values for the hypothesized mean. Thus, interval estimation is recommended as a way of communicating more information about the data.

As another example, suppose that the Environmental Protection Agency (EPA) sets a standard limit of 1.3% total phosphate–phosphorus content in waste water from chemical plants. Over a 3-month period, 200 periodic samples of waste water are taken and analyzed for phosphate–phosphorus content. Results indicate that the mean percentage content is 1.34% with a standard deviation of 0.14%. Is the waste water from this plant within the EPA guidelines? Or, in hypothesis form: Is there a significant difference between the sample mean and the hypothesized population mean? Because the EPA would be only interested in whether the plant exceeded the maximum set percentage, a directional hypothesis should be used. Thus, the hypothesis to be tested would be H_0: $\mu \leq 1.3$. In other words, is the sample mean significantly greater than the set standard taken as the hypothetical mean? The alternative hypothesis would be H_1: $\mu > 1.3$. A one-tailed procedure can be used. Because the sample is considered as a large sample, either the z ratio or the t ratio could be used. For convenience, equation 9.8 will be used. The standard error of the mean would be $0.14 \div \sqrt{200} = 0.0099$. Then,

$$z = \frac{1.34 - 1.30}{0.0099} = 4.04$$

For a one-tailed test of the directional hypothesis, the table value for z would be 1.65 at the 0.05 level of significance and 2.33 at the 0.01 level. Because the computed value of z is greater than the table value for z at both the 0.05 and the 0.01 levels of confidence, the H_0 of no difference, which is the hypothesis that is tested, is rejected, and the phosphate–phosphorus content in the sample is concluded to significantly exceed the EPA maximum requirement of only 1.3%. The 95% confidence level for this problem would be $1.34 \pm 1.65(0.0099) = 1.34 \pm 0.016$, which gives the confidence interval from 1.324 to 1.34 because the EPA is interested in only one direction (i.e., to determine whether the EPA standard of 1.3% could be considered a random variation from the obtained mean of 1.34%). The results are consistent with the test of the hypothesis in that the hypothesized population mean would not be considered as a random sample variation but beyond the chance limits and thus significantly different.

9.4 Standard Error of Frequencies, Percentages, and Proportions

Some situations in research and quality control deal with frequency counts, percentages, or proportions (e.g., the number of defects per 100 for a production run, the percentage of an element in a formulation, or the proportion of cures for a new medication). For each of these instances, the stability of each of these statistics is very important. Estimates of the population values for these statistics can be made in the same manner that the sampling distributions of the mean were made (i.e., by using the sampling distribution of the binomial distribution, which approaches the normal distribution for large samples).

Suppose that for a sample of pills pressed by a certain type of machine, 27 out of 100 are found to be chipped. The question that might be asked is what could be expected for other samples of these pills. To answer this question, the binomial sampling should be used with the mean equal to np and the variance equal to npq, where p is the proportion of chipped cases, and $q = 1 - p$. For this and other problems involving frequency data, the frequencies are converted to proportions by dividing by 100. For the pill problem, $n = 100$, $p = 0.27$, and $q = 1.00 - 0.27 = 0.73$. The mean for this binomial distribution would be equal to np, or $100 \times 0.27 = 27$, and the variance is npq, or $100 \times 0.27 \times 0.73 = 19.71$. The square root of the variance gives the standard deviation, which is 4.44. Because a sampling distribution is being dealt with, this number is also the standard error of the mean frequency. When samples are large, the observed mean frequency can be considered a good estimate of the population mean, and the distribution of frequencies can be interpreted as continuous data. Thus, the normal curve can be used for frequency data in a similar way as for interval data, especially when p is between 0.27 and 0.73, and the SE can be interpreted as other SE values. This approach will hold for samples in which $np \geq 10$ or more. For this problem, the minimum frequency is $100 \times 0.27 = 27$, so the normal approximation can be applied. The 95% confidence limits would be computed by $1.96 \times SE_p$ in the same way as the SE_M, where SE_p is the standard error of the proportion of cases. For the pill problem, $1.96 \times 4.44 = 8.70$. Thus, the mean frequency of chipped pills for similar random samples would be expected to range from 27 ± 8.7, or 18.3–35.7 pills/100 pills. For the 0.99 confidence limits, the interval for the mean frequency would be $\pm 2.58(4.44)$, or in other words, the mean number of chipped pills would be expected to fall between 15.54 and 38.46 pills/100 pills.

For data in the form of frequencies, the general equation for computing the standard error of a frequency (SE_f) is

$$SE_f = \sqrt{npq} \qquad (9.10)$$

where n is the number of elements or cases in the sample, p is the proportion of cases in the category of interest, and $q = 1 - p$. As stated previously, this standard error of a frequency is interpreted in a manner similar to the SE_M by using the normal curve.

For data in the form of proportions, the standard error (SE_p) is computed by

$$SE_p = \sqrt{\frac{pq}{n}} \qquad (9.11)$$

where the terms are the same as in equation 9.10. Suppose that a random sample of 200 persons are asked if they would buy a product containing a controversial chemical ingredient, and only 70 indicate that they would buy that product, which is a proportion of 0.35 of the sample. The question of concern is what proportion of the target population could be expected to buy this product. From equation 9.11,

$$SE_p = \sqrt{\frac{0.35(0.65)}{200}} = 0.0337$$

The 95% confidence limit would then be $\pm 1.96(0.034) = \pm 0.067$, or from $p = 0.283$ to 0.417. Thus, for 95% of other random samples of $n = 200$, the proportion that would be expected to buy the product would likely be from 28% to 42%, or for each sample of $n = 200$, from 56 to 84 persons would likely buy the product. The same results would have been obtained by using equation 9.10.

If the data were in the form of percentages, the standard error would simply be 100 times the value obtained by equation 9.11, or

$$SE_{\%} = 100\sqrt{\frac{p(q)}{n}} = \sqrt{\frac{P(Q)}{n}} \qquad (9.12)$$

where $P = 100(p)$, and $Q = 100 - P$.

The standard error would be interpreted in the same way. For the same problem in terms of percentages, $P = 35$ and $Q = 65$. Then,

$$SE_{\%} = \sqrt{\frac{35(65)}{200}} = 3.37$$

which is 100 times the value obtained for the standard error of the proportion.

9.5 Standard Error of a Standard Deviation

In a number of situations, the variation or variance in products is important, especially with regard to products for which the tolerances are critical. Often in such situations, the variation expected for a total production run needs to be determined on the basis of a sample of items. To make such estimates, researchers can use the same general procedure as for means and frequencies. The standard error of a standard deviation (σ_S) is calculated as

$$SE_S = \frac{\sigma}{\sqrt{2(n)}} \qquad (9.13)$$

and is used in the same way as other standard errors (*2,* p. 235).

Suppose that for a steel alloy, the yield point is specified as 68,000 psi with an allowable tolerance of only 2000 psi. From a production run, a random sample of $n = 150$ segments of steel are tested, and the results indicate a mean yield point of 68,500 psi with a standard deviation of 1800. For the total production run, what proportion of other samples of production of this steel alloy would be likely to exceed the allowable variance? To answer this question, the standard error of the standard deviation has to be calculated, then this standard error is used to calculate a z ratio, and then the table of the normal curve is examined to see what proportion of cases will exceed this z value. From equation 9.13,

$$\mathrm{SE}_s = \frac{1800}{\sqrt{2(150)}} = 103.93$$

The z ratio would then be

$$z = \frac{\sigma_s - \sigma_p}{\mathrm{SE}_s} = \frac{1800 - 2000}{103.93} = -1.92$$

where σ_s is the standard deviation of the sample, σ_p is the specified or hypothesized standard deviation, and σ_S is the standard error of the standard deviation.

For $z = -1.92$, Table B shows that 2.74% of the area of the normal curve will fall at and above that point in the distribution. Thus, 2.74% of the samples are likely to exceed the specified allowable variation of 2000 pound per square inch. Because the researcher is only interested in the proportion of cases likely to exceed the specified range, a directional or one-tailed test would be called for. If the researcher was merely interested in how variable the total production of steel might be in terms of per square inch, then the 95% or 99% confidence limits would be calculated. For the 95% confidence level, the limits would be $\pm 1.96(103.93) = \pm 203.70$, or the standard deviation for other samples will probably be between 1596.30 and 2003.70.

The confidence interval for the standard deviation as given by Freund (*2*, p. 235) is to be used only for large samples, because equation 9.13 should be used only for large-size samples ($n > 100$). Kirk (*5*) and Hays and Winkler (*4*) gave an alternate procedure to be used for determining the confidence interval for the standard deviation. This procedure uses the chi-square statistic and can be used when sample sizes are less than $n = 30$.

9.6 Inferences from Correlation Coefficients

Like other statistics, the Pearson correlation coefficient will vary from sample to sample under the same influence of random sampling variation. As with other statistics, the correlation between two variables on one sample of data is usually computed, and researchers want to generalize to the total population. This generalization can be done by using either of the two basic techniques of statistical inference: point estimation and interval estimation. The basic underlying procedure is the same as for other statistics. That is, if an infinite number of samples of the same size are drawn at random from a given population, and r_{xy} is calculated for each sample, the tabulation of those r values will form a random sampling distribution of r for that particular size sample. Although the sample correlation is a slightly biased estimator of the population correlation rho (ρ), the mean of those correlations will be very close to the population mean. The amount of bias is slight and involves terms of the order of $1/n$. Thus, for large samples, the amount of bias can be ignored for practical purposes. The standard deviation of the distribution of sample correlations, which is the standard error of r (SE_r), is

$$SE_r = \frac{1}{\sqrt{n - 1}} \tag{9.14}$$

where n is the sample size.

Equation 9.14 can be used for correlation coefficients that are near 0, because then the distribution of the sample correlation is approximately normal. However, this equation is of little use when the absolute value of r is considerably greater than 0 (i.e., greater than $r = 0.50$) because as ρ departs from $\rho = 0$, the sampling distribution of r becomes increasingly skewed as r approaches ± 1.00. To show the reason for this skewness, an $r = +0.95$ can vary in one direction only from $+0.95$ to $+1.00$ but theoretically could vary in the other direction from 0.95 through 0 to -1.00. Thus, a negatively skewed distribution results. When $\rho = 0$ and the sample size is greater than $n > 30$, the sampling distribution is symmetrical and approximately normal. Thus, equation 9.14 would apply. However, for other values of ρ, other procedures are recommended.

One procedure for testing the significance of r is to use the t distribution. This procedure is usually used for point estimation or hypothesis testing. The general hypothesis tested is usually in the null form: No correlation is found between the two variables X and Y, or, H_0: $\rho = 0$. The t ratio can then be used to test this hypothesis by using the equation

$$t = r\sqrt{\frac{n - 2}{1 - r^2}} \tag{9.15}$$

The number of degrees of freedom for this t test is given in the numerator as $n - 2$, or two less than the number of pairs of values used to compute r.

Suppose that an engineer wants to determine if a significant correlation is found between tensile strength of plastic and length of time that each piece of plastic was baked at a uniform temperature. For a sample of 20 different time periods, the correlation between these two variables is $r = 0.45$. Does this represent a correlation significantly different from 0 ? The hypothesis to be tested would be H_0: $\rho = 0$. From equation 9.14,

$$t = 0.45 \sqrt{\frac{20 - 2}{1 - 0.45^2}} = 2.138$$

From Table C, for df = 18, at the 0.05 level of significance, $t \geq 2.10$ is required for significance and to reject the null hypothesis. The computed value, 2.138, is greater than the required table value. Thus, the engineer would reject the null hypothesis and conclude that a significant correlation is found between tensile strength and length of baking time. Had the hypothesis at the 0.01 level of significance been tested, the t required at df = 18 would be 2.878. The obtained $t = 2.138$ is less than this required or critical value of t. Thus, at the 0.01 level, the engineer would fail to reject the null hypothesis and would conclude that the correlation is merely chance that these two variables are not significantly related.

For convenience, statisticians have provided tables for testing the null hypothesis $\rho = 0$. The critical values of r for the 0.05 through the 0.005 levels for a one-tailed test are given in Table F in the Appendix. If the obtained r is equal to or greater than the critical value in the table for the alpha level set and for the appropriate degrees of freedom, the hypothesis of $\rho = 0$ can be rejected. For the last problem, the obtained correlation was $r = 0.45$. The degrees of freedom would equal $n - 2$, or $20 - 2 = 18$. From Table F, under the column for the 0.05 level of significance, a correlation of 0.444 is necessary to be considered significant for 18 degrees of freedom, and at the 0.01 level, a correlation of 0.561 would be required for a two-tailed test. Thus, the engineer would come to the same conclusion by using the t ratio. This conclusion is that the correlation obtained, $r = 0.45$, is significantly different from 0 at the 0.05 level of significance but not significantly different at the 0.01 level.

For interval estimation as to where ρ might fall based upon a sample r, Table F can also be used for levels of confidence for a two-tailed test from 90%–99%. For the example cited previously, at the 95% confidence level, for df = 18, any correlation coefficient between −0.443 and +0.443 would be considered a chance correlation, and those would be the limits within which the population correlation is likely to fall if $\rho = 0$. For confidence intervals of r when $r \neq 0$, because the correlation coefficient tends to be increasingly skewed as r approaches ±1.00, a transformation procedure that normalizes the distribution of r must be used. Such a procedure was developed by Fisher and is known as Fisher's z transformation. Values of z_r for each value of r are given by looking up the respective values from Table G in the Appendix. For example, for $r = 0.55$, $z_r = 0.618$, and for $r = 0.92$, $z_r = 1.589$. For negative values of r, z_r values are given a negative sign. The advantage of using the z_r distribution is that this distribution is approximately normal and is independent of the value of rho. Thus, z_r can be interpreted by using the normal curve distribution. The standard error of z_r (SE_{z_r}) is given by

$$SE_{z_r} = \frac{1}{\sqrt{n - 3}} \qquad (9.16)$$

Thus, the standard error of z_r depends only upon the sample size. The standard error of z_r can then be used to determine the confidence limits for r for any level of confidence desired.

For example, suppose that for a given correlation study, $r = 0.84$ was computed from a sample of $n = 165$. For $r = 0.84$, from Table G, $z_r = 1.221$. The standard error for z_r would be $1/\sqrt{162} = 0.079$. As with other standard errors, the confidence limits are computed by multiplying the standard error by the z value for the selected level of confidence. For the 95% confidence interval, $z_r \pm 1.96(SE_{z_r})$, or for this problem, $1.221 \pm 1.96\ (0.079)$, gives the interval for z_r from 1.066 to 1.376. These two z values can then be converted back to r values by using Table G, where in terms of r an interval would be obtained from $r = 0.79$ to 0.88. Thus, the engineer would be confident that 95% of the time the population correlation, ρ, would fall within that range for similar samples of 165 pairs of values.

If the Spearman rank-difference correlation coefficient was used, then the significance of this coefficient can be determined from Table

H in the Appendix. For the nondirectional two-tailed test of the null hypothesis that $\rho = 0$, Table H is entered with n equal to the number of pairs of ranked values in the sample. For example, suppose $\rho = 0.60$ for a sample of 18 individuals. To reject the null hypothesis that $\rho = 0$ at the 0.05 level of significance, $\rho \geq 0.474$. Thus, if $\rho = 0.60$, the null hypothesis would be rejected and the conclusion would be that a significant relationship exists between the two variables. However, at the 0.01 level of significance, $\rho \geq 0.600$ is needed; thus, at this level, the null hypothesis would not be rejected.

References

1. Shewhart, W. A. *Economic Control of the Quality of Manufactured Product;* Van Nostrand: New York, 1931.
2. Freund, J. *Modern Elementary Statistics;* Prentice–Hall: Englewood Cliffs, NJ, 1960.
3. Freund, J. *Mathematical Statistics;* Prentice–Hall: Englewood Cliffs, NJ, 1962.
4. Hays, W. L.; Winkler, R. L. *Statistics;* Holt, Rinehart and Winston: New York, 1971.
5. Kirk, R. E. *Elementary Statistics;* Wadsworth: Belmont CA, 1984.
6. Natrella, M. G. *Experimental Statistics;* U. S. Government Printing Office: Washington, 1963.

Problems

1. The sulfur content of diesel fuel provided by one manufacturer is claimed to be only 0.15%. The sulfur content of 40 random samples of fuel taken over a period of time was analyzed. The mean sulfur content was 0.162% with a standard deviation of 0.040. Is the manufacturer's claim substantiated?

2. The strength of 120 filaments taken at random from randomly spaced intervals of time during a production run was analyzed, and the mean strength was found to be 20.5 oz with a standard deviation of 1.1 oz. Estimate the strength for all production runs for this type of filament.

3. A corporation estimates that approximately 400,000 doctors might be able to use a product. From a random sample of 100 doctors taken from this population, 45% indicated that they would

definitely purchase this product. What would be the 95% fiducial limits of possible sales of this product?

4. In problem 2 in Chapter 6, a correlation of 0.535 was found between refluxing time and the amount of tin extracted from a product for 10 time samples. Is this a significant correlation? Test the hypothesis that $r = 0$ at the 0.01 level of significance.

Chapter 10

Inferences from Two Data Sets

Comparing two procedures or conditions to determine which might be the most productive or better in some way is often important in chemical studies and related areas. For example, two catalysts might be compared to determine which is the most productive, the effects of two temperatures on the tensile strength of plastic could be compared, the productivity rates for two different machines might be compared, or two measurement procedures might be compared to determine if they give the same readings.

Each of these problems can be handled by applying statistical procedures similar to those presented in Chapter 9. As with making inferences from one group, statisticians can make either point or interval estimates as to what the differences are likely to be between two groups. As with inferences from one data set, the main concern is estimating population values from samples. The general hypothesis tested is that no difference exists between the two groups, which would then imply that both could be considered as random samples from the same population with regard to the characteristic being studied. The most frequent comparison involves the means of two groups on some variable (e.g., mean productivity yield, mean reaction time, or mean concentration level). Other comparisons might involve two proportions, two correlations, or two variances. For each of these statistics, one can set up a nondirectional null hypothesis, H_0: $\mu_1 \geq \mu_2 = 0$, or a directional null hypothesis, H_0: $\mu_1 - \mu_2 > 0$, where μ_1 and μ_2 are the means for groups 1 and 2, respectively.

One way to consider these comparisons is in terms of the variance ratio (i.e., the ratio between two independent estimates of a

1453–0/88/0187$10.25/1

single population parameter). Essentially, each comparison is made by forming a ratio of the observed difference to the standard error of that difference, or in simple terms, forming a ratio of the observed difference to the difference that would probably occur if nothing but chance was operating. If nothing but chance were operating, the obtained difference would probably be equal to the difference obtained by chance variations between the two observations, and the ratio would be equal to 1. If the obtained difference is less than what might occur by chance, then the ratio is less than 1. If the obtained difference is larger than what would be expected on the basis of chance alone, then the ratio is greater than 1, and if larger than the critical value for a given level of significance, the ratio indicates a significant difference or a difference that is larger than that expected on the basis of chance alone. This difference might then be due to the factors or treatments being compared. Two ratios are commonly used for comparing statistics from two groups. One is the z ratio, which can be used when the population variances are known, and the other is the t ratio, which has to be used when the population variances are not known.

10.1 Comparing Means

Several ways are available to set up a study to compare the means for two groups. Several considerations must be taken into account. The first consideration is that the two groups can be taken as two independent random samples or can be matched or related in some way. Most often, samples are taken on a random sampling basis and thus are considered as independent samples (e.g., when a random sampling of one production run is compared to a random sampling of another production run). Samples would be considered as related or matched if they are paired off on a one-to-one basis for equivalence on some characteristic (e.g., each item in each pair having the same percentage of an element, or if two different measures are taken on the same items and then these measures are compared). The second consideration is whether the variances are equivalent for the two sets of data. One of the assumptions for comparing two means is that the variances should not be significantly different. Research by Havlicek and Peterson (*1*) indicated that violation of this assumption does not effect the t test or the analysis of variance (ANOVA) if the sample sizes are equal in number of cases, and has little effect until sample sizes become very disproportionate (e.g., $n = 10$ and $n = 40$). If sample sizes are disproportionate, an alternate equation for t can be used when unequal variances and unequal numbers of cases occur in the two groups being compared. The final consideration is whether the population variances are known or not, or if large- or small-size samples are involved. If σ is known, or if the sample size is large (e.g., $n > 100$), then the z distribution can be used because the result can usually be considered as distributed as the normal distribution. If σ is not known or if the sample size is small, then the t test should be used. Many statisticians prefer the t test regardless of sample size because the population variance is seldom, if ever, known.

The differences between two population means can be compared by two general methods: point estimation, in which the focus is on hypothesis testing, and interval estimation.

10.1.1 Point Estimation

Consider a situation in which the means are compared for two large, independent random samples for which the variances are considered equivalent. An engineer wants to determine whether the viscosities of two different batches of oil are the same. The hypothesis tested would be that no difference exists in the mean viscosity of oil from batch A compared to the mean viscosity of oil from batch B, or H_0: $\mu_A - \mu_B = 0$. The alternative hypothesis would be H_1: $\mu_A - \mu_B \neq 0$. For this comparison, 100 random samples of oil are taken from each batch so that $n_A = n_B = 100$, where n_A and n_B are the number of samples in batches A and B, respectively. Hypothetical results for these samples are presented in Table 10.1.

For large samples or when the population variances are known, the distribution of the differences between means for independent random samples can be considered normally distributed with a mean

Table 10.1. Hypothetical Descriptive Statistics for Oil Viscosity Problem

Statistic	Batch A	Batch B
n	100	100
Mean	10.18	10.42
S	2.10	1.80
SE_M	0.21	0.18

of zero and a standard error that can be estimated by

$$\mathrm{SE}_{D_M} = \sqrt{\sigma_{M_1}^2 + \sigma_{M_2}^2} \tag{10.1}$$

where SE_{D_M} is the standard error of the difference between means, σ_{M_1} is the standard error of the first mean, and σ_{M_2} is the standard error the second mean.

The standard error of difference between means is really an estimate of what the standard deviation of the differences $M_1 - M_2$ would be for a large number of samples. This value indicates what kind of differences would be expected to happen purely on the basis of random sampling. To determine if the obtained difference is greater than what would be expected to happen by chance (i.e., to test the null hypothesis that $\mu_1 - \mu_2 = 0$), a z ratio is computed by dividing the obtained difference by the standard error of the mean difference, or

$$z = \frac{M_1 - M_2}{\mathrm{SE}_{D_M}} \tag{10.2}$$

The obtained z is then compared to the values of z in the table of the normal curve to determine the probability that the obtained z would occur by chance.

For the oil viscosity problem, from the data in Table 10.1, the SE_{D_M} would be

$$\mathrm{SE}_{D_M} = \sqrt{0.21^2 + 0.18^2} = 0.276$$

The obtained difference between the two means would then be divided by 0.276 to form the z ratio,

$$z = \frac{10.18 - 10.42}{0.276} = -0.87$$

Table B in the Appendix, shows that the z ratio has to be greater than or equal to 1.96 to be significant at the 0.05 level of significance. Because the computed z ratio is less than this value, no significant difference in these two means is concluded, and the engineer would not reject the null hypothesis. The obtained difference is similar to that expected to occur solely on the basis of random sampling. Thus, from the data available, no significant difference is found in the viscosities of the oil from the two batches sampled.

Suppose that the data presented in Table 10.1 are from two different analyses of viscosity, A and B, taken on the same sample of oil, and the engineer wants to determine whether the two measurement analyses give the same results. In this case, each sample of oil would be analyzed by procedure A and then by procedure B. Thus, the engineer would end up with 100 pairs of data, and the two measures taken for each sample would be considered as related or matched data.

For related or matched samples, the standard error of the difference would be

$$SE_{D_M} = \sqrt{\sigma_{M_1}^2 + \sigma_{M_2}^2 - 2(r_{12})(\sigma_{M_1})(\sigma_{M_2})} \tag{10.3}$$

in which this equation is the same as equation 10.1 with the addition of a correction for the correlation, r_{12}, between the two sets of measures. Suppose for the viscosity problem, $r_{AB} = 0.90$. Then,

$$SE_{D_M} = \sqrt{0.21^2 + 0.18^2 - 2(0.90)(0.21)(0.18)} = 0.092$$

and

$$z = \frac{10.18 - 10.42}{0.092} = -2.609$$

As shown in Table B in the appendix, this value is significant at both the 0.05 and the 0.01 levels of significance. Thus, the difference is beyond what would be expected by chance alone, and the engineer would conclude that a significant difference exists at the 0.01 level of significance because $z > 2.58$, and thus $p < 0.01$.

For such problems, both the correlation coefficient and the test for significant differences between the means might be very useful. The correlation provides information with regard to how consistently each pair of samples is analyzed or measured (i.e., whether or not the sample having a high viscosity measured by procedure A also has a high viscosity measured by procedure B). The test for a significant difference between means provides information with regard to whether the overall average level of viscosity as measured by one procedure is the same, lower, or higher than the overall average level as measured by the other procedure. Thus, two types of information are provided, each of which could be very important for such studies.

Now consider testing the difference between means for small samples, where $n < 100$, or as some statisticians indicate, when

$n < 30$ or 50, and when the population variances are not known. For such situations, Fisher (*2*) provided several equations for computing t ratios to test the hypothesis of no difference between two means. For independent samples, the means are said to be uncorrelated or independent, and the t equation for testing the hypothesis of no difference between the two means is

$$t = \frac{M_1 - M_2}{\sqrt{\left(\frac{\sum x_1^2 + \sum x_2^2}{n_1 + n_2 - 2}\right)\left(\frac{n_1 + n_2}{n_1 n_2}\right)}} \qquad (10.4)$$

where M_1 is the mean for the first group, M_2 is the mean for the second group, n_1 is the number of cases in the first group, n_2 is the number of cases in the second group, $\Sigma x_1{}^2$ is the sums of squares for the first group, and $\Sigma x_2{}^2$ is the sums of squares for the second group.

This equation combines the sums of squares for both groups and is sometimes referred to as the pooled estimate of the population variance. This equation assumes that both samples came from the same population; thus, there should be only one estimate of the population variance. Therefore, homogeneity of variance is assumed when using this equation. When the two samples are equal in size (i.e., $n_1 = n_2$), then

$$t = \frac{M_1 - M_2}{\sqrt{\frac{\sum x_1^2 + \sum x_2^2}{n_1(n_1 - 1)}}} \qquad (10.5)$$

Suppose a botanist wants to compare the percentage of potassium oxide (K_2O) in fertilizer produced in two production plants, A and B. The null hypothesis tested is that no difference in mean percentage of K_2O in fertilizer occurs from plant A compared to the mean percentage of K_2O in fertilizer from plant B, or H_0: $M_A - M_B = 0$. Ten samples are taken at random from plant A's production and eight are taken from plant B. The hypothetical data are presented in Table 10.2. To use equation 10.4, the means and sums of squares have to be computed, and these are summarized in Table 10.3 as well as the standard deviations, which are usually provided as part

Table 10.2. Percentage K_2O in Fertilizer from Two Plants

n	Plant A	Plant B
1	14.8	14.2
2	15.2	14.8
3	14.7	13.9
4	14.3	14.2
5	15.0	13.6
6	14.6	14.3
7	14.9	14.0
8	14.4	13.8
9	15.4	
10	14.5	

NOTE: All values are in percent.

of the descriptive statistics for such studies. The computed t would be

$$t = \frac{14.78 - 14.10}{\sqrt{\left(\frac{1.116 + 0.94}{10 + 8 - 2}\right)\left(\frac{10 + 8}{10(8)}\right)}} = \frac{0.68}{0.17} = 4.00$$

The obtained t now has to be compared to the theoretical distribution of t to determine if it is significant (i.e., to determine the probability that this t ratio could have been generated by chance alone). The t values for the 0.05 and the 0.01 levels of significance are given in Table C in the Appendix. However, the appropriate degrees of freedom (df) must be known to use Table C. For the independent t test, the number of degrees of freedom is equal to $n_A + n_B - 2$, or the total number of cases in both samples minus 2. For this problem, df = 10 + 8 − 2 = 16. With 16 degrees of freedom, the obtained t would have to equal or be greater than 2.120 to be significant at the 0.05 level and be equal to or greater than 2.921 to be significant at the 0.01 level. Because $t = 4.00$, the botanist can conclude that a significant difference exists at the 0.01 level of significance. The null hypothesis would be rejected, and the botanist would conclude that the mean percentage of K_2O in the fertilizer from plant A is significantly greater than the mean percentage of K_2O in the fertilizer from plant B.

The t test is said to be robust against most violations of the basic assumptions of equal measurement scales, normally distributed var-

Table 10.3. Summary Statistics for the Data in Table 10.2

Statistic	Plant A	Plant B
n	10	8
Mean	14.780	14.100
S	0.352	0.366
ΣX	147.800	112.800
ΣX^2	2185.600	1591.420
Σx^2	1.116	0.940

iables, and homogeneity of variances, but not all combinations of these violations. When $n_1 = n_2$, the violation of homogeneity of variances does not affect the t test. However, when both $n_1 n_2$ and $S_1{}^2\ S_2{}^2$, where $S_1{}^2$ and $S_2{}^2$ are the variances for groups 1 and 2, respectively, Cochran and Cox (*3*) and Edwards (*4*) suggested an approach based on separate variance estimates. Their equation is

$$t = \frac{M_1 - M_2}{\sqrt{\dfrac{S_1^2}{n_1} + \dfrac{S_2^2}{n_2}}} \qquad (10.6)$$

The resulting t is then compared to a modified t value rather than the t value in Table C because this t is not exactly distributed like Fisher's t. The modified t (t') is

$$t' = \frac{w_1 t_1 + w_2 t_2}{w_1 + w_2} \qquad (10.7)$$

where t_1 and t_2 are the values of t from Table C for $n - 1$ degrees of freedom for each sample size, $w_1 = S_1{}^2/n_1$, and $w_2 = S_2{}^2/n_2$.

If the sample sizes are equal, then $t' = t$, there is no need to compute t', and the values of t from Table C can be used. However, as pointed out before, when the sample sizes are equal, the regular t test using equation 10.4 is not influenced by lack of homogeneity of variance and thus can be used. The t test of equation 10.6 is interpreted by comparing it to a calculated t based upon only $n_1 -$ 1 degrees of freedom, which requires a larger t for significance than the t obtained by using equation 10.4 based upon $n_1 + n_2 - 2$ degrees of freedom. Thus, using equation 10.6 when an equal number of cases in each group is involved results in a conservative test, in that larger differences are required to obtain a significant t. Thus, when $n_1 = n_2$, equation 10.6 is not recommended. This equation should be used only when unequal groups of n and unequal variances are involved.

Consider an example of comparing the yield of whole catalytic naphtha from two production plants, A and B. The hypothetical data are given and summarized in Table 10.4 for the percentage yield from each plant. For this data, equation 10.6 yields

$$t = \frac{40.4 - 44.429}{\sqrt{\frac{33.977}{15} + \frac{5.617}{7}}} = -2.302$$

For this problem, the obtained $t = -2.302$ would be compared to t' computed from equation 10.7, which is

$$t' = \frac{2.265(2.145) + 0.802(2.447)}{2.265 + 0.802} = 2.224$$

where $w_A = 33.977/15 = 2.265$ and $w_B = 5.617/7 = 0.802$.

Table 10.4. Percentage Yield of Naphtha from Two Plants

Statistic	Plant A	Plant B
n		
1	40	46
2	36	42
3	46	48
4	48	45
5	34	44
6	32	45
7	41	41
8	42	
9	40	
10	35	
11	43	
12	51	
13	38	
14	33	
15	47	
n_T	15	7
M	40.400	44.429
S	5.829	2.37
S^2	33.977	5.617
Σx^2	475.600	33.714

NOTE: All values are in percent.

Because the obtained t is greater than the t' required for significance, the conclusion would be that a significant difference occurs in percentage yield of naphtha between the two plants at the 0.05 level of significance. If homogeneity of variance was assumed and the regular Fisher t test used as in equation 10.4, the result would have been

$$t = \frac{40.4 - 44.429}{\sqrt{\left(\frac{475.60 + 33.714}{15 + 7 - 2}\right)\left(\frac{15 + 7}{15(7)}\right)}} = 1.742$$

which would not have been significant at the 0.05 level of significance. Thus, the difference in variance plus unequal sample sizes caused t to be lower and nonsignificant for this example. Generally, research has shown that when the sample with both the larger n and larger variance occurs together, the t computed from equation 10.4 is smaller because the larger variance is weighted proportionately more than the other variance. This weighting causes the error term to be greater. If the larger variance is for the sample with the smaller n, then this variance will be weighted proportionately less, and the error term will be smaller. This weighting results in a spuriously inflated t ratio. When the larger variance is weighted more, more Type II errors would be made, whereas when the larger variance is weighted less, more Type I errors would be made. Thus, the equation provided by Cochran and Cox (*3*) and Edwards (*4*) should be used for situations in which there are significant differences in variances and differences in the number of cases in each sample. A test for determining if the variances are significantly different is given in Section 10.2.

Consider the situation in which the data are correlated or matched (e.g., pairing individuals on some trait, taking pairs of samples of products produced by two machines matched on pressure at which they are produced, or measuring the reaction time of the same individuals with and without a drug). Suppose that the reaction time for 12 men is taken before and after taking a drug and results in the data presented in Table 10.5. A pharmacologist wants to know whether the drug has an effect on reducing reaction time. Because the main interest is on reducing reaction time, a directional hypothesis could be used. The statistical hypotheses corresponding to this scientific hypothesis are H_0: $\mu_w \geq \mu_{w/o}$ and H_1: $\mu_w < \mu_{w/o}$, where μ_w

Table 10.5. Reaction Time With and Without a Drug

Statistic	Without Drug	With Drug	Difference	D^2
n				
1	420	402	18	324
2	310	298	12	144
3	450	455	−5	25
4	480	450	30	900
5	400	388	12	144
6	380	390	−10	100
7	420	398	22	484
8	460	445	15	225
9	390	390	0	0
10	410	401	9	81
11	320	310	10	100
12	435	405	30	900
ΣX	4875	4732	143	
ΣX^2	2009625	1892032	3427	3427
M	406.25	394.33	11.917	
S	51.48	48.66	12.515	

NOTE: The symbol D^2 denotes the squared difference between reaction times.

is the mean for the group with the drug, and $\mu_{w/o}$ is the mean for the group without the drug.

For related samples, the equation for t for paired comparisons or correlated data is

$$t = \frac{M_D}{\sqrt{\dfrac{\sum x_D^2}{n(n-1)}}} \qquad (10.8)$$

where M_D is the mean difference between the two measures, Σx^2_D is the sum of squared differences, and n is the number of pairs in the sample.

From equation 5.4, using the data in Table 10.5,

$$\sum x_D^2 = 3427 - (143^2/12) = 1722.92$$

and then

$$t = \frac{11.917}{\sqrt{\dfrac{1722.92}{12(11)}}} = \frac{11.917}{3.613} = 3.298$$

The number of degrees of freedom for the correlated t test is given by the number of pairs of values minus 1, or df $= n - 1$. For this problem, df $= 12 - 1 = 11$. With 11 degrees of freedom, Table C shows that $t = 1.796$ is required at the 0.05 level for a one-tailed test. Because the computed t is greater than this value, the pharmacologist would conclude that the difference is significant and that the reaction time is significantly reduced by taking the drug. Statisticians always test the null hypothesis, which in this problem would be rejected, and the alternative directional hypothesis, which was stated, would be supported.

Equations 10.8 and 10.3 provide the same results. Thus, for large samples, either procedure could be used. However, as noted before, often the correlation coefficient computed for equation 10.3 is useful in indicating consistency of the changes for the pairs of scores. If this use is of no concern, then equation 10.8 would be easier to use.

10.1.2 Interval Estimation

Although the main focus when comparing the differences between means is on hypothesis testing, which involves point estimates, interval estimates of the differences between means can be computed and are often useful in determining what differences could be expected between two population means. The procedure and logic for making these interval estimates are the same as that used for interval estimates of the population mean. Because the standard error of the difference between means is the same as the standard deviation of the distribution of those differences, the standard deviation is used to compute the interval in which the differences between the two means for random samples from that population are likely to fall.

The general procedure for computing this interval estimate is to add and subtract from the difference obtained between the two sample means the product of the standard error of the difference between means multiplied by the z or t value for the level of significance desired. The standard error is computed in the usual way for the statistic being used (e.g., z ratio or t test). As with the interval estimate of one mean, this procedure is based upon the assumption that the difference between the two sample means is a good estimate of what the difference between the two population means would be, and that the means and standard deviations of the samples are also good estimates of what the population parameters are. For random sampling from approximately normal distributions and when the size of each sample is equal to or exceeds 30 for independent samples, or exceeds 60 for dependent samples, these estimates are probably fairly close to the population values. Thus, the interval estimate of the population difference between means will probably be representative of the population.

As with estimates of any population parameter, the probability of an interval including a particular parameter is the confidence coefficient. The confidence coefficient of an interval estimate is the probability that the sampled interval includes the population parameter being estimated, and is usually expressed as a percentage. This percentage is usually calculated by $100(1 - \alpha)$. The probability that the interval defined by the 95% confidence limit includes the population parameter is 0.95.

Confidence intervals for estimating the differences between various population parameters are interpreted in the same way as the confidence level for estimating any population parameter. That is, once the lower and upper limits are specified for the sample being considered, this interval should be interpreted as only one of all possible intervals that could be computed for many random samples of the same size under the same conditions by using the same procedures to estimate this interval. The interval changes from one sample to another, not the difference between the population parameters. Many confidence intervals could be computed, one for each set of samples, whereas the difference between population parameters is fixed and does not vary. Thus, once the interval is calculated, the fixed difference between the population parameters either is in or not in the interval. The intervals vary, not the population parameters. Therefore, statisticians cannot say that at a given level of confidence, the interval will contain the difference between the population parameters, because this difference cannot vary. Rather, the intervals vary from sample to sample, and thus a correct interpretation is that a 95% confidence interval is calculated so that 95% of the intervals that could be computed from all possible random samples would contain the estimated difference between population parameters. Or, one of the 95% confidence limits for the difference between population parameters is as computed, or the probability is 0.95 that the interval computed will include the true difference between population parameters.

The standard error of the difference between means was computed to be 0.276 by applying the procedure for determining the interval estimate of the difference between means to the oil viscosity problem. If the confidence interval for the difference between means was set at the 0.05 level, then

$$\begin{aligned} CI_{95\%} &= -0.24 \pm 1.96(0.276) = -0.24 \pm 0.54 \\ &= -0.78 \text{ to } +0.30 \end{aligned}$$

Thus, a 95% confidence interval ($CI_{95\%}$) for the difference between these two population means, $\mu_A - \mu_B$, is from -0.78 to $+0.30$. Therefore, 95% of the intervals that could be calculated would include the hypothesized zero difference between the population means, which is estimated to be a difference of -0.24 from the samples, if the means are compared in the same order (i.e., $\mu_A - \mu_B$).

For the problem of comparing two different measures of the same samples of oil, the z ratio for related samples was used, and the standard error of the difference between means for this problem was 0.092. The resulting significant z ratio is 2.609. Computing the confidence interval for the difference between means at the 0.01 level yields

$$\begin{aligned} CI_{99\%} &= -0.24 \pm 2.58(0.089) = -0.24 \pm 0.23 \\ &= -0.48 \text{ to } -0.00 \end{aligned}$$

Thus, a 99% confidence interval ($CI_{99\%}$) for the differences between population means is from -0.48 to -0.00. Therefore, 99% of other intervals constructed under the same conditions would not be expected to contain the hypothesized zero population difference.

When the procedure is applied to a problem for which the t test has been used, the same procedure is followed with the exception that the t value rather than the z value is used. For the t test comparing the percentage of K_2O in fertilizer from two plants (A and B), the standard error of the difference between means was computed to be 0.17 and the difference between means was 0.68. These values yield a t ratio of 4.00, which is significant at the 0.01 level. For this problem, the confidence interval for the difference between means would be

$$\begin{aligned} CI_{99\%} &= 0.68 \pm 2.921(0.17) = 0.68 \pm 0.50 \\ &= 0.18 \text{ to } 1.18 \end{aligned}$$

Thus, a 99% confidence interval for the difference between population means, $\mu_A - \mu_B$, is 0.18–1.18. For the 95% confidence interval, the standard error of the difference between means, 0.17, would have been multiplied by 2.12, which is the t value at the 0.05 level for 16 degrees of freedom. A 95% confidence level would then be from 0.32 to 1.04.

For the problem of reducing the reaction time by a drug using matched or related samples, the standard error of the mean for the related t test was 3.613. The difference between means was equal to 11.917. At the 0.05 level of significance with 11 degrees of freedom, $t = 2.201$ is required for a two-tailed test because both smaller and larger differences that could occur between the two means are to be estimated. Thus,

$$\begin{aligned} \text{CI}_{95\%} &= 11.917 \pm 2.201(3.613) = 11.\,917 \pm 7.95 \\ &= 3.967 \text{ to } 19.867 \end{aligned}$$

A $100(1 - 0.05)\%$ confidence interval for $\mu_w - \mu_{w/o}$ is 3.967–19.867. Because this interval does not include 0, the researcher can be confident that less reaction time would occur with the drug than without the drug.

Computing the confidence interval for the difference between population means gives not only an estimate of what these intervals would be for a large number of similar samples, but also helps to understand the results of the test of the hypothesis that no difference between those two means occurs. For example, in the problem comparing oil viscosities, the z ratio was not significant, indicating no significant difference between the two means; thus, the null hypothesis, H_0: $\mu_A - \mu_B = 0$, was not rejected. When the confidence interval was computed, the confidence interval was from a negative difference of -0.78 to a positive difference of 0.30. That is, one mean could be lower or higher than the other. Compare this to the significant z ratio for the second problem for the matched samples of oil viscosity for which a significant difference was found between the means for the matched samples of measurements of oil viscosity. The confidence interval for the difference between means was from -0.48 to -0.00. In this case, all of the estimated differences are likely to be in one direction, or all negative. With repeated sampling, all of the means for measure B are likely to be higher than the means for measure A, and no reversals occur in the direction of the difference.

This rationale follows the testing of null hypotheses (i.e., the average difference will be equal to zero if nothing but chance is operating). Thus, if the null hypothesis were true, then some of the differences between means would be negative and some would be positive. If the majority of the differences between means are either negative or positive, the difference between means is probably significantly different from H_0: $\mu_A - \mu_B = 0$.

As indicated previously for the standard error of the mean, interval estimates of the difference between means are useful in quality control and in other situations in which tolerances in products have to be within specified limits.

10.2 Comparing Variances

Often researchers are interested in determining whether a significant difference occurs in the variability of products produced under different conditions, or whether a change in procedures results in more or less variability in the product. Reducing variability may be an important goal, especially if tolerances are reduced or if a demand arises for greater consistency in a product. The general hypothesis tested is that no significant difference occurs in the variance of one set of data compared to another set of data, or $\sigma^2_1 - \sigma^2_2 = 0$. To test these hypotheses, point estimates are usually used because little interest is raised in interval estimates of these differences. As with the comparison of means, the two sets of data can be independent or can be related or matched in some way.

The tests for comparing variances are often referred to as tests of homogeneity of variance. If the test is not significant, the two sample variances are said to be homogeneous and can be considered as random estimates of the same population variance. If the test is significant, then the variances would be said to be heterogeneous, and the two samples would be considered as coming from two different populations with regard to variance. A number of tests can be used, and two tests for comparing two variances will be described in this chapter. If several variances are to be compared and the sample sizes are the same, then Hartley's F_{max} test could be used (*5*).

For independent samples, the difference between the two variances is tested by forming a ratio by dividing the larger variance by the smaller variance, and this procedure is used for both small- and large-size samples. If the two variances are equal, then this ratio is equal to 1.00, and the two variances can be considered as two random samples from the same population, or from two populations having the same variance. The ratio is distributed as the *F* ratio, and the table of the distribution of *F*, in Table D in the Appendix, is used to interpret the results. The equation for the *F* ratio is

$$F = S_1^2/S_2^2 \qquad (10.9)$$

where S_1^2 is the larger variance, and S_2^2 is the smaller variance. The number of degrees of freedom associated with each variance is calculated by df $= n - 1$.

Applying this test to the variances given in Table 10.4 shows that the variance of the percentage yield of naphtha for plant A is 33.977, whereas the variance for plant B is 5.617. Is the variance for plant A greater than the variance for plant B at the 0.10 level of significance? For this data,

$$F = 33.977/5.617 = 6.049$$

The number of degrees of freedom for the larger variance is 15 − 1 = 14, and df = 7 − 1 = 6 for the smaller variance. With df = 14 for the larger variance and df = 6 for the smaller variance, Table D shows that $F \geq 3.96$ is required to be considered significantly different at the 0.05 level of significance. Because the obtained *F* ratio of 6.049 is greater than the critical value in the table, the directional hypothesis that the variance for plant A is greater than the variance for plant B is supported. Had the 0.02 level been used, the critical value needed for a significant *F* at this level of significance is 7.60, and at this level, the null hypothesis would not have been rejected, and the conclusion would be that the variances were equivalent.

As will be shown in the next chapter, the *F* distribution in Table D is set up in terms of two-tailed tests for the general null hypothesis. The two-tailed test will be applicable for most *F* ratios. However, for this *F* ratio, the larger variance is always divided by the smaller variance, thus making a one-tailed test. Consequently, the probabilities listed for Table D must be doubled, and for this test, the critical values are for the 0.10 and 0.02 levels, respectively, rather than the 0.05 and 0.01 levels. For most other uses of the *F* distribution, the larger variance is not arbitrarily placed in the numerator, and thus the two-tailed test probabilities as given are applicable. Other probability levels are given in other *F* tables such as those in Beyer (*6*) or Owen (*7*).

Situations occur in which two variances are to be compared from samples that are related or matched in some way. For example, suppose that for the study on reaction time with and without a drug, the main interest was in reducing the variation in reaction time rather than the average reaction time. That is, does administration of the drug change the variation of reaction times for the persons included in the sample from the variation of reaction times without the drug? The null hypothesis would be that no difference exists in the variance in reaction time between the no-drug and drug conditions, or $\sigma^2_{w/o} - \sigma^2_{w} = 0$. Because the same group of persons is tested under both conditions, the data are considered to be related or matched. The F ratio does not apply to related data, but a t test can be used for testing the difference between variances for related data. Because a positive correlation is likely between the two sets of data, the t ratio uses the correlation between the pairs of values to adjust the error term. The equation for t is

$$t = \frac{(S_1^2 - S_2^2)\sqrt{n - 2}}{2(S_1)(S_2)(\sqrt{1 - r_{12}^2})} \qquad (10.10)$$

where S_1 and S_2 are the two estimates of the population standard deviation, n is the number of pairs of values in the data set, and r_{12} is the correlation between the two sets of values.

For the data presented in Table 10.5,

$$t = \frac{(51.48^2 - 48.66^2)\sqrt{12 - 2}}{2(51.48)(48.66)\sqrt{1 - 0.97^2}} = 0.728$$

The number of degrees of freedom for this t test is equal to $n - 2$. For this problem, df $= 12 - 2 = 10$, and the critical values of t for the 0.05 and 0.01 levels, respectively, are 2.228 and 3.169. The computed t does not approach either of these two values; thus the null hypothesis is not rejected, and the conclusion is that no significant difference occurs between the variances of the values with and without the drug.

10.3 Comparing Correlation Coefficients

In prediction studies, where the correlation coefficient is used to estimate or predict a future quantity, often a choice exists between two or more predictors. In such cases, the logical question is whether there is a significant difference between the correlation of one predictor (a) and the predicted variable (c) and another predictor (b) and the same predicted variable (c). The null hypothesis that would be tested is that no difference exists between the two population correlation coefficients, or $\rho_{ac} - \rho_{bc} = 0$.

As noted in Chapter 9, the sampling distribution of the correlation coefficient is usually skewed (i.e., the sampling distribution is usually not normally distributed) and does not represent an interval scale. Thus, in testing the significance of the correlation or in averaging correlations, Fisher's z' transformation should be used. For these same reasons, Fisher's transformation of r to z' should be used when testing the difference between two correlation coefficients because the standard error of z' is only a function of n and not of r. When two samples are independent, the standard error of the difference between the respective Fisher's z' values ($D_z{}'$) can be computed by

$$\mathrm{SE}_{D_{z'}} = \sqrt{\frac{1}{n_1 - 3} + \frac{1}{n_2 - 3}} \qquad (10.11)$$

where n_1 and n_2 are the number of pairs of values in the first and second sets of data, respectively. The two correlation coefficients to be compared are first transformed to their respective Fisher z' values by using Table G in the Appendix. Then, a z ratio is computed by

$$z = \frac{z_1' - z_2'}{\mathrm{SE}_{D_{z'}}} \qquad (10.12)$$

The resulting z is then interpreted by using the table of the normal curve, which is Table B.

Suppose that a water treatment operator is interested in comparing the correlation between the percent of waste solids removed (c) and the flow rate of the effluent being fed into the system for two filtration systems, A and B, where the flow rate into A is a, and the flow rate into B is b. For system A, the correlation between these two variables is $r_{ac} = 0.78$ for 40 random samples taken, whereas for system B, $r_{bc} = 0.88$ for 50 random samples taken. The corresponding Fisher's z' values for these two correlation coefficients are

1.05 and 1.38, respectively. Combining equations 10.11 and 10.12 would yield

$$z = \frac{1.05 - 1.38}{\sqrt{\dfrac{1}{40 - 3} + \dfrac{1}{50 - 3}}} = -1.507$$

A $z = 1.96$ is required for significance at the 0.05 level of significance. Because the obtained value of $z = 1.507$ does not equal or exceed this value, the water treatment operator would conclude that the two correlation coefficients are not significantly different. Thus, the correlation between these two variables for the two filtration systems can be considered as equivalent.

In most studies comparing correlation coefficients involving chemical reactions, the two sets of data will usually be independent because, if combined, a compounding effect would arise. Thus, the procedure just described for testing the significance of the difference between two correlation coefficients will most often be used. However, for some studies, the correlations are based on data that are related, such as in personnel studies in which one wants to determine whether one selection test has a higher correlation with job performance than another selection test. Such problems usually have three correlation coefficients: the correlation between job performance (c) and the first selection test (a), r_{ac}; the correlation between job performance (c) and the second selection test (b), r_{bc}; and the correlation between the two selection tests, r_{ab}. For one group of persons who have taken both selection tests, their job performance is correlated with each of the selection tests. The test that has the highest correlation with job performance would provide the best estimate of job performance for future groups of candidates. For such situations, Hotelling (*8*) developed a t test that takes into account the relationship between the two selection tests. Hotelling's equation is

$$t = (r_{ac} - r_{bc}) \frac{(n - 3)(1 + r_{ab})}{\sqrt{2\{1 - r_{ab}^2 - r_{ac}^2 - r_{bc}^2 + [2(r_{ac})(r_{bc})(r_{ab})]\}}} \qquad (10.13)$$

where n is the number of cases in the group. Suppose that the correlation between job performance and the first test is $r_{ac} = 0.50$, and the correlation between these two variables for the second test is $r_{bc} = 0.62$. Is there a significant difference between these two correlation coefficients? The hypothesis that would be tested is that no difference exists between the two correlation coefficients, or H_0: $\rho_{ac} - \rho_{bc} = 0$. The alternative hypothesis is H_1: $\rho_{ac} - \rho_{bc}$ 0. To test this hypothesis by using equation 10.13, the correlation between the two tests has to be computed, which for this problem is given as $r_{ab} = 0.75$. If a sample of 100 persons were tested by both tests and had job performance ratings, then

$$t = (0.50 - 0.62)$$

$$= -2.136$$

The number of degrees of freedom for this test is $n - 3 = 97$. With the closest degrees of freedom to 97, df = 100, from Table C at the 0.05 level, $t = 1.980$ is required. Thus, the null hypothesis that no difference occurs between these two correlation coefficients would be rejected, and the conclusion would be that the second test can predict job performance more accurately than the first test. Often in such studies, a test that is usually used is compared with a new test to determine if the new test is superior to the old test. If this is the situation, a directional hypothesis could be used, and a one-tailed test could be made that would then require $t = 1.66$ to be significant at the 0.05 level of significance.

10.4 Comparing Proportions

In studies in which the data are in the form of proportions or percentages, often the proportions of two groups are compared. For example, the proportion of rejects of one production procedure might be compared to the proportion of rejects from another production procedure, or the proportion of persons reacting favorably to two products might be compared. When the proportions to be compared come from two independent random samples, the method that is recommended uses a weighted mean of the two sample proportions to provide an estimate of the population proportion in the computation of the standard error of the difference between uncorrelated proportions. The equation for computing the weighted mean for the two proportions ($\bar{p}_w$) is

$$\bar{p}_w = \frac{n_1 p_1 + n_2 p_2}{n_1 + n_2} = \frac{f_1 + f_2}{n_1 + n_2} \qquad (10.14)$$

where p_1 and p_1 are the proportions in the two samples that are favorable or represent the targeted event, f_1 and f_2 are the frequencies favoring the event in each sample, and n_1 and n_2 are the number of cases in each sample. For samples in which both n_1 and n_2 equal 5 or more in either sample, the sampling distribution of the differences $p_1 - p_2$ approaches the normal distribution; thus, the z ratio can be used to determine the probability that the difference could have happened by chance. The z ratio for comparing uncorrelated proportions is

$$z = \frac{p_1 - p_2}{\sqrt{\bar{p}_w(\bar{q}_w)\left(\frac{n_1 + n_2}{n_1 n_2}\right)}} \tag{10.15}$$

where p is the proportion of favorable responses in each group, $\bar{p}_w$ is the weighted average proportion for the two groups, and $\bar{q}_w = 1 - \bar{p}_w$.

When dealing with samples for which the product np or nq is between 5 and 10, a correction for continuity is recommended because frequencies increase by increments as discrete data, whereas the z value is continuous. This correction can be made by reducing the absolute size of $p_1 - p_2$ by $0.5[(n_1 + n_2)/n_1n_2]$. As will be shown in Section 10.5, when either of the products np or nq is less than 5, the chi-square (χ^2) test can be used to determine if a significant difference occurs in the two proportions.

Suppose that the proportion of defective products from two suppliers, A and B, are to be compared to determine if a significant difference occurs in the proportions of defective products. The null hypothesis to be tested would be that there is no difference in the proportion of defective parts between supplier A compared to supplier B, or H_0: $p_A - p_B = 0$. For a random sample of 80 parts from supplier A, 12 parts, or $p_A = 0.15$, are found to be defective, whereas for a random sample of 90 parts from supplier B, 10 parts, or $p_B = 0.111$, are found to be defective. The weighted-mean proportion obtained by using equation 10.14 would be

$$\bar{p}_w = \frac{80(0.15) + 90(0.111)}{80 + 90} = 0.129$$

Applying equation 10.15,

$$z = \frac{0.150 - 0.111}{\sqrt{(0.129)(0.871)\dfrac{(80 + 90)}{80(90)}}} = 0.757$$

The result of this test is compared to the values of 1.96 and 2.58, which are the usual values for the z ratio at the 0.05 and 0.01 levels of significance. Because the obtained $z = 0.757$ does not equal or exceed either of these values, the null hypothesis would not be rejected, and the conclusion would be that no significant difference occurs between these two proportions. If the parts from supplier A are less expensive and not significantly more defective than the parts from supplier B, then buying from supplier A would be advantageous.

10.4.1 McNemar's Test

In some situations, the proportions are based on samples that are related or correlated in some way. For example, the proportion of rejects before and after modification of a production procedure could be compared, or the same group of individuals may be asked to give their preference for some product on two different occasions to determine if their responses are consistent. Thus, the proportion of rejects or the proportion of responses would be considered to be correlated.

To compare correlated proportions, McNemar (*9*) provided a procedure that does not require the computation of the standard error of the estimated population proportion or the correlation coefficient. As an example, suppose that a group of individuals are asked whether they would purchase a product on two different occasions, before and after another product was introduced by a competitor. This survey could be a test of loyalty for a particular brand. Thus, paired responses are available for each person in the form of yes/no preferences on both occasions. Thus, one person might respond "yes" on both occasions, another "yes" on the first occasion and "no" on the second occasion, and another person might respond "no" on both occasions, or "no" first and then "yes." The paired responses are tabulated into a fourfold, or 2 × 2, table such as in Figure 10.1, and a tally is made in each cell dependent on the paired responses. For example, the individual who responded "yes" on the first occa-

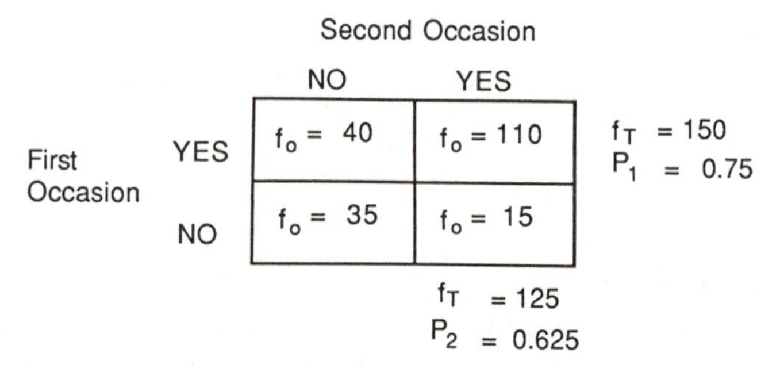

Figure 10.1. Tally of "yes" and "no" responses on two occasions for a group of 200 persons.

sion and "no" on the second occasion would be tallied into cell A. The individual who responded "no" on the first occasion and "no" on the second occasion would be tallied into cell C, and the individual who first responded "no" and then responded "yes" would be tallied into cell D. The person who responded "yes" both times would be tallied into cell B.

The individuals who fall into cells A and D are those who have changed their preference, and these are the individuals whose change represents the difference between the two preferences being tested. The difference in the proportions of individuals in cells A–D divided by n is equal to the difference between the proportion responding "yes" on the first occasion and the proportion responding "yes" on the second occasion (i.e., $p_1 - p_2$). Thus, the difference between the two frequencies in cells A and D is tested in McNemar's equation as the test of $p_1 - p_2$. McNemar's equation is

$$z = \frac{f_A - f_D}{\sqrt{f_A + f_D}} \qquad (10.16)$$

where f_A and f_D are the number of individuals falling in the two cells A and D, respectively.

Suppose that 200 individuals were asked if they would purchase a product on two different occasions, and each was given the opportunity to respond "yes" or "no" on each occasion. The paired responses are tabulated into the fourfold table in Figure 10.1, which indicates that 110 individuals responded "yes" on both occasions, 40 individuals responded "yes" and then changed to "no", 15 first responded "no" then changed to "yes," and 35 responded "no" on both occasions. Equation 10.16 yields

$$z = \frac{40 - 15}{\sqrt{40 + 15}} = 3.37$$

The obtained $z = 3.37$ is significant at the 0.01 level of significance. Thus, the conclusion would be that a significant change occurred in the proportion responding "yes" on the two occasions. When asked their preference on the first occasion, a total of 150 individuals responded "yes," or $p_1 = 0.75$. On the second occasion, a total of 125 individuals responded "yes," or $p_2 = 0.625$. The difference between these two proportions is significant at the 0.01 level; thus, the conclusion would be that a significant change occurred in the proportion of individuals responding "yes" on the two occasions.

10.4.2 Cochran's *Q* Test

The McNemar test for two related groups presented can be extended to three or more groups, and this extension is known as the Cochran *Q* test (*10*). As with the McNemar test, the data can be from the same individuals, machines, or whatever is measured or tested at three or more time periods or over three or more conditions. Or the data can be from three or more groups of individuals or other things that have been matched on some relevant variable or variables. As with the McNemar test, the data are nominal and are in the form of frequencies, and each value can take on only two values, 0 or 1. These values usually represent some dichotomous variable such as yes or no, hard or soft, easy or difficult, pleasant or unpleasant, or positive or negative.

Suppose that six chemical engineers are asked to rate three microcomputers with regard to which computer should be purchased for laboratory experiments. Each engineer is to rate each microcomputer as to whether that computer is acceptable, a "1" rating, or unacceptable, a "0" rating. Because each engineer rates all computers, the data have to be considered related. The hypothetical results are presented in Table 10.6. The null hypothesis tested is that no difference exists in the preferences of these six engineers for the three computers.

The general equation for the Cochran *Q* test is

$$Q = \frac{(k - 1)[(k \sum C^2) - T^2]}{k(T) - \sum R^2} \qquad (10.17)$$

Table 10.6. Hypothetical Choices of Three Computers by Six Engineers

Statistic	Computer Rating 1	2	3	R	R^2
n					
1	1	0	1	2	4
2	0	0	1	1	1
3	1	1	0	2	4
4	1	0	1	2	4
5	0	1	1	2	4
6	1	0	1	2	4
C	4	2	5		
C^2	16	4	25		
T				11	
T^2				121	
ΣR^2					21

NOTE: The symbol C denotes the sum.

where k is the number of groups; R is the row totals; C is the column totals; and T is the grand total, or $C + R$. Applying equation 10.17 to the data in Table 10.6,

$$Q = \frac{(3 - 1)[(3 \times 45) - 11^2]}{3(11) - 21} = 2.33$$

The number of degrees of freedom for the Cochran Q test are equal to $k - 1$, or for this example, $3 - 1 = 2$. The Cochran Q test is interpreted by using the chi-square (χ^2) table with the appropriate degrees of freedom. At the 0.05 level with df $= 2$, from Table E in the Appendix, $\chi^2 = 5.991$ is required for significance. Because the computed Q is less than the table value, the null hypothesis is not rejected, and the conclusion would be that no significant difference exists in the preferences for these three computers by the six engineers.

10.5 Comparing Frequency Distributions

In many research studies, the data are in the form of frequencies, and a need arises to make inferences about a population distribution from the sample frequency distribution. In the majority of such cases, a set of observed frequencies is compared to a theoretical distribution to determine if a significant difference exists between the two frequencies. The chi-square (χ^2) analysis can be used for this test, and two general types of tests used are (1) tests of goodness of fit and (2) tests of independence.

10.5.1 Tests of Goodness of Fit

In tests of goodness of fit, a set of observed frequencies is compared with a corresponding set of expected or theoretical frequencies. For example, suppose that a random sample of 100 consumers indicated their preference for four different brands of instant coffee, and 38 expressed a preference for brand A, 21 for brand B, 20 for brand C, and 21 for brand D. Do these observed frequencies indicate a preference for brand A, or could this set of observed frequencies be a chance deviation from what would have been expected if no real preference existed? If nothing were operating but chance, each brand would have received 25 preferences. Or suppose that defective products produced by three shifts of workers over a designated period of time were tabulated, for which 27 occurred during the first shift, 26 during the second shift, and 37 during the third shift. Do these frequencies indicate that more defective products are produced during the third shift? To answer these questions, χ^2 can be applied to the data. For the second example, if nothing is operating but chance, one-third of the defective products should occur during each shift. The data can be set up in terms of observed and expected frequencies as in Table 10.7. The null hypothesis would be $H_0: f_o - f_e = 0$, and $H_1: f_o - f_e\ 0$, where f_o is the observed frequencies, and f_e is the expected frequencies.

The general equation for χ^2 is

$$\chi^2 = \sum \left(\frac{(f_o - f_e)^2}{f_e} \right) \qquad (10.18)$$

Table 10.7. Distribution of Defective Products over Three Shifts

Shift	f_o	f_e	$(f_o - f_e)$	$(f_o - f_e)^2$	$(f_o - f_e)^2/f_e$
1	27	30	3	9	0.30
2	26	30	4	16	0.53
3	37	30	7	49	1.63
Total	90	90			2.46

The values are summed for all categories in the table. For the defective products problem, $\chi^2 = 2.46$ as computed and shown in Table 10.7. To interpret this result, the computed χ^2 has to be compared to the critical values of χ^2 as given in Table E, which is the theoretical sampling distribution of χ^2. As with the table of t, a different sampling distribution of χ^2 results for each of the degrees of freedom. For this use of χ^2, the number of degrees of freedom is equal to the number of categories minus 1, or df $= k - 1$, where k is the number of categories. For this problem, df $= 3 - 1 = 2$. With two degrees of freedom, Table E indicates that χ^2 has to be equal to or greater than 5.99 to be significant at the 0.05 level of significance. Thus, because the computed χ^2 does not equal or exceed this value, the conclusion is that no significant difference occurs between the observed frequencies and those that would have happened by chance alone. Thus, researchers would fail to reject the null hypothesis of no difference between the expected and the observed frequencies and conclude that no significant difference occurs in the number of defective products produced by any of the three shifts.

Applying the χ^2 test to the preferences for the four brands of coffee, the data would be set up as in Table 10.8. The computed χ^2 for this data is 9.04, and with df $= 4 - 1 = 3$, the result is significant at the 0.05 level of significance because at this level, $\chi^2 \geq 7.815$ is significant. Thus, the null hypothesis, $f_o - f_e = 0$, is rejected, and the conclusion is that a significant difference occurs in the obtained distribution of preferences from what would have been obtained by chance alone. By inspection of the distribution of frequencies, a significant preference exists for brand A.

The goodness of fit χ^2 can be used when expected frequencies are distributed by chance, or the expected frequencies could follow any known or theoretical distribution (e.g., the normal curve distribution). For example, if the distribution of defective products for

Table 10.8. Preference for Four Brands of Coffee

Shift	f_o	f_e	$(f_o - f_e)$	$(f_o - f_e)^2$	$(f_o - f_e)^2/f_e$
A	38	25	13	169	6.76
B	21	25	4	16	0.64
C	20	25	5	25	1.00
D	21	25	4	16	0.64
Total	100	100			9.04

several plants or sections is distributed such that 25% occur during shift 1, 30% during shift 2, and 45% during shift 3, then this proportional distribution could serve as the expected frequency distribution, and an observed distribution of frequencies from another plant or section would be tested against this known distribution.

10.5.2 Tests of Independence

If two sets of variables are involved (e.g., sex and preference, or sex and frequency of defective products), then χ^2 can be used to determine whether the two variables are independent of each other or are associated in some way. For this type of problem, both variables are usually *categorical* or *nominal data*; that is, data that can be reduced to categories. The data are arranged in the form of a bivariate frequency distribution, also referred to as a contingency table. The major question to be answered for such bivariate distributions is whether the classifications of responses are independent of each other or are associated in some way. As an example of this type of problem, suppose that in the preference test for the four brands of coffee, a difference is thought to arise in preferences for the four brands between men and women. Thus, the two variables would be sex and brand of coffee and would be set up as in Table 10.9. The null hypothesis tested would be that no difference exists in the distribution of preferences for men as compared to the distribution of preferences for women.

First, expected frequencies have to be obtained. These frequencies could be computed by using the multiplication theorem of probability, but a more convenient method is available. For each cell in such tables, the expected frequency can be obtained by multiplying the row total by the column total and then dividing this product by the total number of frequencies for the total table. For example, for the 38 men who preferred brand A, the row total for brand A is 64, and the column total for men is 100. Thus, the expected frequency

Table 10.9. Coffee Preference for Men and Women

Brand	Men	Women	Total
A	38	26	64
B	21	15	36
C	20	18	38
D	21	31	52

for men who preferred brand A is $(64 \times 100) \div 190 = 33.68$. For women who prefer brand A, the expected frequency is $(64 \times 90) \div 190 = 30.32$. The expected frequencies for each of the eight cells are presented in Table 10.10.

Then, χ^2 is computed in the usual way by squaring the difference between expected and observed frequencies, dividing the result by the expected frequency, and then summing over all cells. This procedure is done for each cell in the table, and for the frequencies in Tables 10.9 and 10.10, χ^2 would be as shown in Table 10.11.

For this problem, the computed χ^2 is 4.7679, and this value has to be compared to the table value of χ^2 to determine if the two sets of frequencies are independent of each other. The degrees of freedom for this use of χ^2 is computed by multiplying the number of rows (R) minus 1 times the number of columns (C) minus 1, or df = $(R - 1)(C - 1)$. For this problem, there are four rows, one for each brand of coffee, and two columns. Thus, the number of degrees of freedom is equal to $(4 - 1)(2 - 1) = 3$. With df = 3, the χ^2 needed for significance at the 0.05 level of significance is 7.815. Because the computed χ^2 is only 4.7679, the null hypothesis is not rejected, the observed frequencies are not different from how they would be expected to fall by chance alone, and the two distributions are concluded to be independent. Thus, the preferences of men and women are not significantly different.

In interpreting χ^2 analyses, converting the obtained frequencies to proportions is often helpful, especially when the total number of frequencies is different for the two variables being compared, because this conversion puts each category on a common basis. For the example of coffee preference, because different numbers of men and women were included in the study, the frequencies in Table 10.9 are difficult to compare on an absolute basis. However, if these frequencies are converted to proportions based upon row totals, then

Table 10.10. Expected Frequencies for Coffee Preference Data

Brand	Men	Women
A	33.68	30.32
B	18.95	17.05
C	20.00	18.00
D	27.37	24.63

Table 10.11. Computation of χ^2 for Coffee Preferences for Men and Women

Brand	Sex	f_o	f_e	$(f_o - f_e)$	$(f_o - f_e)^2$	$(f_o - f_e)^2/f_e$
A	M	38	33.68	4.32	18.6624	0.5541
A	F	26	30.32	4.32	18.6624	0.6155
B	M	21	18.95	2.05	4.2025	0.2218
B	F	15	17.05	2.05	4.2025	0.2465
C	M	20	20.00	0.00	0.0000	0.0000
C	F	18	18.00	0.00	0.0000	0.0000
D	M	21	27.37	6.37	40.5769	1.4825
D	F	31	24.63	6.37	40.5769	1.6475
Totals						4.7679

NOTE: M denotes males, and F denotes females.

the proportion of men and women selecting each brand of coffee is easier to compare. Each frequency in Table 10.9 was divided by the appropriate row total to obtain the proportions shown in Table 10.12.

For comparison purposes, the column totals can also be divided by the total number of cases included in the analysis. Such column proportions can serve as an indication of the expected proportions for each cell in that column if the two variables are independent. From Table 10.12, the proportions are fairly consistent up and down each column with an exception for brand D. However, as indicated by the nonsignificant χ^2, this deviation could happen by chance alone and thus not reflect a real preference for this brand of coffee. Had these frequencies been significantly associated, then the proportions would have been different to a greater extent, reflecting real differences between men and women. However, when interpreting a significant χ^2, researchers must remember that the χ^2 is computed by using all discrepancies taken as a whole, and that all discrepancies make a contribution to the significant χ^2. However, the relative contributions of χ^2 computed for each cell can be taken into consideration, and by inspection, the cells making the most contribution to the overall χ^2 can be seen.

10.5.3 Assumptions Underlying Chi-Square

Although χ^2 is usually considered a nonparametric statistic, some basic assumptions should be taken into consideration when it is used. The most important assumption is that the observations have

Table 10.12. Coffee Preferences Expressed as Proportion of Row Totals

Brand	Men	Women	Row Totals
A	0.59	0.41	64
B	0.58	0.42	36
C	0.53	0.47	38
D	0.40	0.60	52
Total	0.526	0.474	190

to be independent. Generally, the total number of observed frequencies tallied has to equal the total number of cases included in the sample. For example, if each person in the coffee preference example were given three occasions to taste the four brands of coffee, then 300 tallies would be made for the 100 persons tallied in Table 10.8. The preferences for each person are unlikely to be independent. Thus, the data would not have met the assumption of independent observations, and χ^2 would not be appropriate to use as the statistical analysis.

The second assumption is that, as with all inferential statistics, the sample has been drawn at random. To the extent that the sample is biased, the result will also be biased.

The final assumption is that the observed frequencies are normally distributed about the expected frequencies. With random sampling and large numbers of cases included in the sample, this assumption is usually met. However, when the data are such that the expected frequencies are small (e.g., less than 5 for 2×2 tables, or less than 3 for larger tables), then the distribution of observed frequencies around the expected frequencies tends to be positively skewed. As will be discussed in this section, there are ways small expected frequencies can be accommodated in 2×2 tables. However, recent studies on what the expected cell frequency should be as a minimum indicate that χ^2 will be distributed as expected even when the expected cell frequencies are less than 5 in 2×2 tables (*11*).

Another consideration has to be made when using χ^2 and that involves the size of the sample. The size of the computed χ^2 is directly related to the size of the sample. If a χ^2 is computed for a 2×2 table with only 20 frequencies compared to a 2×2 table with 200 frequencies, the table with 200 frequencies will be 10 times as large because of the sample size, even though the number of categories remains the same. Thus, in interpreting χ^2, statisticians must consider the sample size and remember that large samples may produce significant differences that are so small that they have no practical importance. Likewise, if the sample size is very small, a high degree of relationship may exist between the two variables, but the χ^2 will not be sensitive to revealing the significance of this relationship.

The final consideration is that because χ^2 is computed on nominal data, the obtained values of χ^2 can increase only by increments of whole values, whereas the theoretical distribution of χ^2 represents a continuous distribution. For large sample sizes, this discrepancy poses no problem, but for small sample sizes, a discrepancy may result between the obtained χ^2 and the theoretical distribution of χ^2 that would have an influence on the exact probabilities associated with each obtained χ^2. However, for most problems involving both large sample sizes and for tables greater than 2×2, the discrepancy is minimal and usually not taken into consideration. For tables larger than 2×2, most statisticians indicate that if 80% of the cells have expected frequencies greater than 5, then no distortion occurs in the interpretation of (*11*, p. 194).

10.5.4 Yates' Correction

A correction can be applied to χ^2. This correction is known as Yates' correction for discontinuity. This procedure involves reducing the discrepancies between observed and expected frequencies by 0.5 before squaring. For a single classification χ^2, the correction is made as in the following example. Suppose that 50 persons were asked to state their preference for a new product, and 32 stated a preference for product A and 18 stated a preference for product B. Does this distribution indicate a significant difference in preference for these two products? If there is no preference, then half of the persons should select product A and half product B. To answer this question, the χ^2 analysis with Yates' correction for discontinuity can be applied to yield

$$\chi^2 = \frac{(|32 - 25| - 0.5)^2}{25} + \frac{(|18 - 25| - 0.5)^2}{25} = 3.38$$

The vertical bars indicate that 0.5 is to be subtracted from the absolute discrepancy without regard to sign before squaring. For this problem, the obtained $\chi^2 = 3.38$ is compared to the table value of χ^2 for df $= 1$, which is 3.84. Because the obtained χ^2 is less than this value, this difference between observed and expected frequencies can be concluded to happen more than 5 times out of 100 trials if nothing but chance were operating, and thus the results probably represent only a chance difference. Without Yates' correction, $\chi^2 =$ 3.92, which is significant, would have been obtained. Thus, without Yates' correction, a Type I statistical error would probably have been made (i.e., the null hypothesis would have been rejected when in fact it probably should have been retained).

Yates' correction can also be applied to 2×2 tables. For example, if the preferences of men and women for two products were to be compared, then

$$\chi^2 = \frac{n[f_A(f_D) - f_B(f_C)]^2}{(f_A + f_B)(f_A + f_C)(f_B + f_D)(f_C + f_D)} \qquad (10.19)$$

where f_A, f_B, f_C, and f_D represent the frequencies in cells A,B,C, and D, respectively. Equation 10.19 is the general equation for χ^2 for a 2×2 table, where the data would be set up as in Figure 10.2, which also shows how the marginal row and column totals would be computed.

Suppose that the preferences for 10 men and 10 women for the two products are as in Figure 10.2. Is there a significant difference in the preference of men and women for products A and B? To answer this question, the χ^2 without Yates' correction would be

$$\chi^2 = \frac{20[2(3) - 7(8)]^2}{(9)(10)(10)(11)} = 5.05$$

	MEN	WOMEN	
Product A	A $f_o = 2$	B $f_o = 7$	A + B
Product B	C $f_o = 8$	D $f_o = 3$	C + D
	A + C	B + D	

Figure 10.2. Data layout for a 2×2 table.

which would indicate a significant difference in the preferences of men and women at the 0.05 level of significance because the computed χ^2 is greater than the χ^2 required for one degree of freedom ($\chi^2 = 3.84$). However, because the expected frequencies for the A and B cells are less than 5, (i.e., $(9 \times 10)/20 = 4.5$), Yates' correction should be applied by using equation 10.20. The correction is made to reduce the absolute difference in the numerator by $n/2$.

$$\chi^2 = \frac{n[|f_A(f_D) - f_B(f_C)| - (n/2)]^2}{(f_A + f_B)(f_A + f_C)(f_B + f_D)(f_C + f_D)} \qquad (10.20)$$

With Yates' correction,

$$\chi^2 = \frac{20(|6 - 56| - 10)^2}{(9)(10)(10)(11)} = 3.23$$

which is not significant at the 0.05 level. Thus, because the expected frequency was less than 5 for two of the cells, Yates' correction should have been used, and the conclusion should be that no significant difference exists in the preference of men compared to women for these two products. Camilli and Hopkins (*12, 13*) showed that Yates' correction should be applied to 2×2 tables only when the expected frequency is less than 8 for any of the cells and should not be applied when the expected frequency is greater than 8 for all of the cells. When the expected frequency is greater than 8, the correction results in a too conservative test of the significance of the differences. Thus, fewer statistically significant results and more Type II statistical errors result.

References

1. Havlicek, L. L.; Peterson, N. L. *Psych. Bull.* **1976,** *84,* 373–377.
2. Fisher, R. A. *Statistical Methods for Research Workers;* Hafner: New York, 1970.
3. Cochran, W. G.; Cox, G. M. *Experimental Designs;* Wiley: New York, 1957.
4. Edwards, A. L. *Experimental Design in Psychological Research;* Holt, Rinehart and Winston: New York, 1968.

5. Dayton, C. M. *The Design of Educational Experiments;* McGraw–Hill: New York, 1970.
6. Beyer, W. H. *Handbook of Tables for Probability and Statistics;* The Chemical Rubber Company: Cleveland, OH, 1966.
7. Owen, D. B. *Handbook of Statistical Tables;* Addison–Wesley: Reading, MA, 1962.
8. Hotelling, H. *Ann. Math. Stat.* **1940,** *11,* 217–283.
9. McNemar, Q. *Psychological Statistics;* Wiley: New York, 1962.
10. Cochran, W. G. *Biometrika* **1950,** *37,* 256–266.
11. Roscoe, J. T. *Fundamental Research Statistics;* Holt, Rinehart and Winston: New York, 1969.
12. Camilli, G.; Hopkins, K. D. *Psych. Bull.* **1978,** *85,* 163–167.
13. Camilli, G.; Hopkins, K. D. *Psych. Bull.* **1979,** *86,* 1011–1014.

Problems

1. Per-unit production costs for two plants that produce the same product are being compared to determine cost differences. The mean per-unit production cost per item at plant A was $0.80 for a random sample of 400 items with a standard deviation of $0.10. For plant B, a random sample of 300 units was taken, and the mean production cost per item was found to be $0.92 with a standard deviation of $0.22. Is there a significant difference at the 0.01 level between these two plants with regard to mean per-unit production costs? Test the null hypothesis H_0: $\mu_A - \mu_B = 0$.

2. For the data in problem 1, is there a significant difference in the variance of per-unit production costs between these two plants? Test the hypothesis H_0: $\sigma_A^2 - \sigma_B^2 = 0$ at the 0.01 level of significance.

3. A testing laboratory is trying to decide whether new equipment would reduce the time required to analyze water samples. Ten technicians were randomly selected to run analyses on the old equipment and then the same analyses with the new equipment. The results in number of minutes per analysis are given for both the old and new equipment. Is there a significant reduction in the amount of time required for these analyses? Set up and test the directional hypothesis H_0: $\mu_{old} - \mu_{new} = 0$ at the 0.01 level of significance.

Technician	Old Analysis	New Analysis
1	18	15
2	20	19
3	12	14
4	16	15
5	15	15
6	22	20
7	14	16
8	19	17
9	13	15
10	18	16

4. For the data in problem 3, is there a significant reduction in the variation in the amount of time needed for each analysis? Set up and test the hypothesis H_0: $\sigma^2_{old} - \sigma^2_{new} = 0$ at the 0.01 level of significance.

5. For the data in problem 3, is the correlation between the time required with the old compared to the new equipment significant at the 0.01 level?

6. In testing the effectiveness of a new medicine compared to a standard product, 160 patients out of a random sample of 200 patients showed improvement of their condition with the new product, compared to 130 patients out of another random sample of 200 patients showing improvement with the standard product. Is the new product more effective than the standard product in bringing about an improved condition in the number of patients with this disease?

Chapter 11

Analysis of Variance

Methods to determine whether two means are significantly different represent some of the most frequently used statistical procedures because they provide statistical models for comparing the results of two-group investigations. However, the relative effects of three or more treatments or conditions are frequently compared. For instance, the yield of a chemical obtained by using three or more different catalytic methods might be investigated, the effectiveness of four types of fertilizer on the yields of corn grown in standard-sized plots might be evaluated, or the amount of sulfur dioxide removed from three towers varying in height might be compared. The *analysis of variance* is the statistical procedure used to test the significance of the difference between the means for two or more sets of data. For the two-sample situation, the analysis of variance is mathematically equivalent to the independent *t* test.

The one-way or simple analysis of variance, or *ANOVA*, is used to test the null hypothesis that two or more independent samples were drawn from the same population with regard to mean level on the dependent variable, or specifically that no significant difference occurs among the sample means with regard to one dependent variable. In an experimental design, the samples are usually drawn at random from the same population, subjected to different experimental conditions, and then compared on a common criterion variable. Or the samples may be randomly drawn from different populations, and then the means of these samples compared to determine whether they represent the same population with regard to a specific criterion. As an example of samples drawn from different populations, suppose that yields from four production plants were compared to determine if any significant differences occur among the four plants with regard to mean yields of a chemical or product.

1453–0/88/0225$08.50/1

If no significant differences are found, then these four plants can be considered as random samples from a common population of plants with regard to mean yield.

11.1 Basic Concepts

The basic rationale underlying the analysis of variance is that any observed value is made up of many parts and is influenced by many factors. For example, the yield of a chemical is determined by a number of factors including, but not limited to, the production process; catalysts used; impurities; and variations in the equipment, operators, temperature, and pressure. In statistical terminology, statisticians speak of the observed values as composites of these factors and try to partition the various sources of variance that influence the observed value so that some are high and others are low. The goal of analysis of variance is to account for as much of the variance as possible in the sets of observed values. In the simple or one-way analysis of variance, only one factor or source of variation is investigated. This factor is defined as the independent variable. In the example of the mean yields of a chemical or product, the production process used in the four plants would be the independent variable. For other studies, the independent variable might be various catalysts that could be compared, or different temperature or pressure levels could be studied with regard to their effect on the yield of a chemical. In Chapters 12 and 13, more complex analyses will be presented in which two or more independent factors are investigated simultaneously in a given study. Regardless of the type of analysis of variance used, the procedure is to partition the variation of observed values into sources that can be accounted for and sources that cannot be accounted for, which are considered error variance.

For one set of data, the variance of the values from the mean of that set of values was shown in Chapter 9 to provide an estimate of the population variance. The standard error of the mean provided the basis for computing upper and lower limits where the means from random samples would be likely to fall. If this standard error of the mean is squared and multiplied by the number of cases included in the sample, another estimate of the population variance is obtained. If nothing but chance is operating and all of the factors listed that make up each value are combined together as random variations, then with a large number of random samples taken from a given population, these two estimates of the population variance should be equal because only random sampling errors are operating for both estimates. If a ratio of these two estimates is formed and only chance is operating, then the ratio should be equal to 1. In conceptual form, a ratio of the sampling error for means divided by the sampling error for individual cases for a large number of random samples should average to be approximately 1.

Suppose that some of the random samples are treated differently in a systematic way so that instead of purely random variation operating, a systematic variation is introduced, such as changing the production procedure or a catalyst for some groups. The means for these samples or groups are going to be influenced by both the systematic treatment plus the same errors that were operating when considered as random samples. The numerator of the ratio of these two estimates of the population variance is composed of two components, reflecting both the treatment and the sampling error, and the ratio for these two estimates becomes the sampling error for means plus the treatment effect divided by the sampling error for individual cases. If the treatment has an effect on the group means, then this ratio would be expected to be greater than 1. If this ratio is greater than 1, some of the sampling variance has been partitioned into variance due to treatments. In analysis of variance terms, the estimate of the population variance based on the deviations of means from the grand mean is defined as the *between-groups variance* or, if more than two groups, *among-groups variance*, and the other estimate based on deviations of scores from group means is defined as the *within-groups variance*. To determine whether the treatment has produced an effect, the ratio of these two estimates (F) is

$$F = \frac{\text{between–groups variance}}{\text{within–groups variance}} \qquad (11.1)$$

and then this F is compared to the distribution of F to determine the probability that the obtained F ratio could have occurred by chance alone.

The partitioning of variance is the basis of the analysis of variance procedure. However, for computations, the basis is the sum of squares, which is actually divided into additive parts. For any analysis of variance, the deviation of any single value from the total grand mean, $X_i - \overline{X}_G$, can be divided into two parts. As shown in Figure 11.1, one part of this deviation, which is defined as $X_i - \overline{X}$, is the deviation of the value from the subgroup mean. The other part consists of the deviation of the subgroup mean from the grand mean, or

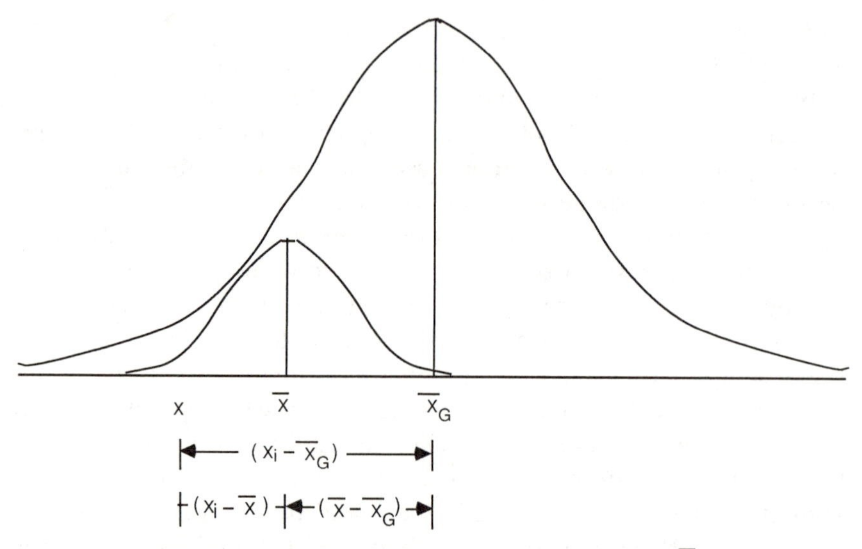

Figure 11.1. Components of deviation $(X_i - \overline{X}_G)$.

$\overline{X} - \overline{X}_G$. If this partitioning is done for all individual values in all groups, the total deviations from the grand mean equal

$$\Sigma(X_i - \overline{X}_G) = \Sigma(\overline{X} - \overline{X}_G) + \Sigma(X_i - \overline{X}) \quad (11.2)$$

This same additive relationship holds for the squared deviations. Thus,

$$\Sigma(X_i - \overline{X}_G)^2 = \Sigma(\overline{X} - \overline{X}_G)^2 + \Sigma(X_i - \overline{X})^2 \quad (11.3)$$

$$SS_T = SS_b + SS_w \quad (11.4)$$

where SS_T is the total sum of squares, SS_b is the sum of squares between groups, and SS_w is the sum of squares within groups. The mathematical derivations for this relationship are given in advanced statistics textbooks (*1–4*). Once the sums of squares are obtained, dividing each sum of squares by its respective degrees of freedom gives the two estimates of the population variance that are compared in the analysis of variance.

As a way of demonstrating the concept that the total sum of squares is equal to the sum of squared deviations about the individual group means plus the sum of the individual group means from the grand mean for all groups, hypothetical values are given for three groups in Table 11.1 and shown in Figure 11.2. The means for the

Table 11.1. Components of Sums of Squares

X	T	W	B
	Group A		
5	1	4	9
4	4	1	9
4	4	1	9
3	9	0	9
3	9	0	9
3	9	0	9
2	16	1	9
2	16	1	9
1	25	4	9
Total	93	12	81
	Group B		
9	9	4	1
8	4	1	1
8	4	1	1
7	1	0	1
7	1	0	1
7	1	0	1
6	0	1	1
6	0	1	1
5	1	4	1
Total	21	12	9
	Group C		
10	16	4	4
9	9	1	4
9	9	1	4
8	4	0	4
8	4	0	4
8	4	0	4
7	1	1	4
7	1	1	4
6	0	4	4
Total	48	12	36
$\bar{X}_A = 3$	$\bar{X}_B = 7$	$\bar{X}_C = 8$	$\bar{X}_G = 6$

$$SS_T = 93 + 21 + 48 = 162$$
$$SS_w = 12 + 12 + 12 = 36$$
$$SS_b = 81 + 9 + 36 = 126$$

NOTE: Abbreviations are as follows: X is a raw value for each observation, T is the squared deviation of each raw value from the grand mean, or $(X_i - \bar{X}_G)^2$, W is the squared deviation of each raw value from each group mean, or $(X_i - \bar{X})^2$, and B is the squared deviation of each group mean from the grand mean, or $(\bar{X} - \bar{X}_G)^2$.

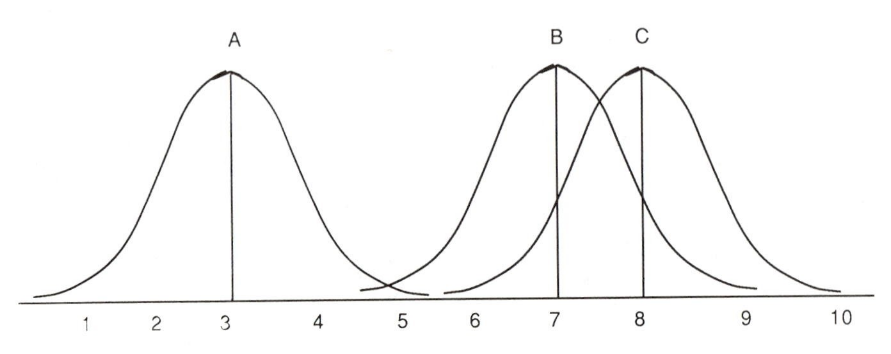

Figure 11.2. Distribution of values from Table 11.1.

three groups A, B, and C, are, respectively, 3, 7, and 8, and the overall grand mean for all 27 cases is 6. For each group, the squared deviation of each value from the grand mean is given under column T for $(X_i - \overline{X}_G)^2$, and the totals for the three groups or for the 27 cases will equal the total sum of squared deviations for all cases, or SS_T. The sum of squared deviations for within groups, which is the squared deviation of each value from its respective group mean, is given under column W for $(X_i - \overline{X})^2$, and the sum of the squared deviations of each group mean from the grand mean is given under column B for $(\overline{X} - \overline{X}_G)^2$. As can be noted from Table 11.1, the total sum of squares is equal to 162, and this is composed of the within sum of squares, which is 36, and the between sum of squares, which is 126.

When the respective sums of squares are divided by their degrees of freedom, then the two estimates of the population variance to be compared are obtained. In the analysis of variance, the sums of squares and the corresponding degrees of freedom are put in a summary table such as shown in Table 11.2. As shown in this table, the source for each sum of squares is given along with the appropriate degrees of freedom. Then each sum of squares is divided by its degrees of freedom to obtain the estimate of the population variance, which is usually labeled mean square, or ms, in analysis of variance tables. The F ratio is then formed by dividing the mean square between by the mean square within.

The number of degrees of freedom for the one-way analysis of variance is obtained as follows: For the total number of cases in all groups, the total number of cases minus 1 is the total degrees of freedom (df_T). The within degrees of freedom (df_w) is the total number of cases minus the number of groups, or $df_w = n_T - k$, where n_T is the total number of cases in all groups, and k is equal to the

Table 11.2. Summary Table of Analysis of Variance

Source	SS	df	ms	F
Between	126	2	63	42.00
Within	36	24	1.5	
Total	162	26		

number of groups. The between degrees of freedom (df_b) equals the number of groups minus 1, or $k - 1$. As with the sums of squares, the degrees of freedom are also additive, and the between plus the within degrees of freedom equal the total degrees of freedom.

To evaluate the F ratio, the obtained F is compared to the table value of F for the number of degrees of freedom for the numerator, which is equal to the between degrees of freedom, and the number of degrees of freedom for the denominator, which is equal to the within degrees of freedom. Thus, for the hypothetical example, the degrees of freedom are 2 and 24, respectively. From Table D in the Appendix, with 2 and 24 degrees of freedom, $F = 3.40$ is required to be significant at the 0.05 level, and $F = 5.61$ is required for significance at the 0.01 level. Because the obtained F is greater than either of these two values, one would then conclude that a significant difference exists somewhere between pairs of means. In Section 11.3, methods to determine which pairs of means are significantly different will be presented.

As with other statistics, the deviation approach as presented in Table 11.1 is satisfactory for small samples and for computer applications. However, more convenient equations are available for computing the sums of squares necessary for the analysis of variance. The total sum of squares can be computed by the general equation 11.5, which was first presented for computing the standard deviation.

$$SS_T = \Sigma X_T^2 - \frac{(\Sigma X_T)^2}{n_T} \qquad (11.5)$$

where $\Sigma X^2{}_T$ is the sum of the squared values for all cases, ΣX_T is the sum of the values for all cases, and n_T is the total number of cases.

To compute the sum of squares within groups,

$$SS_w = \Sigma X_T^2 - \sum^{k} \left(\frac{(\Sigma X_i)^2}{n_i} \right) \qquad (11.6)$$

where $(\Sigma X_i)^2$ is the squared sum of the values for each group, and n_i is the number of cases in each group, and ΣX_T^2 os the sum of the squared values for all cases.

The sum of squares between groups is

$$SS_b = \sum^{k} \left(\frac{(\Sigma X_i)^2}{n_i}\right) - \left(\frac{(\Sigma X_T)^2}{n_T}\right) \tag{11.7}$$

where ΣX_i is the sum of values for each group, and ΣX_T is the sum of values for all cases.

Only three quantities have to be computed to obtain the three sums of squares. The first quantity is $(\Sigma X_T)^2/n_T$, which is the second term in equations 11.5 and 11.7. This quantity is calculated by adding all of the values for all groups, or for the total sample size, by squaring this sum and then dividing by n_T, the total number of cases. The second quantity, which is the first term in equations 11.5 and 11.6, is obtained by simply squaring each value for all cases in all groups and adding these squared values together. The last quantity needed, which is the second term in equation 11.6 and the first term in equation 11.7, is obtained by summing the values in each group, squaring this sum for each group, dividing each of these squared values by the number of cases in each group, and then summing the quotients for all groups.

To show the computations obtained by using these three equations, the sums of squares for the data in Table 11.1 will be recalculated. The sums and sums of squares are presented in Table 11.3.

Using the three equations for the respective sums of squares yields

$$SS_T = 1134 - \frac{162^2}{27} = 162$$

$$SS_w = 1134 - \left(\frac{27^2}{9} + \frac{63^2}{9} + \frac{72^2}{9}\right) = 36$$

$$SS_b = \left(\frac{27^2}{9} + \frac{63^2}{9} + \frac{72^2}{9}\right) - \frac{162^2}{27} = 126$$

which is exactly what was obtained by using the deviations shown in Table 11.1. As a computational check,

$$SS_T = SS_w + SS_b$$

$$162 = 36 + 126$$

Table 11.3. Values and Squared Values for the Data in Table 11.1

Group A		Group B		Group C	
X	X^2	X	X^2	X	X^2
5	25	9	81	10	100
4	16	8	64	9	81
4	16	8	64	9	81
3	9	7	49	8	64
3	9	7	49	8	64
3	9	7	49	8	64
2	4	6	36	7	49
2	4	6	36	7	49
1	1	5	25	6	36
Total = 27	Total = 93	Total = 63	Total = 453	Total = 72	Total = 588

$\sum X_T = 27 + 63 + 72 = 162$

$\sum X_T^2 = 93 + 453 + 588 = 1134$

11.2 Analysis of Variance Example

Suppose the percentage of solids in each of four batches of wet brewer's yeast (A, B, C, and D), each from a different supplier, was to be determined. The researcher wants to determine whether the suppliers differ so that decisions can be made regarding future orders. The null hypothesis would be that no differences occur in the mean percentage of solids among the four batches of wet brewer's yeast. Then the level of significance desired to test this hypothesis would be determined. If the percentage level was very important, and if there were cost differences among the suppliers, then possibly the 0.01 level should be chosen. After the significance level has been set, then random samples from each batch would be collected. Hypothetical values for four batches are presented in Table 11.4. The null hypothesis to be tested would be H_0: $\overline{X}_A = \overline{X}_B = \overline{X}_C = \overline{X}_D$ at the 0.01 level of significance.

Equations 11.5, 11.6, and 11.7 would yield, respectively,

$$SS_T = 7247 - \frac{495^2}{37} = 7247 - 6622.30 = 624.70$$

$$SS_w = 7247 - \left(\frac{184^2}{10} + \frac{130^2}{9} + \frac{107^2}{10} + \frac{74^2}{8}\right)$$

$$= 7247 - 7092.77 = 154.23$$

$$SS_b = 7092.77 - 6622.30 = 470.47$$

Table 11.4. Percentages of Total Solids in Four Batches of Brewer's Yeast

Statistic	Batch A	Batch B	Batch C	Batch D	Totals
n					
1	20	19	11	10	
2	18	14	12	7	
3	16	17	14	11	
4	21	13	10	9	
5	19	10	8	6	
6	17	16	10	11	
7	20	14	13	8	
8	16	12	9	12	
9	19	15	12		
10	18		8		
n_T	10	9	10	8	37
$\sum X$	184	130	107	74	495
$\sum X^2$	3412	1936	1183	716	7247
$\overline{X}$	18.40	14.44	10.70	9.25	13.88
S.D.	1.71	2.70	2.06	2.12	4.17

As a computational check,

$$SS_T = SS_w + SS_b$$

$$624.70 = 154.23 + 470.47$$

$$624.70 = 624.70$$

The summary analysis of variance table would now be set up as shown in Table 11.5. The degrees of freedom are as follows:

$$df_T = n_t - 1 = 37 - 1 = 36$$

$$df_b = k - 1 = 4 - 1 = 3$$

$$df_w = n_t - k = 37 - 4 = 33$$

The significance of the obtained F ratio can be determined by referring to Table D with 3 and 33 degrees of freedom. Table D shows that $F \geq 4.44$ would be required for significance at the 0.01 level. Because the obtained $F = 33.58$ is greater than this table value, the null hypothesis of no significant difference among the means would be rejected, and the conclusion would be that a difference or differ-

Table 11.5. Analysis of Variance Table for Total Solids Data

Source	SS	df	ms	F
Between	470.47	3	156.82	33.58
Within	154.23	33	4.67	
Total	624.70	36		

ences among the means occur. When there are three or more groups, all that the F ratio conveys is that somewhere a difference between or among the means occurs. To show where these differences are, a post hoc test, which will be described in Section 11.3, is then run. If the F ratio is not significant, there is no need to go any further. The null hypothesis would not be rejected, and the conclusion would be that no significant differences occur between or among the means for the groups studied.

As will be presented in Section 11.5, an additional measure can be used after obtaining a significant F ratio to provide information about the strength of the relationships between the independent and dependent variables. This index, which is called omega squared (ω^2), indicates the amount of the variance in the dependent variable accounted for by the independent variable.

11.3 Comparisons among Treatment Means

The analysis of variance F ratio provides an omnibus or overall test for comparing means from two or more random samples. If this F ratio is not significant, the conclusion would be that no real differences occur among the sample means and that any observed differences can be accounted for by random errors. Thus, no need arises to do further analyses with regard to the differences among the means. However, when the overall test of equal means is rejected, further analysis is necessary to determine which means are significantly different. The significant F ratio indicates only significant differences somewhere among the means. Further analyses are necessary to identify which pairs of means or combinations of pairs of means are significantly different.

Comparing more than two sample means involves a number of considerations, and a number of procedures have been suggested, some of which are more widely accepted than others. The main problem is that multiple comparisons violate some of the basic principles of probability theory. For example, the *t* test is appropriate for comparing one pair of sample means but is not appropriate when comparing all possible pairs of differences when more than two means are being compared. To illustrate the problem involved, a review of basic probability shows that if two coins are tossed simultaneously, the probability of getting two heads by using the multiplication theorem is 1/4. If three coins are tossed, the probability of getting two or more heads becomes 1/2, and if four coins are tossed, the probability becomes 11/16. The same probability changes apply to making repeated *t* tests, and thus the probability of making Type I errors changes as the number of *t* tests computed increases.

Tukey (5) was one of the first statisticians to discuss this problem, and he made a distinction between *per-comparison error rate* and *experimentwise error rate*. Tukey defined the number of comparisons between means falsely declared significant divided by the total number of comparisons as the per-comparison error rate, or increase in Type I errors. He defined the experimentwise error rate as the number of experiments with at least one difference falsely identified as being significant divided by the total number of experiments. The relationship between these two types of errors is given by

$$\alpha_{ew} = 1 - (1 - \alpha)^c \qquad (11.8)$$

where α_{ew} is the probability of making an experimentwise error, α is the probability level set for the study, and c is the number of independent comparisons that are made. As an example, if three comparisons are to be made (i.e., when comparing three means), and with an alpha level set at 0.05 for each comparison, the probability of making a Type I error for these three comparisons, or an experimentwise error, is

$$\alpha_{ew} = 1 - (1 - 0.05)^3 = 0.143$$

If six comparisons are to be made as when comparing four means, the experimentwise error rate increases to

$$\alpha_{ew} = 1 - (1 - 0.05)^6 = 0.265$$

Thus, the experimentwise error rate increases directly with the number of comparisons made. The number of comparisons for all pairs of means is given by $k(k - 1)/2$, where k is the number of sample means to be compared. With five treatments, the total possible combination of pairs of means (e.g., A vs. B, A vs. C, A vs. D, and B vs. C) would be $(5 \times 4)/2 = 10$. With 10 treatments, a total of $(10 \times 9)/2 = 45$ pairs of means could be tested. Thus, any procedure for comparing all possible pairs of means has to take this basic consideration of probability into consideration to keep the probability of Type I errors constant.

The references at the end of this chapter, especially references 3 and 4, discuss the problems involved in making multiple comparisons. Three basic considerations have to be taken into account when deciding what procedure should be used when making multiple comparisons of means. The first consideration is whether specific comparisons are planned before the analyses are begun. These comparisons are referred to as *a priori* or planned comparisons. Usually, a priori comparisons are made instead of computing the omnibus F test and involve only comparisons of means that have been specifically identified as having either theoretical or experimental relevance. Unplanned, *post hoc*, or a posteriori comparisons usually involve all comparisons between means and are sometimes referred to as data sifting (i.e., making all comparisons possible to find differences that are significant). In chemical research, studies are set up so that all comparisons between means are important to test. Thus, this textbook will emphasize the post hoc procedures commonly used. However, two types of planned comparisons will be considered in this chapter: those dealing with comparing a standard procedure with each of several treatment means, and, when the independent variable is continuous rather than nominal or categorical, trend analysis.

The second consideration is what comparisons are meaningful and are important to run. Often, a researcher will want to test all combinations of pairs of means, whereas situations arise in which only specific contrasts are important.

The third consideration is when the independent variable is categorical in nature (e.g., different plants or procedures). Then, the researcher may want to compare selected pairs of means or to compare all possible pairs of means. If the independent variable is continuous in nature, such as an incremental percentage of a catalyst, then the researcher may want to compare two means at a time or combine sets of means in an incremental way (or run a trend analysis).

11.3.1 Post Hoc Tests

A number of post hoc methods can be used after obtaining a significant F ratio, which indicates significant differences among the means. These methods usually use the F test or a statistic known as Student's t and vary in terms of the degree to which they control per-comparison error rates. The Scheffe′ method (*6*) is the most conservative with respect to Type I errors, in that this method will lead to the smallest number of significant differences. In fact, Winer (*4*) pointed out that using the Scheffe′ procedure will result in fewer significant results than theoretically expected. The next most conservative test is the Tukey A test, which is recommended by Winer (*4*, p. 201) because it is easy to apply and is applicable to many situations. Next in order of conservative tests is the Tukey B test, followed by the Newman–Keuls (*7, 8*) test, and then the Duncan (*9*) test. Discussions and comparisons of these post hoc tests are given in Winer (*4*) and Keppel (*3*). The Scheffe′ and Tukey A tests will be described in this Section. The Scheffe′ test is suggested because the Type I error rate is at a maximum equal to the significance level set for all possible comparisons, even though too few significant differences may be identified. The Tukey A procedure is less conservative than the Scheffe′ method, but like the Scheffe′ method uses a single critical range (CR) for the testing of all differences between pairs of sample or treatment means.

For the Scheffe′ test, the critical range for comparing all possible pairs of means (CR_S) is given by

$$CR_S = \sqrt{(a-1)F}\sqrt{\frac{2(ms_w)}{n}} \qquad (11.9)$$

where a is the number of sample means being compared, F is the critical F value from Table D for the degrees of freedom between and within, n is the number of cases in each group, and ms_w is the mean square value within from the analysis of variance summary table.

The general equation for the Tukey A test, often referred to as the *honestly significant difference* test or HSD, is

$$CR_T = (q_{k,df_w})\sqrt{\frac{ms_w}{n}} \qquad (11.10)$$

where CR_T is the critical range for the Tukey test, q represents an entry from the table of Student's t statistic (Table I in the Appendix) for the maximum number of treatment groups being compared (k) and the degrees of freedom for within groups (df_w), n is the number of cases in each sample, and ms_w is the mean square value within from the analysis of variance table.

Applying the Scheffe′ test to the data in Table 11.3 and the summary analysis of variance data in Table 11.2 yields

$$CR_S = \sqrt{(3 - 1)(3.40)}\sqrt{\frac{2(1.5)}{9}}$$

$$= 2.608 \times 0.577 = 1.50$$

Thus, any difference between means equal to or greater than 1.50 would be significantly different at the 0.05 level of significance. Applying the Tukey A test to the same data yields

$$CR_T = 3.53\sqrt{\frac{1.5}{9}} = 1.44$$

where 3.53 is taken from Table I for $k = 3$ and $df_w = 24$.

Thus, the Tukey test results in a slightly smaller critical range than the Scheffe′ test. Applying these critical ranges to the means for the three groups as shown in Table 11.1 shows that the mean for group A is significantly different from the means for both group B and group C. However, the means for groups B and C are not significantly different.

The Tukey A equation that uses Student's t test should be used only for comparing groups of equal size. However, if there are slight differences in values of n such as for the four batches in Table 11.4, the harmonic mean of the n values can be used. To apply the Tukey A procedure to the data in Table 11.4, the harmonic mean (HM) of the numbers in each sample must first be computed.

$$HM = \frac{\text{number of groups (batches)}}{\frac{1}{n_A} + \frac{1}{n_B} + \frac{1}{n_C} + \frac{1}{n_D}} \quad (11.11)$$

$$= \frac{4}{\frac{1}{10} + \frac{1}{9} + \frac{1}{10} + \frac{1}{8}}$$

$$= \frac{4}{0.10 + 0.111 + 0.10 + 0.125} = 9.174$$

Then, using the data from the summary analysis of variance table (Table 11.5) yields

$$\mathrm{CR}_T = 3.84\sqrt{\frac{4.67}{9.174}} = 2.738$$

Thus, any means that differ by 2.74 or more would be considered significantly different at the 0.05 level of significance. In this computation, with four batches and 33 degrees of freedom for the within sum of squares, the closest value from Table I is for $k = 4$ and 30 degrees of freedom. Thus, $q = 3.84$.

When a number of means are to be compared, it is often convenient to put them in order of their magnitude, which is how they happened to fall in Table 11.4, and to show the differences between each pair of means as follows:

	Batch		
	A	**B**	**C**
Batch B	3.96*		
Batch C	7.70*	3.74*	
Batch D	9.15*	5.19*	1.45

Thus, all means are significantly different from each other as denoted by * except for the means for batch C compared to batch D.

The same procedure of computing the harmonic mean could be used for applying the Scheffe′ procedure for these four groups or batches with unequal values of n. However, an alternative Scheffe′ procedure can be used. This procedure involves computing a t test for each comparison (t_S) by using the following equation:

$$t_S = \frac{\bar{X}_1 - \bar{X}_2}{\sqrt{\mathrm{ms}_w\left(\frac{1}{n_1} + \frac{1}{n_2}\right)}} \qquad (11.12)$$

and then comparing this t to a computed K value, which is the critical value to be used for these comparisons rather than the t values from Table C in the Appendix.

$$K = \sqrt{\mathrm{df}_b(F_{\mathrm{df}_{b,w}})} \qquad (11.13)$$

where df_b is the degrees of freedom between groups from the analysis of variance table, and $F_{df_{b,w}}$ is the table value of F for the degrees of freedom between and within. Applying this procedure for comparing means for batches B and C yields

$$t_S = \frac{14.44 - 10.70}{4.67\left(\frac{1}{9} + \frac{1}{10}\right)} = 3.76$$

This t would then be compared to

$$K_{0.05} = \sqrt{3(2.90)} = 2.95$$

Thus, at the 0.05 level of significance, these two means would be considered to be significantly different. At the 0.01 level,

$$K_{0.01} = \sqrt{3(4.46)} = 3.658$$

and these two means would also be significantly different because the computed $t = 3.766$ is greater than the K value needed for this comparison at this level of significance.

The Scheffe′ test can also be used to compare combinations of means (e.g., A and B compared to C and D, or A, B, and C compared to D). The procedure uses the weighted mean, and the general equation is

$$t_S = \frac{\left(\frac{n_1\bar{X}_1 + n_2\bar{X}_2}{n_1 + n_2}\right) - \left(\frac{n_3\bar{X}_3 + n_4\bar{X}_4}{n_3 + n_4}\right)}{\sqrt{ms_w\left[\left(\frac{1}{n_1 + n_2}\right) + \left(\frac{1}{n_3 + n_4}\right)\right]}} \qquad (11.14)$$

where 1, 2, 3, and 4 denote the different groups.

For comparing means for batches A and B compared to batches C and D at the 0.01 level of significance.

$$t_S = \frac{\left(\frac{10(18.40) + 9(14.44)}{10 + 9}\right) - \left(\frac{10(10.7) + 8(9.25)}{10 + 8}\right)}{\sqrt{4.67\left[\left(\frac{1}{10 + 9}\right) + \left(\frac{1}{10 + 8}\right)\right]}} = 9.072$$

The obtained t would be compared to $K = 3.658$, which was obtained previously for this data, and because the obtained value of 9.072 is larger than the required K value, the conclusion would be that the combined mean for batches A and B is significantly different than the combined mean for batches C and D at the 0.01 level of significance.

11.3.2 A Priori Tests

As discussed in the beginning to this section, comparisons among treatment means can be made on an a priori basis or on a post hoc basis. Comparisons made on a post hoc basis were discussed in the last section and can be made only if the omnibus F test is significant. A priori or planned comparisons, which will be presented in this section, can be made whether the omnibus test is significant or not and usually involve only comparisons that the researcher is interested in rather than making all possible comparisons between pairs of means.

One of the a priori or planned tests that can be made is to compare all treatment means with the mean for a control condition. Because this is planned, these comparisons can be made regardless of the outcome of the omnibus F test. Dunnett (*10*) provided a test for such comparisons. This test is

$$C_{\text{diff}} = d_k \sqrt{\frac{2(\text{ms}_w)}{n}} \qquad (11.15)$$

where C_{diff} is the critical difference between needed treatment means and the control mean, d_k is the value from Table J in the Appendix at the intersection of the column headed by the number of means to be compared and the row for the df for the mean square within groups, ms_w is the mean square within groups, and n is the number of cases in each group. As with the Tukey test, if the sample sizes vary, then the harmonic mean of the sample sizes can be used. If batch A in Table 11.4 was the control group to which the other batches would be compared, the harmonic mean of the n values would be

$$\text{HM} = \frac{4}{\frac{1}{10} + \frac{1}{9} + \frac{1}{10} + \frac{1}{8}} = 9.174$$

Then, equation 11.15 can be used for the data in Table 11.5 for a two-tailed test at the 0.05 level of significance:

$$C_{\text{diff}} = 2.47\sqrt{\frac{2(4.67)}{9.174}} = 2.49$$

Any difference between the mean of the control group (in this example, batch A) and the other batches that is greater than this critical difference of 2.49 would be considered significantly different at the 0.05 level. Because all of these differences are larger than 2.49, all of the batches would be considered significantly different from the control group. If the n values are considerably different, an alternate procedure to computing the harmonic means of the values of n would be to compute the standard error of the difference separately for each comparison as

$$t_D = \frac{\overline{X}_c - \overline{X}_i}{\sqrt{\text{ms}_w\left(\frac{1}{n_1} + \frac{1}{n_2}\right)}} \qquad (11.16)$$

where $\overline{X}_c$ is the control mean and $\overline{X}_i$ is the mean for a specific treatment. If the computed value of t_D is greater than the critical value for k and df_w in Table J, those two means are considered to be significantly different.

As pointed out by Winer (*4*, p. 203), the level of significance for each individual comparison with the control will be less than the level of significance chosen for all tests (e.g., 0.02 rather than 0.05 in the previous example). However, for all of the possible comparisons, the significance level will be equal to the level set for all comparisons. This condition is due to the fact that the tables set up by Dunnett (*11*) are for all comparisons that can be made with the control, not for each specific contrast.

Another a priori test that can be made is to plan to compare selected pairs of means or combination of means for any given study by using orthogonal comparisons. The basis for these comparisons is that the degrees of freedom for the between-groups sum of squares can be partitioned into a number of independent parts equal to $k - 1$ degrees of freedom. Essentially this means that the between-groups sum of squares can also be divided into as many parts or bits of information as there are degrees of freedom. This test can be done whether the overall F test is significant or not and is based on the additive nature of the sum of squares. That is, the total sum of squares can be broken down into a number of additive parts.

Orthogonal, or independent, comparisons are based on weighting the sums of squares for each group using coefficients selected by the researcher. The only restrictions are that the number of comparisons cannot exceed the $k - 1$ degrees of freedom associated with the between-groups sum of squares, the comparisons have to be independent or orthogonal, and the sum of the weights has to equal zero. Because the sum of squares is dealt with, the number of cases, n, in each group has to be equal (i.e., $n_1 = n_2 = n_3 = n_4$). Two comparisons are orthogonal when the sum of the cross products of their weights is equal to zero. Consider an example of four experimental treatments, and before the study is begun, the researcher states that the only treatment comparisons that are important are for comparing A with B (comparision 1), C with D (comparision 2), and the combinations of A and B with C and D (comparision 3). For each comparison, the means would be weighted as follows:

	A	B	C	D
Comparison 1	1	−1	0	0
Comparison 2	0	0	1	−1
Comparison 3	1	1	−1	−1

Comparisons 1 and 2 are orthogonal because the sum of their cross-products is zero, or

$$\text{comparisons 1 and 2: } (1)(0) + (-1)(0) + (0)(1) + (0)(-1) = 0$$

Likewise, comparisons 1 and 3 and 2 and 3 are also orthogonal because

$$\text{comparisons 1 and 3: } (1)(1) + (-1)(1) + (0)(-1) + (0)(-1) = 0$$

$$\text{comparisons 2 and 3: } (0)(1) + (0)(1) + (1)(-1) + (-1)(-1) = 0$$

The orthogonal-weighted sums of squares are additive (i.e., the sum of the weighted parts equals the total sum of squares for between groups).

As an example of making a priori comparisons, consider the data in Table 11.6 for four experimental treatments for which only the

Table 11.6. Yields for Four Experimental Treatments

Statistic or Comparison	A	B	C	D	Totals
	8	7	11	12	
	7	12	9	10	
	4	8	13	16	
	6	13	12	11	
	3	10	10	13	
	5	9	14	9	
$\sum X$	33	59	69	71	232
$\sum X^2$	199	607	811	871	2488
$\overline{X}$	5.50	9.83	11.50	11.83	9.67
S.D.	1.87	2.32	1.87	2.48	3.26
Comparison 1	1	−1	0	0	
Comparison 2	0	0	1	−1	
Comparison 3	1	1	−1	−1	

comparisons of treatments A with B (comparison 1), C with D (comparison 2), and A and B with C and D (comparison 3) are considered important. The coefficients for each of these three comparisons are given in the table as well as the sum of values (ΣX), sum of squared values (ΣX^2), means and standard deviations for each treatment, and the sum of values and sum of squared values for all cases in the four treatments. The sums of squares for the total, within, and between components can be calculated by using equations 11.5, 11.6, and 11.7, respectively. For the data,

$$SS_T = 2488 - \frac{232^2}{24} = 245.33$$

$$SS_w = 2488 - \left(\frac{33^2}{6} + \frac{59^2}{6} + \frac{69^2}{6} + \frac{71^2}{6}\right)$$

$$= 2488 - 2395.34 = 92.66$$

$$SS_b = 2395.34 - \frac{232^2}{24} = 152.67$$

The treatment sum of squares, SS_b, is now partitioned into orthogonal components by obtaining the weighted sum of the values for each comparison. This partitioning is done by multiplying each group total by coefficient of the comparison squaring the sum of those products, and then dividing by n times the sum of the squared coefficients in the comparison, as in the following equation:

$$\mathrm{SS}_c = \frac{(c_1 t_1 + c_2 t_2 + c_3 t_3 + c_4 t_4)^2}{n(c_1^2 + c_2^2 + c_3^2 + c_4^2)} \qquad (11.17)$$

where t is the ΣX for that specific treatment, and c is the coefficient of the comparison for that treatment.

For the three comparisons comp 1, comp 2, and comp 3, respectively,

$$\mathrm{SS}_{comp_1} = \frac{[1(33) + -1(59) + 0(69) + 0(71)]^2}{6(1^2 + -1^2 + 0^2 + 0^2)} = 56.33$$

$$\mathrm{SS}_{comp_2} = \frac{[0(33) + 0(59) + 1(69) + -1(71)]^2}{6(0^2 + 0^2 + 1^2 + -1^2)} = 0.33$$

$$\mathrm{SS}_{comp_3} = \frac{[1(33) + 1(59) + -1(69) + -1(71)]^2}{6(1^2 + 1^2 + -1^2 + -1^2)} = 96.00$$

As a computational check, the between sum of squares has been partitioned into its three orthogonal components, and these should add up to the computed SS_b, or $56.33 + 0.33 + 96.00 = 152.66$, which is equivalent to $\mathrm{SS}_b = 152.67$ as computed. Thus, of the total variation among the four methods, much of this variation is accounted for by the first and third comparisons and very little by the second comparison. An analysis of variance table for orthogonal comparisons can then be set up as in Table 11.7. Each comparison has one degree of freedom, and the F ratios are computed by dividing each mean square by the mean square within. Computing the overall F ratio is not necessary because a priori tests can be made whether or not the overall F ratio is significant. For this example, with 1 and 20 degrees of freedom for each comparison, at the 0.01 level of significance, $F \geq 8.10$ is needed for significance. Thus, both comparison 1 and comparison 3 are significant at the 0.01 level, and the null hypothesis of no significant difference between means for these two comparisons would be rejected. Thus, the means for treatments A and B are significantly different as are the combined means for treatments A and B compared with treatments C and D. However, treatment C is not significantly different from treatment D.

In setting up orthogonal comparisons, any set of coefficients can be used as long as they add up to zero and the sum of their cross products also adds up to zero. For example, the coefficients could

Table 11.7. Analysis of Variance for Orthogonal Comparisons

Source	SS	df	ms	F
Comparison 1	56.33	1	56.33	12.17
Comparison 2	0.33	1	0.33	0.07
Comparison 3	96.00	1	96.00	20.73
Within	92.66	20	4.63	
Total	245.33	23		

be as follows:

	A	B	C	D
Comparison 1	3	−1	−1	−1
Comparison 2	0	0	1	−1
Comparison 3	0	2	−1	−1

These comparisons would be considered orthogonal because the sum of the cross products for each comparison (comparison 1 vs. comparison 2, comparison 1 vs. comparison 3, and comparison 2 vs. comparison 3) all equal zero. As pointed out by both Winer (*4*) and Keppel (*3*), statisticians may not be interested in all $k - 1$ comparisons that account for all of the independent information in their data. Situations may arise in which the between sum of squares is partitioned into incomplete sets of orthogonal comparisons and some nonorthogonal comparisons, as long as the total sum of squares for these comparisons does not exceed the between sum of squares that must be partitioned. With a priori tests, statisticians usually are concerned only with comparisons that have specific experimental relevance, whether these comparisons are orthogonal or not, and these comparisons may or may not account for all of the between variance. The only restriction is that the sums of squares for these contrasts do not exceed the sum of squares for between-group variation.

11.4 Trend Analysis

If the independent variable in a study represents a quantitative interval or ratio scale that indicates increasing amounts of that variable (e.g., increasing amounts of a catalyst, pressure, temperature, or dosage), then a *trend analysis* may be done that will show the relationship between the independent and dependent variables. In trend analysis, statisticians are usually interested in determining whether means for treatment groups increase significantly in a linear or nonlinear way, or if any significant trend occurs at all with the independent variable rather than determining if the difference between any contiguous pair of means is significant. For example, a hypothesis may be that a relationship between the independent and dependent variables fits one of the trends depicted in Figure 11.3 representing

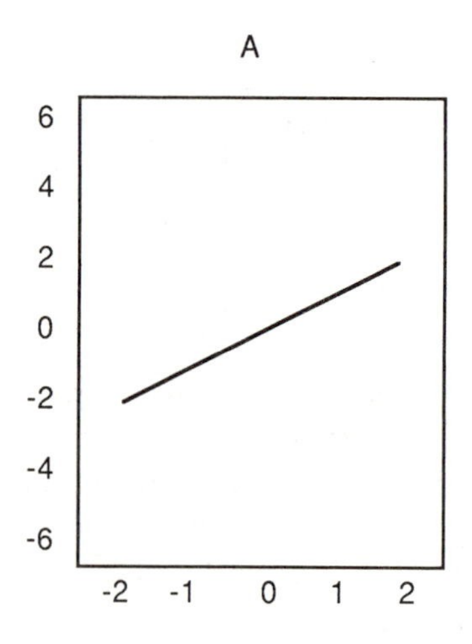

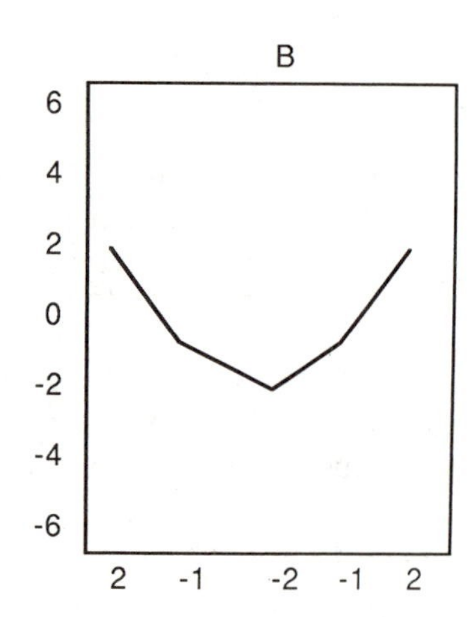

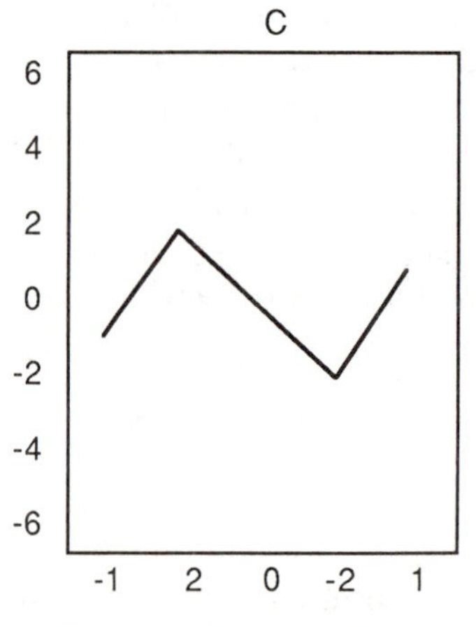

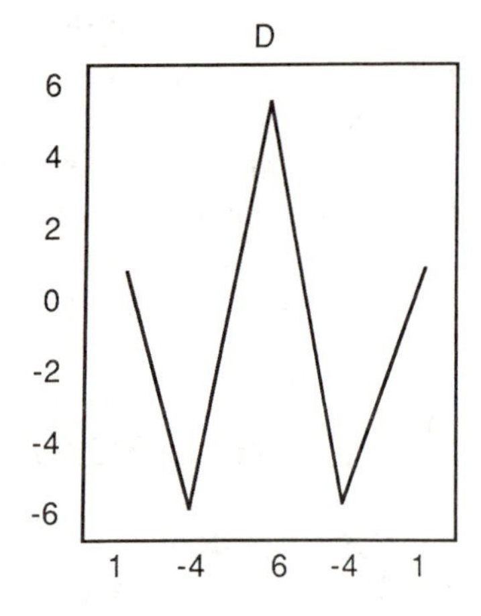

Figure 11.3. Graphs of coefficients of (A) linear, (B) quadric, (C) cubic, and (D) quartic trend components.

a linear, quadric, cubic, or quartic trend. Such a trend might be important in determining the optimal conditions for productivity, or for predicting productivity if a linear relationship exists.

Trend analysis represents an extension of the orthogonal tests and is usually applied in studies in which the levels of the treatment variable are equally spaced and when the sample size is the same for all treatment groups. When these conditions are not met, other procedures can be used, as suggested by Robson (*12*). The procedure described in this section using orthogonal polynomials requires equal n values and equally spaced quantitative intervals for the independent variable. For example, in Table 11.6, the four treatments might represent different amounts of a catalyst that might vary in increments of five percentage points (i.e., 5%, 10%, 15%, and 20%).

Trend analysis uses coefficients for orthogonal polynomials that show the general form of relationships such as in Figure 11.3. Actually, trend analysis correlates the treatment means with the various sets of orthogonal polynomials and then determines which correlations are significant. If the correlation is significant for a specific trend, then that trend fits that specific relationship or shape portrayed by one of the coefficients shown in Figure 11.3. The polynomial coefficients for various numbers of treatment conditions are given in Table K in the Appendix. More extensive tables are provided by Anderson and Houseman (*13*) and Fisher and Yates (*14*).

As an example, suppose that a researcher is interested in determining the form of the relationship between the increments of percentage of a catalyst and yield of a chemical. After deciding upon four percentage levels of the catalyst, the researcher then collects six sample yields for each level. The data in Table 11.6 can be used for this example, where the concentration of the catalyst is 5%, 10%, 15%, and 20% for treatments A, B, C, and D, respectively. The yields for each of the six samples are given for each treatment. Because there are $k = 4$ treatment groups, three possible relationships—linear, quadric, and cubic—are possible. From Table K, the coefficients of orthogonal polynomials for these three shapes are as follows:

	A	**B**	**C**	**D**
Linear	−3	−1	1	3
Quadric	1	−1	−1	1
Cubic	−1	3	−3	1

The computational procedures are similar to those used for orthogonal comparisons in Section 11.6 with the only change being the use of the orthogonal polynomials as the coefficients. For this problem, the total, within, and between sums of squares have been calculated and are, respectively, 245.33, 92.66, and 152.67.

The treatment sum of squares is partitioned into the components by using equation 11.17. For the linear, quadric, and cubic components, respectively,

$$SS_{\text{linear}} = \frac{[-3(33) + -1(59) + 1(69) + 3(71)]^2}{6(-3^2 + -1^2 + 1^2 + 3^2)} = 128.13$$

$$SS_{\text{quadric}} = \frac{[1(33) + -1(59) + -1(69) + 1(71)]^2}{6(-1^2 + -1^2 + -1^2 + 1^2)} = 24.00$$

$$SS_{\text{cubic}} = \frac{[-1(33) + 3(59) + -3(69) + 1(71)]^2}{6(-1^2 + 3^2 + -3^2 + 1^2)} = 0.53$$

As a computational check, these three sums of squares should equal the between sum of squares, which they do within 0.01, which may be due to rounding errors. The analysis of variance table would be set up as in Table 11.8. From Table D with df = 1 and 20, an $F \geq 8.10$ is significant at the 0.01 level. Thus, the F ratio for the linear relationship is significant, which indicates that the relationship between the independent and dependent variables is linear.

Several points and limitations should be mentioned for trend analysis. First, the underlying trend in the population to which one is making inferences is based upon a limited number of data points, and how these incremental points are selected may have an influence on the trend that is found. As pointed out by Winer (*4*), Keppel (*3*), and Hays (*2*), any trends that are found are limited to both the data points selected by the researcher and the range of the treatment variables. The trend may be of one form for one range of treatment variables and of another form for a different range of that variable.

Table 11.8. Trend Analysis of Variance Table

Source	SS	df	ms	*F*
Linear	128.13	1	128.13	27.67
Quadric	24.00	1	24.00	5.18
Cubic	0.53	1	0.53	0.11
Within	92.66	20	4.63	
Total	245.33	23		

Thus, the nature of the data, the increments for the treatments, and the general purpose must all be considered when planning trend analyses. Any trend that is found is limited to the range of the treatment variable that is analyzed. To establish a trend, several incremental points should be selected, possibly four or five at a minimum, and within each level, a sufficient number of cases should be available for establishing a reliable trend. The trend analysis presented in this section requires equal increments for the independent variable and an equal number of cases at each incremental level. Procedures can be adopted for situations having unequal increments and unequal number of cases at each increment (*3, 12*).

Second, the signs of the orthogonal polynomial coefficients can be reversed to fit trends that represent a negative rather than a positive correlation (i.e., inverted shapes of those depicted in Figure 11.3). The sums of squares will be identical because the numerator in equation 11.17 is squared. In selecting the coefficients of orthogonal polynomials, plotting out the obtained means for each treatment condition is always a good idea to see what the general trend is.

Finally, trend analysis is based on the assumption that the orthogonal polynomials will appropriately fit a set of data. However, some data are more accurately described by an exponential function or a logarithmic function. But as pointed out by many statisticians [e.g., Keppel (*3*) and Winer (*4*)], the polynomials provided in Table K approximate these other functions quite well, and thus can be used for the majority of data that might be dealt with.

Trend analysis is closely related to regression analysis, and once the degree of the best fitting polynomial is found, regression coefficients for that curve can be calculated (*13, 14*).

11.5 Strength of Relationship

As pointed out by Keppel (*3*) and Winer (*4*), other factors being equal, the level of significance is a function of sample size and does not give an indication of the magnitude of the treatment effect. For example, given the same difference among means for two studies, the study with the larger sample size is more likely to be significant than the study with a smaller size sample. Or if both are significant, the study with the larger sample size will likely be significant at a much higher level of significance, say the 0.001 level compared to the 0.05 level. Thus, the overall F test merely provides information with regard to the probability of obtaining that size of an F ratio by chance, but gives no information about the strength or magnitude of the treatment effect.

Several measures can be used to estimate the strength of association between the independent and dependent variables. Hays (*2*) introduced the omega-squared (ω^2) index as a measure of the strength of the association between the independent and the dependent variables. This index is basically a correlation ratio contrasting the variance between groups to the total variance in the study. Two equations are most commonly used to compute this index. The first is

$$\omega^2 = \frac{SS_b - (k - 1)(ms_w)}{SS_T + ms_w} \qquad (11.18)$$

which is based upon the the mean square within and the between and total sums of squares from the analysis of variance table. For the data in Table 11.2,

$$\omega^2 = \frac{126 - (3 - 1)(1.5)}{162 + 1.5} = 0.752$$

Another equation based on the F ratio and mean square within is

$$\omega^2 = \frac{(k - 1)(F - 1)}{(k - 1)(F - 1) + k(s)} \qquad (11.19)$$

where F is the computed F ratio, k is the number of treatment groups, and s is the number of subjects in each group. For the same data in Table 11.2,

$$\omega^2 = \frac{(3 - 1)(42.0 - 1)}{(3 - 1)(42 - 1) + 3(9)} = 0.752$$

The value of ω^2 indicates that 75% of the total variance is due to treatment effects, and 25% is due to error variance. Omega squared is interpreted in a similar way as r^2 or the coefficient of determination.

However, as pointed out by Keppel (*3*), some points have to be considered when using ω^2. The main factor is that the relative size of ω^2 is directly related to the levels of the treatment factor selected by the researcher. Thus, if larger differences are selected by the researcher, other things being equal, the value of ω^2 will also be relatively larger. Only in situations in which the levels of the independent variable are selected at random and represent the total population of levels that could be included would ω^2 truly represent the percentage of the total variance accounted for. Also, Keppel (*3*) pointed out that in exploratory work, researchers are more interested in large treatment effects, but as subsequent research becomes more refined, a major factor in the research is not the size of the treatment effect, but to find small differences that have theoretical interest. Thus, ω^2 is another index that can aid the researcher to interpret data completely as possible. However, as with other types of analyses, the interpretation of the results have to be made with a consideration of all factors involved in the particular study.

References

1. Box, G. E. P.; Hunter, W. G.; Hunter, J. S. *Statistics for Experimenters;* Wiley: New York, 1978.
2. Hays, W. L. *Statistics for Psychologists;* Holt, Rinehart and Winston: New York, 1963.
3. Keppel, G. *Design and Analysis: A Researcher's Handbook;* Prentice–Hall: Englewood Cliffs, NJ, 1973.
4. Winer, B. J. *Statistical Principles in Experimental Design;* McGraw–Hill: New York, 1971.
5. Tukey, J. W. *Biometrics* **1949,** *5,* 99–114.
6. Scheffe′, H. *Biometrika* **1953,** *40,* 87–104.
7. Keuls, M. *Euphytica* **1952,** 112–122.
8. Newman, D. *Biometrika* **1939,** *31,* 20–30.
9. Duncan, D. B. *Biometrics* **1955,** *11,* 1–41.
10. Dunnett, C. W. *J. Am. Stat. Assoc.* **1955,** *50,* 1096–1121.
11. Dunnett, C. W. *Biometrics* **1964,** *20,* 482–491.
12. Robson, D. S. *Biometrics* **1959,** *15,* 187–191.
13. Anderson, R. L.; Houseman, E. E. *Research Bulletin 297*, Iowa State University, Ames, Iowa, 1942.
14. Fisher, R. A.; Yates, F. *Statistical Tables for Biological Agricultural and Medical Research;* Hafner: New York, 1963.

Problems

1. A researcher wants to investigate the corrosion resistance of different metal coatings. One sheet of tin is divided into 40 pieces, which are then randomly assigned to five sets. Each set is then

coated with the same amount of the different coatings A, B, C, D, and E, and then subjected to the same corrosion elements for the same amount of time. Test the hypothesis at the 0.05 level of significance of no difference among the means in amount of corrosion for these five coatings, which are as follows:

A	B	C	D	E
8	6	9	7	8
7	7	8	9	7
5	7	6	6	8
6	8	7	8	7
7	6	8	7	9
6	6	8	9	8
7	8	7	8	7
5	7	6	9	8

2. If the result is significant for problem 1, which means are significantly different from each other at the 0.05 level of significance?
3. Suppose that set A in problem 1 was a standard to which the other coatings were to be compared. Are any of the coatings significantly different in their corrosion resistance compared to this standard at the 0.05 level of significance?
4. Suppose that for the data in problem 1, sets A and B were of one type of coating and were to be compared to sets C, D, and E, which are of a second type of coating. Is there a significant difference between these two types of coatings at the 0.05 level of significance?

Chapter 12

Introduction to Factorial Designs

12.1 Interaction between Two Factors

The simple or one-way analysis of variance is suitable for investigating differences among groups or treatments divided by a single dimension or factor, such as differences among several plants, production runs, or varying amounts of a catalyst. For these analyses, only one principle or basis for variation occurs. However, for most chemical studies, many factors must be considered, and designing a study so that the influence of two or more factors can be investigated simultaneously is often advisable. For example, factors such as pressure, temperature, amount of a catalyst, when the catalyst is added, and proportions of ingredients all may have individual as well as combined influences on the yield of a chemical product.

The advantages of the factorial design can be shown by a simple illustration involving only two factors—pressure and temperature—each of which is investigated at two levels. The two levels of each of these two factors can be designated by P_1 and P_2 for the two levels of pressure, and T_1 and T_2 for the two levels of temperature. The design of the study is usually displayed in a table such as shown in Figure 12.1, which shows the two levels of each factor along the margin and the mean values for each cell (denoted by A, B, C, and D), which in an investigation would usually be based on a number of samples taken for each condition (i.e., for T_1 and P_1, T_1 and P_2, T_2 and P_1, and T_2 and P_2).

The effect of changing only temperature can be shown by comparing the combined means of A and C with the combined means

1453–0/88/0255$10.75/1

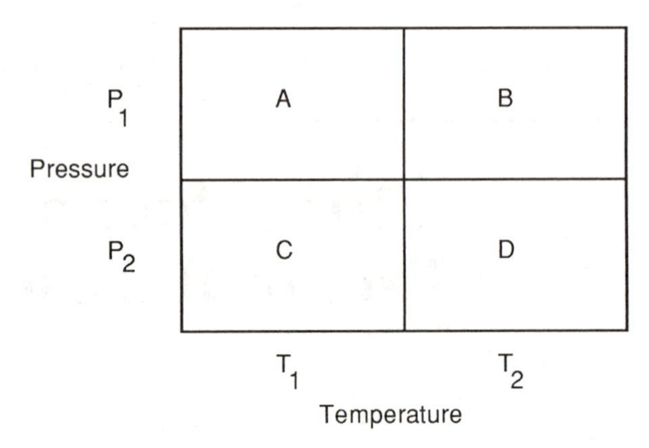

Figure 12.1. Table layout for a 2 × 2 factorial design.

of B and D, disregarding the two pressure levels. Likewise, the effect of changing only pressure can be shown by comparing the combined means of A and B with the combined means of C and D, disregarding the two temperature levels. These comparisons are called *tests of main effects* and are the same kind of tests that would be made if separate studies were done by using the one-way analysis of variance. However, often the main concern in the factorial design is whether a differential effect occurs when the two factors are combined. If the effect of one factor is different at the various levels of the other factor, then an interaction occurs between these two factors. If no change or a consistent change occurs in one of the factors over various levels of the other factor, then these two factors are operating independently of each other and no interaction occurs between these two factors. Two factors interact if differences in an effect for different levels of one factor are different at two or more levels of the other factor.

To show the interaction or lack of interaction between two factors, consider the hypothetical outcomes of the 2 × 2 (read "two by two") factorial designs shown in Figure 12.2. For factorial designs, the means in the margins of the table reflect the main effect of each factor taken independently, and the means within the body of the table reflect the presence or absence of an interaction effect. For this discussion of these possible outcomes, all differences are assumed significant, and if one factor changes from one level to another level, then the interaction is significant. As will be shown, three hypotheses can be tested for the 2 × 2 or two-way factorial design. In null form,

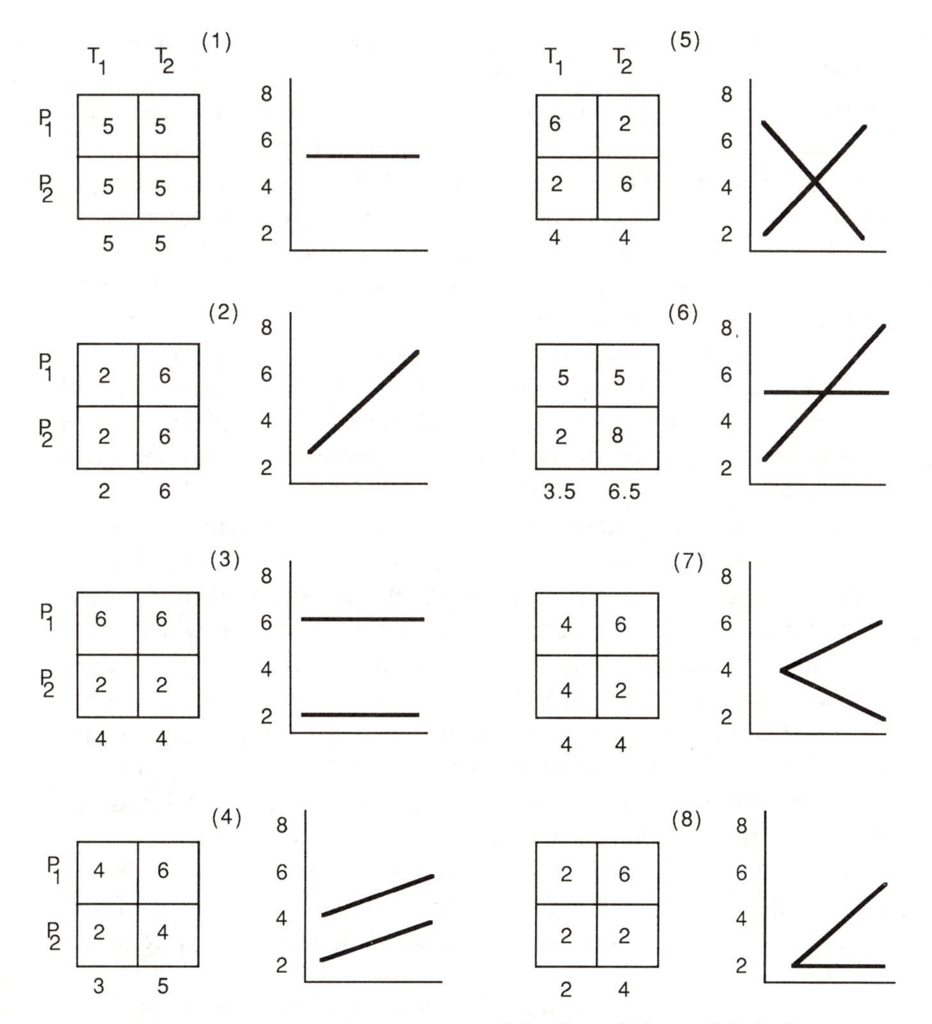

Figure 12.2. Possible outcomes of the 2 × 2 factorial design.

these hypotheses are

1. No significant main effect due to the column variable, or for the example, no significant difference in yield between T_1 compared to T_2 (H_0: $\mu_{T_1} - \mu_{T_2} = 0$).
2. No significant main effect due to the row variable, or for the example, no significant difference in yield between P_1 compared to P_2 (H_0: $\mu_{P_1} - \mu_{P_2} = 0$).
3. No significant interaction between the two factors.

The first example in Figure 12.2 (outcome 1) has no significant main effects or interaction effects. The mean yield is at 5 regardless of which level of pressure or temperature is operating. This yield indicates the absence of both main effects and an interaction effect. For this outcome, yield is not influenced by either pressure or temperature alone or in combination. The plot of this outcome shows parallel and, in this case, overlapping lines for yields of P_1 compared to P_2 over the two levels of temperature. Thus, none of the three hypotheses would be rejected.

For outcome 2, the parallel and overlapping lines in the plot indicate that the effect of temperature changes are the same regardless of the pressure level; thus, no differential effect of temperature is indicated at the two levels of pressure and therefore, no significant interaction occurs. The marginal means of $T_1 = 4$ and $T_2 = 4$ for comparing the two pressure levels indicate no significant main effect due to pressure. However, the marginal means $T_1 = 2$ and $T_2 = 6$ indicate a significant main effect due to this factor, and thus only this hypothesis would be rejected. For outcome 3, the marginal means indicate a significant main effect due to pressure but no main effect due to changes in temperature. Because the yield for P_1 is consistently higher than the yield for P_2, no significant interaction occurs, which is also shown by the parallel lines in the plot for this outcome.

For outcome 4, both of the main effects are significant, indicating that the yield is significantly different for both comparisons of P_1 and P_2 and of T_1 and T_2. However, these data show a consistent trend: the results for P_1 are always two points above P_2, regardless of temperature level. Thus, no differential effect or significant interaction occurs.

For outcome 5, no significant main effects occur as can be noted from the equal marginal means. However, a significant interaction can be noted from the plot of these data. For this outcome, the yield is high for P_1 at T_1 but low at T_2, whereas just the opposite effect happens for the yield for P_2. On the basis of just the main effects, one might conclude that changes in either pressure or temperature had no effect on yield. However, the interaction of these two factors produced very different results. In fact, the interaction is so severe that the simple main effects are cancelled. This interaction is a good example that main effects reflect only treatment averages and may or may not reflect the constituent parts. As will be shown in this chapter, when a significant interaction effect occurs, post hoc analyses have to be done for each factor separately to determine significant differences between or among the levels of each factor. In statistical terminology, the interaction shown for outcome 5 is known as a *disordinal interaction* because the lines on the plot cross, a result indicating that the effects at one level of one factor reverse themselves at different levels of the other factor.

For outcome 6, a significant main effect occurs for temperature but not for pressure, as indicated by the means. However, a significant interaction occurs because the yield for P_1 is not influenced by changes in temperature, whereas changes in temperature have an effect on the yield for P_2. Again, this result would be classified as a disordinal interaction because the lines in the plot cross.

For outcome 7, a significant main effect occurs for pressure but not for temperature. The interaction is significant because the yield is the same for P_1 and P_2 at T_1, but the yield is higher for P_1 at T_2 compared to the yield for P_2 at T_2. This type of interaction is known as an *ordinal interaction* because the lines do not cross, a condition that indicates a larger difference between levels of one factor at one level of the other factor compared to other levels of that second factor.

The last outcome in Figure 12.2 illustrates an outcome where all three effects are present (i.e., where all three hypotheses are rejected). Both main effects are significant, as can be noted from the marginal means. There is also an ordinal interaction, indicating a differential effect due to the combination or interacting effects of the two factors.

In the last four outcomes, the advantage of the factorial design becomes apparent when the marginal means that would be tested in the one-way analysis of variance are compared to the results tested in the factorial design. For outcome 5, the separate analyses of variance would have indicated that neither temperature nor pressure has an influence on yield. However, when these two factors are studied in combination, the interaction effect indicates that the highest yield is obtained with the combination of either P_1 and T_1 or P_2 and T_2. The highest yields for other combinations of pressure and temperature can be obtained by comparing the cells for outcomes 6, 7, and 8.

12.2 Advantages of the Factorial Design

Factorial designs have several advantages over one-way designs. As pointed out by Davies (*1*), experiments done by chemists, engineers, physicists, and others in the physical sciences are intended to determine the effects of many factors on the yield or quality of a product, power consumption of a process, effectiveness of a product, and resistance of a material to corrosion, for example. In most cases, these factors work in combination rather than separately. Thus, all combinations of the different levels of all factors must be examined in order to discover which combination of factors will be most effective.

The main advantage of the factorial design is that it allows one to study the effects of factors both separately and in combination with other factors. As shown in Figure 12.2, situations occur in which the testing for simple main effects would not reveal large differences because these may be averaged out over several levels of an interacting factor. For example, outcome 5 in Figure 12.2 shows marked differences when the two factors are combined, but separately one would conclude no effect due to differences in pressure or temperature if these were studied independently. Often the factorial design can be set up with little extra effort on the part of the researcher, as, for example, when studying productivity of workers. If a researcher thinks that differences may exist between male and female workers, a factorial design can be set up by keeping a record of the sex of each worker and doing an analysis that incorporates sex as one of the factors rather than putting all workers together in one analysis regardless of sex.

The factorial design can also be set up in a way that is economical with regard to the total number of cases needed. For example, suppose that in the example of studying the effects of pressure and temperature on yield of a chemical, two studies were set up separately: one to study the effect of the two levels of pressure, and the other to study the effect of the two levels of temperature. Suppose a sample of $n = 10$ cases was selected for each group being compared. Thus, a grand total of 20 cases would be included for each study, or a total of 40 cases. If a factorial design was set up to study the effects of these two factors in the same study, a total of only 20 cases would be needed to retain the same degree of precision as in the first study. That is, each main effect would be tested with 10 cases. However, because these cases are actually used for both of the main effects as well as the interaction effect, considerable savings result with regard to use of the sample obtained for the study. The designs for the two separate one-way analyses and the one two-way analyses are depicted in Figure 12.3.

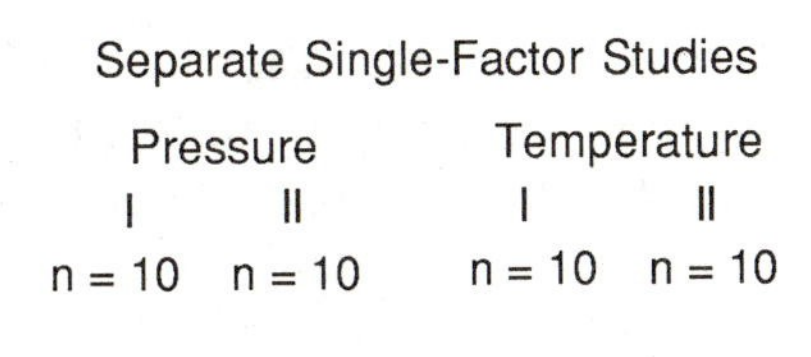

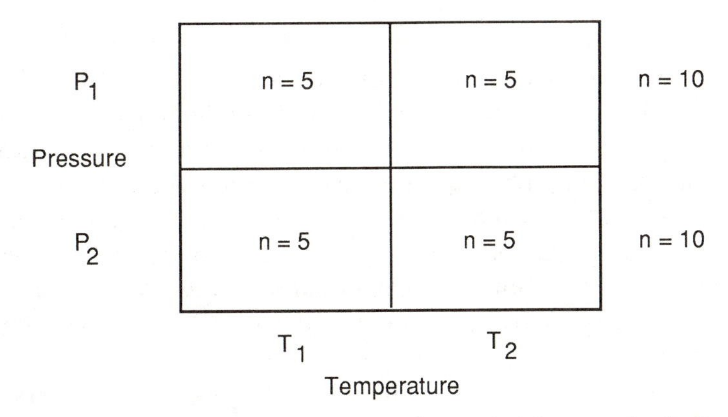

Figure 12.3. Comparison of sample size needed for two single-factor studies and one two-factor study.

Factorial designs can also be used to increase the experimental control in a study, or in statistical terms, to reduce the error variance or within-groups variance. For example, in productivity studies, different equipment or machines are often used, and the results may be influenced by these differences. A design could be set up that would take these differences into account by using each machine with each treatment, which is known as *blocking* and will be covered in Chapter 14. Essentially, this design reduces the variability due to machines and thus makes the study more controlled or accurate.

Finally, wider generalizations may be possible from factorial designs because various levels of one factor are studied over several levels of another factor or several other factors. In the single-factor study, all factors except the experimental factors should be held constant to ensure that any differences among the levels is due to the experimental treatment and not to other factors. This procedure could limit generalizing the results only to other situations in which the other factors were the same as those in the experiment. The factorial design usually permits a wider range of generalization because the results at the various levels of one factor are studied at different levels for one or several other factors. Also, if there is an interaction of the factors included in the study, the factorial design will point out which levels are most effective. This result is often very important.

12.3 Designing Factorial Studies

As with any type of analysis, a number of decisions have to be made with regard to how to set up a factorial experiment. Possibly the first consideration is how many factors should be included in the study. The term "factor" usually is synonymous with the terms "treatment" and "independent" or "experimental variable," and usually consists of a series of related events or procedures. For example, the previous example had two factors: pressure and temperature. Each was set up at only two levels, but in a real situation, there could be as many levels as theory, experience, or the desired thoroughness for the study dictates.

The number of factors included in the study, the levels of each factor, and the spacing of the levels are determined by the researcher and are based upon which combinations of levels of these factors are important to include. The number of factors and levels of each factor determine the dimensions of a factorial experiment, and these are indicated by numbers such as a 4×5 (read "four by five") factorial design for two factors: the first of which has four levels and the second has five levels. Factorial designs are not limited to two factors but can include three or more factors. However, factorial designs involving more than three factors pose problems in the interpretation of the interaction effects. Thus, most factorial designs are usually limited to three factors. Therefore, a $2 \times 4 \times 6$ factorial design has three factors: The first factor has two levels, the second factor has four levels, and the third factor has six levels.

In planning such studies, the number of treatment groups, or *cells*, required for the study is determined by the dimensions of the design. Thus, a 4×5 design will have $4 \times 5 = 20$ cells to consider because 20 treatment combinations are possible. For a $4 \times 5 \times 5$ factorial design, 100 treatment combinations are possible. Thus, researchers must consider the overall dimension of the design and the number of cases required for the total study. Although procedures are available to compute a factorial design for a study having only one case per cell (*2*), there is no within variance and thus no error sum of squares available. For the computational procedures discussed in this chapter, the computations are limited mainly to those situations having at least two independent observations for each cell. For reliability and to estimate the effect in each cell, a minimum of three to five cases should be included for each cell, depending on the degree of homogeneity of the criterion being studied. Also, the sample sizes for each cell must be either equal or proportional to each other. Such designs are referred to as *orthogonal designs*. Orthogonal designs make it easier to separate and test the hypotheses involving both main and interaction effects. Methods are available to handle factorial designs for which the cell frequencies are neither equal nor proportional (a nonorthogonal design) , but these are beyond the scope of this book. These analyses were discussed by Winer (*3*) and Keppel (*4*).

The designs discussed in this book are also completely crossed and balanced. That is, all possible cells in the matrix are considered, and each cell occurs the same number of times. Thus, a 3×4 design would have 12 cells, and each level of the first factor occurs at each level of the other factor. Methods of handling incomplete designs and unbalanced designs are available (*3, 4*).

A number of assumptions have to be made when considering factorial designs. For orthogonal designs, the observations in each cell must be sampled at random and independently from a normal population, and the number of observations in each cell must be the same or proportional to each other. In general, the assumptions for factorial designs are as follows:

- the errors are normally distributed with the sum of the errors expected to equal zero for each treatment combination,
- the errors for each treatment combination have the same variance, and
- the errors are independent both within each treatment combination and across all treatment combinations.

In practice, these three assumptions are usually met by

- selecting cases in a random, independent manner;
- selecting measures of the variables that are normally distributed in the population; and
- selecting samples from populations whose variances of the measures are equal or equivalent.

The factorial designs are fairly robust to violations of normality and, if cell sizes are equal, to violations of the assumption of homogeneity of variances. Thus, for orthogonal designs, the three assumptions are usually met through setting up the design with an equal number of independent observations randomly assigned to each cell.

12.4 Factorial Models: Fixed, Random, and Mixed

Three types of analysis of variance models can be considered for a study. All of these models are particular cases of the general *finite statistical model*, which is the basis of all analyses of variance. This finite model assumes that each value in all groups can be expressed as a deviation from the overall grand mean, and that this deviation can be partitioned into additive components that are based on deviations of each value from its row mean, column mean, the interaction of row and column factors, and an error term. How the levels of each factor are selected determines the type of model used, and the type of model in turn determines which error factor should be used in testing for column and row effects as well as interaction effects.

The choice of the model depends on the nature of the variables used as the basis for the factors of the investigation, and investigators must decide which model best represents their study. The three models are the

- fixed-effects model,
- random-effects model, and
- mixed-effects model.

A *fixed-effects model* is one in which the investigator specifies or fixes the particular levels of each factor. When dealing with categorical data such as sex, manufacturing plants, specific elements, compounds, or different methods of production, researchers must consider the fixed-effects model because the levels of these factors are usually automatically fixed. If the levels of a factor represent a quantitative variable such as amount of a catalyst, operating pressure, or temperature, often the investigator is interested in testing only specific levels of each factor. Thus, the investigator fixes those factors at specified levels. In general, any factor is considered fixed if the investigator would use the same levels of that factor on replications of the study. This model is appropriate when many levels of each factor could be considered, but only specified levels are of interest. Thus, the study would be considered as exhausting all treatment levels of interest, and the only inferences that can be drawn from that study would be to those levels specified by the investigator.

If, for each factor, a large number of levels could be used, and if the investigator is interested in making inferences regarding the entire range of possible levels, then the levels of each factor can be randomly selected from the large range of possible levels. When the levels of each factor are chosen at random, the so-called *random-effects model* is being used. For example, factors such as pressure, temperature, proportion of an element, or concentration of a catalyst could be studied at many levels. Or, if a large number of types of production procedures, variation in catalysts, or treatments are possible, then levels of each of these factors could be randomly selected for a study. Usually, the investigators randomly select the number of levels they feel represents that factor. The basic consideration is whether valid inferences can be made to all possible levels on the basis of those selected. For example, if the range of temperature of interest varies from 20 to 90 °C, then possibly six or seven levels might be randomly selected for study. Even though only a random set of levels is studied, the basic underlying assumption is that the selected levels represent all possible levels of that variable, and if the study were replicated, different levels would be randomly selected.

If, in a factorial design, some of the factors are fixed and other factors are selected at random, then a *mixed-effects model* is used. For example, three production procedures might be selected for a study to determine if an interaction occurs with operating pressure. The study might be limited because only these three production procedures are available (fixed factor), but is not limited to any level of operating pressure. Thus, operating pressure could be randomly determined, and the levels of this factor could be determined at random (random factor).

As shown in the next section, the three models are identical computationally with regard to the sums of squares. The only differences among the models are changes in the error term used for testing row, column, and interaction effects, and the types of inferences that can be made from each model.

Inferences can be made to the entire population of levels considered in the random-effects model but only to those levels included in a study in the fixed-effects model. The choice of model depends upon the nature of the variables being considered and the types of inferences desired. Although the random-effects model has the greatest potential with regard to making inferences, a representative sampling of a number of levels must be considered in order to adequately represent the total population of possible levels of a factor. Often this consideration would require a factorial design with a large number of levels for each factor (e.g., an 8×8 design, or, for a three-factor study, an $8 \times 8 \times 8$ design). The 8×8 design has a total of 64 cells to consider in the complete model; the three-factor design has 512 cells to consider. If only three cases were used for each cell, 192 and 1536 cases would be required for the first and second designs, respectively. The cost for such studies would be considerable. Thus, the fixed-effects design is usually considered for cost considerations as well as the consideration that usually only specific levels of the factors are of interest. For example, if only two types of a product are available (e.g., coarse or fine ammonium chloride), then setting up and studying the two fixed levels of this factor would be more advisable than trying to obtain other levels of this product and then randomly selecting levels to study.

12.5 Computations for the Two-Way ANOVA

In the one-way analysis of variance (ANOVA), the total variance of all values included in the study was partitioned into two independent sources: within-group variability and between-group variability. For factorial designs, the total variability is partitioned into additional sources, and in the two-way ANOVA, the total variance is partitioned into four parts: the sum of squares for rows (SS_r), the sum of squares for columns (SS_c), the sum of squares for interaction between rows and columns ($SS_{r \times c}$), and the sum of squares within groups or cells (SS_w).

Thus, $SS_T = SS_r + SS_c + SS_{r \times c} + SS_w$. As with the one-way analysis of variance, the sums of squares instead of the variance are used when the computations are carried out. Then the sums of squares are divided by their respective degrees of freedom to obtain the mean squares or variances. Finally, F values are calculated by dividing the mean square for each effect by the appropriate error term, which is usually the within-group variance or mean square.

The degrees of freedom (df) for the ANOVA are also additive so that

$$\mathrm{df}_T = \mathrm{df}_r + \mathrm{df}_c + \mathrm{df}_{r\times c} + \mathrm{df}_w \qquad (12.1)$$

where df_T, df_r, df_c, $\mathrm{df}_{r\times c}$, and df_w are the total, rows, columns, interaction, and within degrees of freedom, respectively. For the two-way analysis of variance, df_T = (total number of cases) − 1, df_r = (number of rows) − 1, df_c = (number of columns) − 1, $\mathrm{df}_{r\times c}$ = (number of rows − 1)(number of columns − 1), and df_w = (total number of cases) − (number of groups).

The equations for the necessary sums of squares will be presented by using the raw-values approach (i.e., by using the actual values obtained in an investigation rather than deviation values). The equations are very similar to those used in the one-way analysis of variance. For example, the equation for the total sum of squares (SS_T) is the same.

$$\mathrm{SS}_T = \Sigma X_T^2 - \frac{(\Sigma X_T)^2}{n_T} \qquad (12.2)$$

where $\Sigma X^2{}_T$ is the sum of the squared values summed over all cases included in the study, n_T is the total number of cases, and ΣX_T is the sum of all values for all cases included in the study.

The sum of squares between columns (SS_c) is calculated by

$$\mathrm{SS}_c = \sum^{C} \left(\frac{(\Sigma X_c)^2}{n_c}\right) - \frac{(\Sigma X_T)^2}{n_T} \qquad (12.3)$$

where ΣX_c is the sum of the values for each column, C is the total number of columns, n_c is the number of cases per column, and ΣX_T and n_T are the same as in equation 12.2.

The sum of squares for rows (SS_r) is given by

$$\mathrm{SS}_r = \sum^{R} \left(\frac{(\Sigma X_r)^2}{n_r}\right) - \frac{(\Sigma X_T)^2}{n_T} \qquad (12.4)$$

where ΣX_r is the sum of the values for each row, R is the total number of rows, and n_r is the number of cases for each row.

For the within sum of squares (SS_w),

$$SS_w = \Sigma X_T^2 - \sum^{G} \left(\frac{(\sum X_g)^2}{n_g} \right) \qquad (12.5)$$

where $\Sigma X^2{}_T$ is the sum of squared values summed over all cases included in the study, ΣX_g is the sum of values for each group in the study, and n_g is the number of cases in each group.

The sum of squares for interaction ($SS_{r \times c}$) is then obtained by

$$SS_{r \times c} = SS_T - (SS_c + SS_r + SS_w) \qquad (12.6)$$

Only five terms are needed for the calculation of all sums of squares. Also, because $SS_{r \times c}$ is computed indirectly, no computational check is made for the total sum of squares. Thus, repeating the computations for the first four sums of squares is important to detect computational errors.

12.6 Factorial Design Example

Consider a laboratory investigation of two conditions that affect the yield of a chemical product, where the two factors are temperature and concentration of reactant. For simplicity, only two levels of temperature, T_1 and T_2, and three levels of concentration, C_1, C_2, and C_3, are considered. The 2 × 3 design would be set up as in Table 12.1,

Table 12.1. Hypothetical Data for a Two-Way ANOVA

n	C_1		C_2	C_3
		T_1		
1	3		6	8
2	6		4	6
3	4		7	9
4	2		5	7
5	5		6	8
		T_2		
1	4		4	3
2	2		3	1
3	5		1	4
4	3		4	1
5	4		2	2

NOTE: The abbreviations C_1, C_2, and C_3 denote the concentration levels. The abbreviations T_1 and T_2 denote the temperature levels.

which shows the yield for each sample for all groups (i.e., for all combinations of temperature and concentrations). Because specific levels have been selected and set by the investigator, this study would be a fixed-effects design.

The hypotheses that will be tested for this study are as follows:

1. No difference in mean yield occurs for the T_1 condition compared to the T_2 condition, or H_0: $\mu_{T_1} - \mu_{T_2} = 0$. (This hypothesis tests the main effect for rows.)
2. No differences in mean yield occur among the three levels of concentration, or H_0: $\mu_{C_1} = \mu_{C_2} = \mu_{C_3}$. (This hypothesis tests the main effect for columns).
3. No interaction effect occurs as a result of differential effects of combinations of temperature and concentration.

The necessary sums and sums of squares are given in Table 12.2. Although the sums of the squared values for each group, column,

Table 12.2. Cell Totals, Sums, Means, and Standard Deviations for the Data in Table 12.1

Statistic	C_1	C_2	C_3	Totals
		T_1		
n	5	5	5	15
$\sum X$	20	28	38	86
$\sum X^2$	90	162	294	546
$\bar{X}$	4.00	5.60	7.60	5.73
SD	1.58	1.14	1.14	1.94
		T_2		
n	5	5	5	15
$\sum X$	18	14	11	43
$\sum X^2$	70	46	31	147
$\bar{X}$	3.60	2.80	2.20	2.87
SD	1.14	1.30	1.30	1.30
		Totals		
n	10	10	10	30
$\sum X$	38	42	49	129
$\sum X^2$	160	208	325	693
$\bar{X}$	3.80	4.20	4.90	4.30
SD	1.32	1.87	3.07	2.18

and row are not necessary for computation of the ANOVA, they are given in the table for computations of standard deviations. For most studies, to completely describe the results, the means and standard deviations for each effect as well as for each group should be computed.

The sums of squares for the data in Table 12.1 are calculated as follows:

$$SS_T = 693 - \frac{129^2}{30} = 138.30$$

$$\begin{aligned} SS_r &= \left(\frac{86^2}{15} + \frac{43^2}{15}\right) - \frac{129^2}{30} \\ &= 493.07 + 123.27 - 554.7 = 61.64 \end{aligned}$$

$$\begin{aligned} SS_c &= \left(\frac{38^2}{10} + \frac{42^2}{10} + \frac{49^2}{10}\right) - \frac{129^2}{30} \\ &= 144.4 + 176.4 + 240.1 - 554.7 = 6.20 \end{aligned}$$

$$\begin{aligned} SS_w &= 693 - \left(\frac{20^2}{5} + \frac{28^2}{5} + \frac{38^2}{5} + \frac{18^2}{5} + \frac{14^2}{5} + \frac{11^2}{5}\right) \\ &= 693 - (80 + 156.8 + 288.8 + 64.8 + 39.2 + 24.2) \\ &= 693 - 653.8 = 39.2 \end{aligned}$$

$$SS_{r\times c} = 138.3 - 61.64 - 6.2 - 39.2 = 31.26$$

The degrees of freedom for this problem are

$$df_T = 30 - 1 = 29$$

$$df_r = 2 - 1 = 1$$

$$df_c = 3 - 1 = 2$$

$$df_{r\times c} = (2 - 1)(3 - 1) = 2$$

$$df_w = 30 - (2)(3) = 24$$

To complete the ANOVA, a summary analysis of variance table is set up as Table 12.3. Because this example would be considered as a fixed-effects design, the appropriate error term would be the within mean square. Thus, the mean square for rows, columns, and

Table 12.3. Summary ANOVA Table for the Data in Table 12.2

Source	SS	df	ms	F
Rows (Temperature)	61.64	1	61.64	37.82
Columns (Concentration)	6.20	2	3.10	1.90
$R \times C$ (Interaction)	31.26	2	15.63	9.59
Within (Residual)	39.20	24	1.63	
Total	138.30	29		

interaction would be divided by the within mean square to obtain the F ratios. Thus, the row F ratio (F_r) is 61.64/1.63 = 37.82, the column F ratio (F_c) is 3.10/1.63 = 1.90, and the interaction F ratio ($F_{r \times c}$) is 15.63/1.63 = 9.59. With df = 1 and 24, an F = 7.82 is needed for significance at the 0.01 level. Thus, the main effect for rows is significant at the 0.01 level, and the null hypothesis of no difference in yield between T_1 and T_2 is rejected. The conclusion is that the yield for these two temperatures is significantly different. The value of F_c is not significant at the 0.01 level because with df = 2 and 24, an $F \geq 5.61$ is needed for significance at this level. Thus, the hypothesis of no differences in yield among the three levels of concentration is not rejected, and the conclusion would be that no significant differences occur in yield among the three levels of concentration. The $F_{r \times c}$ value is significant at the 0.01 level of significance, a result indicating a differential effect with different combinations of temperature and concentration.

The differential effect becomes readily apparent when the group or cell means are displayed in a graph such as that shown in Figure 12.4. From Table 12.2, the row effects mean yields for T_1 and T_2 are, respectively, 5.73 and 2.87, which are significantly different as indicated by F = 37.82. However, the column means of 3.80, 4.20, and 4.90 are not significantly different. As can be seen in Figure 12.4, the significant interaction becomes apparent as the differences in yield between T_1 and T_2 increase as the concentration level also increases from C_1 to C_3. Thus, at the first level of concentration, little difference in yield occurs at either temperature. However, at the third level of concentration, the largest difference in yield occurs between the first temperature level and the second level. If the main effect for columns was significant, with a significant interaction effect, a post hoc test would have to be run for each level of temperature. Post hoc tests will be discussed in Section 12.9.

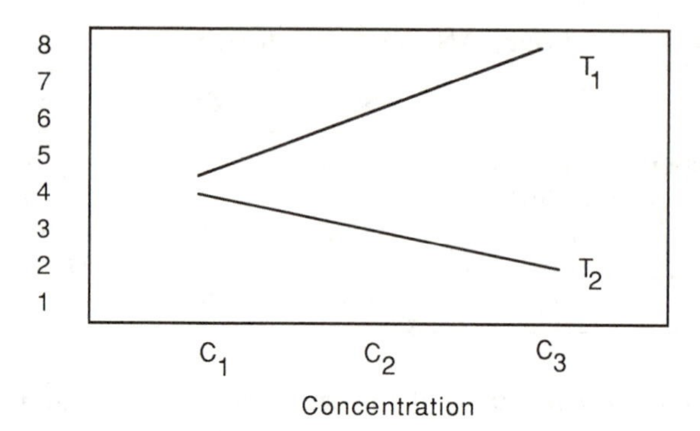

Figure 12.4. Cell means for the 2 × 2 ANOVA problem for the data in Table 12.2. This figure shows an ordinal interaction.

12.7 Selecting the Error Term for Fixed, Random, and Mixed Models

As pointed out in several statistics textbooks (*2–4*), the selection of the appropriate error term for computing the F ratio varies depending on the type of model designed. For the fixed-effects model, the appropriate error term for testing row, column, and interaction effects is the within mean square. For the random-effects model, the proper error term for the test of interaction effects is the within mean square. However, for both the row and column effects, the mean square for the interaction effect is the appropriate error term to use. Thus, for the random model, the F ratios for rows, columns, and interaction would be, respectively,

$$F_r = \frac{\text{ms}_r}{\text{ms}_{r\times c}} \tag{12.7}$$

$$F_c = \frac{\text{ms}_c}{\text{ms}_{r\times c}} \tag{12.8}$$

$$F_{r\times c} = \frac{\text{ms}_{r\times c}}{\text{ms}_w} \tag{12.9}$$

where ms_r, $\text{ms}_{r\times c}$, ms_c, and ms_w are the mean squares for rows, interaction, columns, and within from the ANOVA table, respectively. For the mixed model, where the levels of one factor are fixed and the other levels are selected at random, the proper error term for the interaction effect is the within mean square. For the fixed factor, the appropriate error term is the mean square for the interaction effect. For the random factor, the proper term is the within mean square. Thus, for a mixed design for which the row variable is fixed and the column factor is random, the F ratios would be computed by

$$F_r = \frac{\mathrm{ms}_r}{\mathrm{ms}_{r\times c}} \tag{12.10}$$

$$F_c = \frac{\mathrm{ms}_c}{\mathrm{ms}_w} \tag{12.11}$$

$$F_{r\times c} = \frac{\mathrm{ms}_{r\times c}}{\mathrm{ms}_w} \tag{12.12}$$

For a mixed design for which the row variable is random and the column factor is fixed, the F ratios would be computed by

$$F_r = \frac{\mathrm{ms}_r}{\mathrm{ms}_w} \tag{12.13}$$

$$F_c = \frac{\mathrm{ms}_c}{\mathrm{ms}_{r\times c}} \tag{12.14}$$

$$F_{r\times c} = \frac{\mathrm{ms}_{r\times c}}{\mathrm{ms}_w} \tag{12.15}$$

As already pointed out before, the computation of the sum of squares for the three designs is identical. Thus, the computational procedures are the same with respect to the calculation of the mean squares for these F ratios.

12.8 Strength of Association

For both the one-way analysis of variance and the factorial designs, the overall F test gives the probability that observed differences were likely to happen by chance. However, the F tests in both the one-way ANOVA and the factorial designs do not give information about the strength of each effect. Often researchers compare the level of significance as a means of determining which effect is stronger. However, the level of significance is a function of power and sample size, and the analysis based upon the larger sample size is usually significant at a much higher level than analyses based upon smaller size samples. Thus, all other things being equal, if two experiments are conducted, one with $n = 5$ and the other with $n = 50$, and both are significant at the 0.01 level, which result would indicate the largest effect? Because sample size and power are directly related, the effect of the result with the smaller size would be the most effective. Or take a situation in which two experiments are done. The F test for one experiment is significant at the 0.05 level and the other at the 0.001 level. The second result is not necessarily more effective than the first because the degree of significance is a function of the power of the test and the size of the sample. Thus, an index that indicates the relative strength of the association between the independent factors and the dependent variable independent of sample size should be used. Such an index was introduced by Hays (5) and is called omega squared (ω^2). This index, which was discussed in Chapter 11, can be considered as a correlation ratio contrasting the variability due to the independent factors to the total variability in the study.

For the one-way analysis of variance, the strength of the association is given by

$$\omega^2 = \frac{SS_b - (k - 1)ms_w}{SS_T + ms_w} \tag{12.16}$$

where SS_b is the sum of squares between groups, k is the number of treatment groups, ms_w is the within mean square for the ANOVA table, and SS_T is the total sum of squares. Omega squared indicates the amount of variability accounted for by the independent or treatment variable compared to the total variance of the dependent variable. And as with the correlation coefficient, the larger the index, the stronger the experimental effect.

For factorial designs, the following equations for ω^2 should be used for the fixed-effects model for the row, column, and interaction effects, respectively.

$$\omega_r^2 = \frac{SS_r - df_r(ms_w)}{SS_T + ms_w} \tag{12.17}$$

$$\omega_c^2 = \frac{SS_c - df_c(ms_w)}{SS_T + ms_w} \tag{12.18}$$

$$\omega_{r\times c}^2 = \frac{SS_{r\times c} - df_{r\times c}(ms_w)}{SS_T + ms_w} \tag{12.19}$$

For the random-effects model, the equations are the same except $ms_{r \times c}$ is substituted for ms_w for the main-effects ω^2. Likewise, for the mixed-effects model, the appropriate term to use for the fixed factor would be $ms_{r \times c}$ rather than ms_w. Thus, for the random-effects and the mixed-effects models, the appropriate terms are the same as those used in forming the respective F ratios.

For the analysis in Table 12.3, ω^2 for the row, column, and interaction effects would be, respectively,

$$\omega_r^2 = \frac{61.64 - 1(1.63)}{138.3 + 1.63} = 0.429$$

$$\omega_c^2 = \frac{6.20 - 2(1.63)}{138.3 + 1.63} = 0.021$$

$$\omega_{r \times c}^2 = \frac{31.26 - 2(1.63)}{138.3 + 1.63} = 0.200$$

Thus, the strongest effect is due to the main-row effect, or for this example, temperature, which accounts for 42.9% of the total variance in yield. The interaction effects account for 20% of the variance, whereas the column factor, concentration, accounts for only 2% of the total variance in yield.

12.9 Interpretation of Results

Three hypotheses are tested for the two-way analysis of variance. These hypotheses include

- the test of the main effects for the row variable,
- the test of the main effects for the column variable, and
- the test of the interaction between the two variables.

The first F ratio in the ANOVA table in Table 12.3 is a test of the main effect for the row variable, which for this problem was the effect of temperature. A significant F ratio indicates a significant difference between or among the means for the various levels of temperature represented in the study when these levels are taken into consideration over all levels of the column variable. The second line of the table represents the test of the second main effect, which is for columns, and for this problem is the effect of the various concentrations. Similar to the first F ratio, this test is based upon the variability of the column means independent of the row factor. These two tests are also known as tests or comparisons of marginal means or *tests of main effects*. If no significant interaction occurs between the two factors, then the outcome of the F tests for the main effects can be interpreted without qualification. The F tests for the main effects would be interpreted in the same way as F ratios in the one-way ANOVA. However, if a significant interaction effect occurs, then these main effects have to be interpreted with caution and with a consideration of the type of interaction present.

The third line in Table 12.3 tests for a significant interaction between the two factors being studied and is based upon the variability of the cell means. A significant F ratio indicates a differential effect for various combinations of rows and columns on the dependent variable. The next line in Table 12.3 represents the error factor or residual variability that is not accounted for by the variability among the row means, column means, or cell means. The mean square for this source of variability serves as the error term in the fixed-effects ANOVA.

When interpreting the outcome of the two-way ANOVA, if none of the F ratios is significant, then the statistician would fail to reject all of the null hypotheses and conclude that neither of the factors has an effect on the dependent variable either independently or in combination. If one or both of the F ratios for testing main effects are significant but the interaction F ratio is not significant, then the main effects are interpreted in a manner similar to the one-way analysis of variance. If more than two levels for either main effect occur, then applying an additional comparison would be necessary to identify which levels of these factors differ from each other. As will be shown, these comparisons can be either a priori or post hoc tests and involve similar procedures as for the one-way ANOVA. However, if a significant interaction effect does occur, then the interpretation of main effects has to take this into consideration because the main effects may be influenced by the interaction. The best way to understand if and how the two factors are interacting is to plot the cell means on a chart similar to those appearing in Figures 12.4 and 12.5.

The plotting of the means can be done in a number of ways, depending upon the factor of interest (e.g., temperature or concentration for the example given). However, plotting the cell means in two separate plots is important, for one the row factor is represented on the baseline or abscissa, and the other where the column factor is represented on the baseline. The reason for two plots is that an

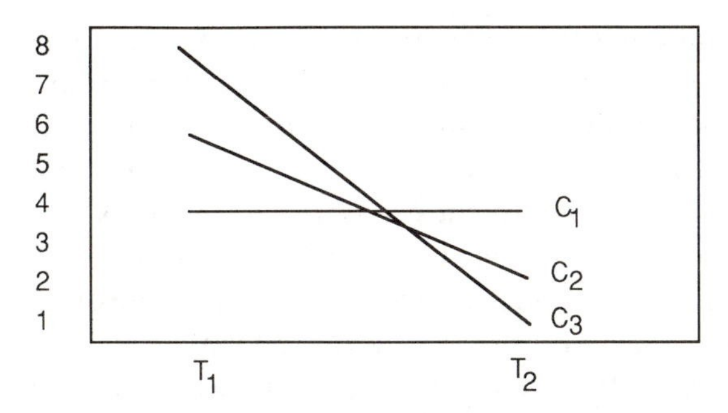

Figure 12.5. Cell means for the 2 × 2 ANOVA problem for the data in Table 12.2. This figure shows a disordinal interaction.

ordinal interaction may occur if the factors are plotted with the levels of the column factor on the baseline, as in Figure 12.4, but a disordinal interaction may occur for the same data if the levels of the row factor are on the baseline, as in Figure 12.5. If an ordinal interaction exists for which one level of the factor is consistently higher than the other level or levels of that factor, such as in Figure 12.4, where T_1 is consistently higher than T_2, then the main effect for that factor can be interpreted as a main effect similar to the one-way ANOVA. However, if a disordinal interaction exists such as in Figure 12.5, then the main effect cannot be interpreted independently of the interaction. For such interactions, testing for differences among levels of that factor for each level of the other factor is recommended. For the example plotted in Figure 12.5, this recommendation would mean testing for simple effects among the three levels of concentration at T_1 and then at T_2.

As with the one-way ANOVA, the tests can be either a priori or planned comparisons that can be made regardless of the significance of the omnibus F test, or post hoc or multiple comparison tests can be made following a significant F ratio if more than two levels of a factor are studied. All of these procedures follow those presented for the one-way ANOVA and are easily adapted to factorial designs.

12.9.1 A Priori Tests

Planned comparisons are usually made on the basis of theoretical or empirical considerations underlying such decisions as well as the basic reason for using a factorial design. That is, when selecting a factorial design, investigators are usually interested in not only the main effects, but also how the factors interact. Thus, rather than comparing all pairs of means in an $R \times C$ matrix, where R denotes the number of rows and C denotes the number of columns, an investigator is mainly interested in comparing means for the main effects or marginal means, or simple effects if a significant interaction occurs. Simple effects are those comparing various levels of either row or column effects at each level of the other factor, which is the recommended procedure when a significant disordinal interaction effect occurs. As with the one-way ANOVA, the comparisons are formulated prior to conducting the study and can be made instead of making the omnibus F tests. For the two-way ANOVA, a maximum of $R \times C - 1$ orthogonal comparisons can be made, where each comparison accounts for one of the total degrees of freedom for row column and interaction effects. Thus, for a 2×3 factorial design, a maximum of $2 \times 3 - 1 = 5$ planned orthogonal comparisons are possible. For a 4×5 design, $4 \times 5 - 1 = 19$ planned orthogonal comparisons are possible. As with the one-way ANOVA, if all or a complete set of orthogonal comparisons are made, these comparisons will account for the total sum of squares for main effects and interaction effects. Of course, a complete set of orthogonal comparisons does not have to be made if all possible comparisons are not desired.

One way to make these planned orthogonal comparisons is to conceptualize the two-way factorial design as a one-way design with $R \times C$ levels. Then the same procedure can be used for making orthogonal comparisons as was used for the one-way ANOVA. This procedure is satisfactory for factorial designs having a small number of levels of each factor, for example, up to a 3×4 design for which there would be 12 levels if each cell was treated as one of the 12 levels of a one-way analysis of variance. For designs with more than three or four levels of each factor, the first procedure that retains the factorial matrix would probably be easier to use.

The data in Table 12.2 can be rearranged into a one-way matrix as shown in Table 12.4. In this table, six cells are arranged as a six-

Table 12.4. Cell Totals Presented As a One-Way ANOVA

Statistic or Planned Comparison	C_1 and T_1	C_2 and T_1	C_3 and T_1	C_1 and T_2	C_2 and T_2	C_3 and T_2
$\sum X$	20	28	38	18	14	11
Comparison 1	1	1	1	−1	−1	−1
Comparison 2	2	−1	−1	2	−1	−1
Comparison 3	0	1	−1	0	0	0

level one-way ANOVA. The first three columns represent the first level of temperature at the three levels of concentration: C_1 with T_1, C_2 with T_1, and C_3 with T_1, respectively. The last three columns represent the second level of temperature at each of the levels of concentration. Either the means for each cell or the sum of values can be used if n is equal for all cells. Using the sums of values to avoid decimal places in the computations may be easier.

Three orthogonal comparisons are presented in Table 12.4. The first compares the first level of temperature with the second level of temperature, disregarding the level of concentration. The second planned comparison involves the first level of concentration compared to the other two levels, disregarding the level of temperature. The third comparison involves comparing the second concentration level with the third concentration level only for the first temperature.

The comparisons are made exactly like those for the one-way ANOVA by using equation 12.20. The computational equation for the sum of squares for each comparison (SS_{comp}) is given by

$$SS_{comp} = \frac{[\Sigma C(\Sigma X)]^2}{n[\Sigma(C^2)]} \qquad (12.20)$$

where C is the coefficient for the contrast, ΣX is the sum of the values for each cell or level, and n is the number of cases in each cell. If means are used rather than the sum of the values in each cell, equation 12.20 is modified to

$$SS_{comp} = \frac{n[\Sigma C(\overline{X})]^2}{\Sigma(C^2)} \qquad (12.21)$$

for which the mean is used rather than the sum of the values. The results are set up in an analysis of variance table the same way as for the one-way ANOVA, and the data in Table 12.4 are summarized in Table 12.5.

Table 12.5. ANOVA for Selected Comparisons in Table 12.4

Source	SS	df	ms	F
Comparison 1	61.63	1	61.63	37.81
Comparison 2	3.75	1	3.75	2.30
Comparison 3	10.00	1	10.00	6.13
Within	39.20	24	1.63	
Total	138.30	29		

The sum of squares for the first comparison (SS_{comp1}) is

$$SS_{comp1} = \frac{[1(20) + 1(28) + 1(38) + -1(18) + -1(14) + -1(11)]^2}{5(1^2 + 1^2 + 1^2 + -1^2 + -1^2 + -1^2)}$$

$$= \frac{(86 - 43)^2}{5(6)} = 61.63$$

which is within rounding errors of what was obtained for the main-row effect sum of squares because these two comparisons should be very close to each other. The sum of squares for the second comparison (SS_{comp2}) is

$$SS_{comp2}$$

$$= \frac{[2(20) + -1(28) + -1(38) + 2(18) + -1(14) + -1(11)]^2}{5(2^2 + -1^2 + -1^2 + 2^2 + -1^2 + -1^2)}$$

$$= \frac{(76 - 91)^2}{5(12)} = 3.75$$

The sum of squares for the third comparison (SS_{comp3}) is

$$SS_{comp3} = \frac{[0(20) + 1(28) + -1(38) + 0(18) + 0(14) + 0(11)]^2}{5(0^2 + 1^2 + -1^2 + 0^2 + 0^2 + 0^2)}$$

$$= \frac{(28 - 38)^2}{5(2)} = 10.00$$

Thus, of the total sums of squares for main and interaction effects, $61.64 + 6.20 + 31.26 = 99.10$, 75.38 has been used or accounted for. The other two possible orthogonal comparisons would use the remaining 23.72 of the sums of squares left unaccounted for. Once computed, the sums of squares are entered in an ANOVA table as in Table 12.5. As with the regular two-way ANOVA displayed in Table 12.3, the main-row effect F ratio is the same because the first comparison was the same as the test for the main effect due to the two selected temperatures. The second comparison is not significant because with df = 1 and 24, an $F = 4.26$ is needed for significance at the 0.05 level. The third comparison is significant, a result indicating that the simple effect of concentration level C_2 is significantly different than the third concentration level C_3 at temperature level T_1. The meaning of these results can be facilitated by using the plots of the results displayed in Figures 12.4 and 12.5.

The sums of squares can also be computed by retaining the two-dimensional matrix as in Table 12.2. The only difference is to modify equations 12.20 and 12.21 to account for the number of cells, or levels, for each column or row. For main effects, or marginal means, the sum of squares for the row main effect is

$$SS_{comp} = \frac{[\Sigma C(\Sigma X_r)]^2}{a(n)[\Sigma(C^2)]} \qquad (12.22)$$

where ΣX_r is the sum of the values for each row over the number of columns, a is the number of subgroups (columns for this example), n is the number of cases at each level, and C is the coefficient for weighting each sum. For comparing the main effect for temperature over the three levels of concentration, the sum of squares for this comparison would be

$$SS_{comp} = \frac{[1(86) + -1(43)]^2}{3(5)(1^2 + -1^2)} = \frac{(86 - 43)^2}{15(2)} = 61.63$$

which is the same as that computed previously. The coefficients for the marginal sums of 86 and 43 taken from Table 12.2 would be 1 and -1, respectively, and for this comparison, $a = 3$ columns or levels of the other factor occur with $n = 5$ cases per cell. Testing for main effects is appropriate when no significant interaction occurs or for an ordinal interaction effect. If a significant disordinal interaction does occur, then the investigator would consider the simple effects of differences in level of one factor at each level of the other factor.

12.9.2 Post Hoc Tests

Post hoc or unplanned comparisons can be made by using the same procedures as for the one-way ANOVA, and the calculations are exactly the same. If no significant interaction effect occurs or for an ordinal interaction, then these comparisons are made on the marginal means. Either the Tukey A or Scheffe′ test presented in Chapter 11 can be used for these comparisons, and the procedure for marginal means is exactly the same. If a significant interaction does occur, then the usual procedure is to test for simple effects (i.e., to test for differences between levels of each row at each level of the other factor or for each column). Essentially, this procedure would be to compute an analysis of variance for the simple effects for each factor separately.

The ANOVA summary table presented in Table 12.3 indicates a significant interaction. Thus, if an investigator were interested in determining if differences occur among the three levels of concentration at either temperature level, then the analysis of variance can be done as a one-way analysis at each level of temperature. The sum of squares for each row can be computed by using the general equation

$$\mathrm{SS}_{\mathrm{comp}} \text{ at row } 1 = \frac{\Sigma(\Sigma X)^2}{n} - \frac{(\Sigma X_r)^2}{a(n)} \qquad (12.23)$$

where ΣX is the sum of the values for each cell or each level of the row factor, ΣX_r is the total sum of values for that row, a is the number of levels in each row, and n is the number of cases in each cell or at each level. For the data in Table 12.2, SS_r for rows 1 and 2 would be, respectively,

$$\mathrm{SS}_{\mathrm{comp}} \text{ at row } 1 = \left(\frac{20^2 + 28^2 + 38^2}{5}\right) - \frac{86^2}{3(5)} = 32.53$$

$$\mathrm{SS}_{\mathrm{comp}} \text{ at row } 2 = \left(\frac{18^2 + 14^2 + 11^2}{5}\right) - \frac{43^2}{3(5)} = 4.93$$

As a computational check of the results, because each of the simple effects is a result of both the interaction effect and the main effect, the total sum of squares for the simple effects should equal the sum of squares for that main effect plus the sum of squares for interaction. Or,

$$\mathrm{SS}_{\mathrm{comp}} \text{ at a row} = \mathrm{SS}_c + \mathrm{SS}_{r \times c} \qquad (12.24)$$

From these calculations and from Table 12.3,

$$32.53 + 4.93 = 6.20 + 31.26$$

$$37.46 = 37.46$$

If these two sums are not equal or within rounding errors, then a computational error has been made.

After the sums of squares have been computed, these are then analyzed as a regular ANOVA as presented in Table 12.6. The degrees of freedom for each column equals $a - 1$, or for this example, three columns minus one is equal to two degrees of freedom. The within

Table 12.6. ANOVA for Simple Column Effects

Source	SS	df	ms	F
Comparison at row 1	32.53	2	16.27	9.98
Comparison at row 2	4.93	2	2.47	1.51
Within	39.20	24	1.63	

sum of squares is from the ANOVA table from Table 12.3. The results indicate that with df = 2 and 24, a significant simple effect occurs at T_1 but not at T_2. The Tukey A or Scheffe′ test could then be calculated to determine which levels of concentration are different. For this problem, the Scheffe′ test yields

$$\begin{aligned} CR_S &= \sqrt{(a-1)F_{df=2,24}}\sqrt{2a(n)(ms_w)} \\ &= \sqrt{(3-1)3.40}\sqrt{2(3)(5)(1.63)} \\ &= 2.61 \times 6.99 \\ &= 18.24 \end{aligned} \tag{12.25}$$

where CR_S is the critical range for the Scheffe′ test. Thus, any difference in the sum of values equal to or greater than 18.24 would be considered significantly different. For this problem, at T_1, C_1 is close to being significantly different than C_3 at the 0.05 level of significance. The only change in the Tukey or Scheffe′ equation from Chapter 11 is that $a(n)$ is used in the second term instead of n.

12.10 Special Considerations

12.10.1 Unequal Cases

The factorial designs in this chapter are considered to be completely crossed factorial designs in which cases are randomly assigned to as many experimental subgroups that are possible for all combinations of the two factors. For example, four levels of the first factor and five levels of the second factor would yield a total of 20 groups of subjects that must be used in the study. For this design, complete crossing of the levels of the factors being considered occurs because each level of one factor occurs at every level of the other factor. An equal number of randomly assigned cases in each of the experimental subgroups is best, and each case can have only one value. Most factorial studies are planned so that the n values are equal in each treatment combination. However, if data are lost for reasons unrelated to the study, (e.g., equipment failure, data being misread, unavailability of subjects during the study, or even subject mortality), then an equal number of cases in each of the subgroups may not be possible. In such situations, the number of cases in each of the groups should be proportional. For example, consider the following 2×3 factorial design for which the number of cases in each subgroup is different.

Row	Frequency in Column		
	1	**2**	**3**
1	4	5	6
2	8	10	12

A matrix is proportional if constant ratios can be set up for all pairs of rows or columns. For the example given, the ratios for rows are $4/8 = 5/10 = 6/12$. For columns, $4/5 = 8/10$ and $5/6 = 10/12$. Other examples of proportionality of cell frequencies are as follows:

Row	Frequency in Column			
	1	**2**	**3**	**4**
1	4	4	4	4
2	6	6	6	6

Row	Frequency in Column			
	1	**2**	**3**	**4**
1	3	4	4	6
2	3	4	4	6

Proportionality is required because the partitioning of the sums of squares into independent components can be done in a relatively simple, straightforward manner. When a proportion of cell frequencies occurs within the matrix, the computations for the sums of squares have to be based on the number of cases used within each cell, otherwise the procedure is the same. Procedures can be used when the number of cases in each subgroup are not equal and are not proportional. These procedures have been described by Winer (*3*), Keppel (*4*), and Dayton (*6*).

If the data being considered in a study are relatively homogeneous within each cell, and if the number of cases in each cell is fairly large, (e.g., 10 or more), then if a value is missing in one of the cells, one procedure that could be used would be to compute the mean for the other values in that particular cell and then substitute the mean of those values for the missing case. For example, if a 3 × 4 complete factorial design had 10 cases per cell, and if 1 or 2 of the 12 cells had a missing case and the other nine values per cell were consistent, then an economical approach would be to use the mean value as a substitute for the missing cases and do the analysis by using 10 cases per cell rather than randomly dropping one case from each of the other cells. Random dropping is another way of handling missing data, but it often wastes valuable data. Thus, the alternate procedures are recommended so that all of the data collected in a study can be used.

One method that is used with unequal cell frequencies is the *unweighted means analysis*. This method is appropriate when a few cases are lost from some of the subgroups in a study. This procedure uses the mean value for each cell rather than the individual values, and then the harmonic mean of the cell frequencies is used as n because the standard error of the mean is proportional to $1/n$ rather than n. This method can also be generalized to higher order factorial designs. Consider a 3 × 3 design that has missing cases. The basic data for this example are given in Table 12.7, which gives the values in terms of n and the means.

The first step is to calculate the harmonic mean (HM) of the cell frequencies.

$$\text{HM} = \frac{9}{(1/6) + (1/5) + (1/6) + (1/5) + (1/6) + (1/6) + (1/5) + (1/6) + (1/6)}$$

$$= 5.62$$

The second step is to calculate the mean of the values for each cell in the matrix. These means have been computed as $\overline{X}$ in Table

Table 12.7. Data for the Example of 3 × 3 Design with Unequal Cell Values of *n*

Summary Statistic	C_1	C_2	C_3	Row Totals
		R_1		
	6	3	5	
	5	5	3	
	6	4	4	
	4	6	2	
	3	3	6	
	6		4	
$\sum X$	30	21	24	
n	6	5	6	
X bar	5.0	4.2	4.0	$\sum X_{R_1} = 13.2$
				$\overline{X}_{R_1} = 4.4$
		R_2		
	4	6	5	
	6	5	6	
	3	7	4	
	5	4	6	
	6	6	7	

		5	5	
$\sum X$	24	33	33	
n	5	6	6	
X bar	4.8	5.5	5.5	$\sum X_{R_2} = 15.8$ $\overline{X}_{R_2} = 5.27$
		R_3		
	8	9	9	
	9	9	6	
	8	8	8	
	6	6	7	
	7	9	8	
		7	9	
$\sum X$	38	48	47	
n	5	6	6	
X bar	7.6	8.0	7.8	$\sum X_{R_3} = 23.4$ $\overline{X}_{R_3} = 7.8$
	$\sum \overline{X}_{C_1} = 17.40$ $\overline{X}_{C_1} = 5.80$	$\sum \overline{X}_{C_2} = 17.70$ $\overline{X}_{C_2} = 5.90$	$\sum \overline{X}_{C_3} = 17.30$ $\overline{X}_{C_3} = 5.77$	$\sum \overline{X}_T = 52.40$ $\overline{X}_T = 5.82$

12.7. The row and column totals are then computed by using these mean values rather than the individual values within each cell. That is, for R_1, the row total of 13.2 is the sum of the three means (5.0 + 4.2 + 4.0), and the row mean of 4.40 is the average of those three means, or 13.2/3 = 4.40. This procedure is used to obtain all of the column and row totals and means plus the total for the entire matrix. Thus, the total sum of the nine means for this matrix is 52.4, and the average of those means is 5.82. For this unweighted means procedure, the analysis proceeds by using the mean in each cell as if only one case occurred per cell and then by adjusting the sums of squares as to what they would have been on the basis of the harmonic mean.

The sums of squares for rows, columns, interaction, and within are calculated, respectively, as follows:

$$SS_r = \text{HM}\left(\frac{\sum^{R}(\Sigma\overline{X}_r^2)}{C} - \frac{(\overline{X}_T)^2}{R(C)}\right) \quad (12.26)$$

$$SS_c = \text{HM}\left(\frac{\sum^{C}(\Sigma\overline{X}_c^2)}{R} - \frac{(\Sigma\overline{X}_T)^2}{R(C)}\right) \quad (12.27)$$

$$SS_{r\times c} = \text{HM}\left(\sum^{R}\sum^{C}(\overline{X}^2) - \frac{\sum^{R}(\overline{X}_r^2)}{C} - \frac{\sum^{C}(\overline{X}_c^2)}{R} - \frac{(\Sigma\overline{X}_T)^2}{R(C)}\right) \quad (12.28)$$

$$SS_w = \Sigma X^2 - \sum^{R}\sum^{C}\left(\frac{(\Sigma X)^2}{n}\right) \quad (12.29)$$

where C is the number of columns, R is the number of rows, $\Sigma\overline{X}_r$ is the sum of the means for each row, $\Sigma\overline{X}_c$ is the sum of the means for each column, $\Sigma\overline{X}_T$ is the sum of all means, ΣX^2 is the sum of each squared value, ΣX is the sum of values in each cell, and n is the number of cases in each cell.

For the data in Table 12.7,

$$SS_r = 5.62\left(\frac{(13.2^2 + 15.8^2 + 23.4^2)}{3} - \frac{52.4^2}{9}\right)$$

$$= 5.62(323.81 - 305.08) = 105.26$$

$$SS_c = 5.62\left(\frac{(17.4^2 + 17.7^2 + 17.3^2)}{3} - \frac{52.4^2}{9}\right)$$

$$= 5.62(305.11 - 305.08) = 0.17$$

$$\text{SS}_{r \times c} = 5.62(5^2 + 4.2^2 + 4^2 + 4.8^2 + 5.5^2 + 5.5^2 + 7.6^2 + 8^2 + 7.8^2) - 323.81 - 305.11 - 305.08 = 5.28$$

$$\text{SS}_w = 6^2 + 5^2 + 4^2 + \ldots + 7^2 + 9^2 + 8^2 - \left(\frac{30^2}{6} + \frac{21^2}{5} + \ldots + \frac{48^2}{6} + \frac{47^2}{6}\right) = 1916 - 1853.37 = 62.63$$

As in other computations, several terms are used in more than one computation [e.g., $(\Sigma \overline{X}_T)^2/R(C)$]. Thus, this procedure reduces the number of total computations that have to be made.

After the sums of squares are computed, the results can be set up as a regular ANOVA table as given in Table 12.8. The degrees of freedom are computed similarly: $\text{df}_r = R - 1$, $\text{df}_c = C - 1$, $\text{df}_{r \times c} = (R - 1)(C - 1)$, $\text{df}_w = n_T - RC$, and $\text{df}_T = n_T - 1$. In Table 12.8, the total sum of squares is not given in the table because, in an unweighted means analysis, the sums of squares for rows, columns, interaction, and within do not add up to the total sum of squares. This inequality is due to the total sum of squares being based on the actual number of cases included, whereas the unweighted means analysis gives equal weight to the data in each cell regardless of the number of cases in each of those cells. Assuming a fixed-effects model, the appropriate error term would be the within mean square. Thus, the F ratios for the three hypotheses tested would be $F = 52.63/1.49 = 35.32$; $F = 0.09/1.49 = 0.06$; and $F = 1.32/1.49 = 0.89$. The only significant effect is a significant difference among the three levels of the row factor. Because no significant interaction effect occurred, the marginal row means can be tested by using the Scheffe′ or Tukey procedure because the frequencies in each row are the same (i.e., $n = 17$). Because the marginal means are given in Table

Table 12.8. ANOVA for the Data in Table 12.7

Source	SS	df	ms	F
Rows	105.26	2	52.63	35.32
Columns	0.17	2	0.09	0.06
Interaction	5.28	4	1.32	0.89
Within	62.63	42	1.49	

12.7, the Scheffe′ procedure for comparing pairs of these means can be given by

$$t_S = \frac{M_1 - M_2}{\sqrt{\text{ms}_w\left(\frac{1}{n_1} + \frac{1}{n_2}\right)}} \tag{12.30}$$

where t_S is the t value for the comparison; M_1 and M_2 are the means for the first and second groups, respectively; ms_w is the within mean square for the ANOVA table; and n_1 and n_2 are the number of cases in each group, respectively. Equation 12.30 can be used for equal or unequal cell frequencies. For the example in Table 12.7, the three marginal comparisons would be as follows: for R_1 and R_2,

$$t_S = \frac{4.40 - 5.27}{\sqrt{1.49\left(\frac{1}{17} + \frac{1}{17}\right)}} = -2.08$$

For R_1 and R_3,

$$t_S = \frac{4.40 - 7.80}{\sqrt{1.49\left(\frac{1}{17} + \frac{1}{17}\right)}} = -8.11$$

and for R_2 and R_3,

$$t_S = \frac{5.27 - 7.80}{\sqrt{1.49\left(\frac{1}{17} + \frac{1}{17}\right)}} = -6.04$$

The calculated t would then be compared to

$$\begin{aligned} K &= \sqrt{(R - 1)F_{\text{df}_{r,w}}} \\ &= \sqrt{2(5.15)} = 3.21 \end{aligned} \tag{12.31}$$

which is the critical value to be used for these comparisons rather than the t values from Table C in the Appendix.

Thus, at the 0.01 level of significance, the mean for level three is significantly greater than the means for levels one and two, but the difference between the means for levels one and two is not significant. As with other t ratios, any ratio equal to or greater than the table or criterion value would indicate a significant difference.

Other procedures are used for estimating the values for cells in which data are lost or not available. A review of these methods was given by Federer (*7*). Winer (*3*) gave a simplified method of estimating missing values that takes into consideration the values in adjacent cells. This procedure is done by forming ratios of the values in adjacent cells. Consider the example in Table 12.9 for a 4×4 factorial design in which one observation occurs per cell. The missing value in the second row, first column ($X_{2,1}$) can be estimated

$$\frac{X_{2,1}}{14} = \frac{(10/15) + (12/16)}{2}$$

$$X_{2,1} = 9.94$$

the missing value in the third row, third column ($X_{3,3}$) can be estimated as

$$\frac{X_{3,3}}{20} = \frac{(18/22) + (19/21)}{2}$$

$$X_{3,3} = 17.2$$

or

$$\frac{X_{3,3}}{16} = \frac{(18/14) + (19/15)}{2}$$

$$X_{3,3} = 20.48$$

Either of these two estimates could be used, and possibly the average of the two estimates, or 18.84, should be used. This procedure assumes that the relationship between the two column values for row 2 is the same as the relationship between the two column values for rows 1 and 3.

Table 12.9. Example of a Matrix with Missing Values

Row	C_1	C_2	C_3	C_4
R_1	10	15	16	19
R_2		14	18	22
R_3	12	16		20
R_4	9	15	19	21

12.10.2 One Observation per Cell

A factorial design can be used for a situation in which only one observation occurs per cell. However, with only one observation per cell, no within-cell variation occurs, and hence, no estimate of the experimental error can be made. With only one observation per cell, the total sum of squares is partitioned into the sum of squares for rows, columns, and interaction. There is no sum of squares for within variance. The computational procedures are the same as for the situation in which two or more observations occur per cell, with the exception that no within sum of squares can be computed. The summary analysis of variance table is set up in the same way with the same exception that no within sum of squares occurs, and thus, no estimate can be made of the error involved in the analysis. For the fixed-effects model, no tests are possible. As suggested by Ferguson (*2*) and Winer (*3*), if the design is a random-effects model, then the two main effects can be tested by dividing the mean square for rows and columns by the mean square for interaction. For this design, the mean square for interaction is used as an estimate of the error or residual variance. No test of the interaction effect can be made when $n = 1$ case per cell. However, Ferguson (*2*) suggested that the ratios calculated previously underestimate what the true ratio would be if the within variance was used as the proper error term. As such, a significant result probably indicates a real difference among the levels of that factor. However, if the result is not significant, then no conclusions should be drawn because this procedure results in a high probability of making a Type II statistical error, (i.e., the probability of accepting a null hypothesis when it is really false may be very high).

Most computer programs for computing factorial analysis of variance permit the selection of the appropriate error term. Thus, the user can select the proper error term for any particular analysis. Also, these programs usually have procedures to handle unequal sample sizes or to make adjustments for missing data.

References

1. *The Design and Analysis of Industrial Experiments;* Davies, O. L., Ed.; Longman Group: London, 1978.
2. Ferguson, G. A. *Statistical Analysis in Psychology and Education;* McGraw–Hill: New York, 1981.

3. Winer, B. J. *Statistical Principles in Experimental Design;* McGraw–Hill: New York, 1971.
4. Keppel, G. *Design and Analysis;* Prentice–Hall: Englewood Cliffs, NJ, 1973.
5. Hays, W. L. *Statistics for Psychologists;* Holt, Rinehart & Winston: New York, 1963.
6. Dayton, C. M. *The Design of Educational Experiments;* McGraw–Hill: New York, 1970.
7. Federer, W. T. *Experimental Design;* Macmillan: New York, 1955.

Problems

1. Suppose that a researcher wants to compare the effectiveness of two different catalysts, A and B, over a range of concentration levels. Differences may exist between the two catalysts depending upon the concentration level used, and the researcher wants to determine whether one catalyst is more or less effective than the other one for all levels of concentration from 10% to 30%. A two-way analysis is decided upon, with the two catalysts as one factor and concentration level as the other factor. Because inferences are to be made with regard to all levels of concentration within the limits of 10%–30%, six levels of concentration were selected at random between these two limits. The yields at various concentration levels for this study are as shown. Are there differences in yield of the two catalysts? Are there differences in yield at the various levels of concentration? Finally, are the two catalysts' effect the same for the levels of concentration selected?

	Yield at Concentration Level					
Catalyst	**13**	**16**	**17**	**22**	**26**	**27**
A	9	8	10	12	12	13
A	6	9	8	9	11	11
A	7	7	9	10	12	12
B	8	9	10	11	12	12
B	5	7	9	9	10	10
B	7	9	7	11	10	11

Chapter 13

Higher Order Factorial Designs

13.1 Basic Concepts

Procedures and principles underlying the two-way analysis of variance can be generalized and expanded to studies involving more than two factors, each of which has no restrictions on the number of levels. As with two-way designs, higher order designs can be described in terms of the number of levels of each factor in the design. For example, a $2 \times 3 \times 4$ design involves three factors, for which there are two levels of the first factor, three levels of the second factor, and four levels of the third factor. For this design, a total of $2 \times 3 \times 4 = 24$ cells or combinations of treatments would be analyzed in the completely crossed design. For a four-way $2 \times 3 \times 4 \times 4$ design, a total of 96 cells or treatment combinations are possible. Higher order designs, such as a five-way or six-way analysis, are also possible. However, interpreting the complex interactions becomes very difficult, and the number of subgroups needed makes the analyses difficult to implement.

The number of hypotheses that can be tested by these higher order designs can be found by combinatorial equations based on the number of factors included in the analysis. For the three-way factorial, three main-effects hypotheses, three two-factor interactions, and one three-way interaction can be tested. Four hypotheses for main effects and 11 interaction hypotheses can be tested in the four-way analysis. These are shown in Table 13.1. The calculations and interpretation of the results higher order designs are a direct expansion of the procedures for the two-way design. The total sum of squares is di-

1453–0/88/0295$10.25/1

Table 13.1. Main and Interaction Effects for Three-Way and Four-Way Factorial Designs

	Description	
Effects	**Three-Way**	**Four-Way**
Main effects	A	A
	B	B
	C	C
		D
Two-way interactions	A × B	A × B
	A × C	A × C
	B × C	A × D
		B × C
		B × D
		C × D
Three-way interactions	A × B × C	A × B × C
		A × B × D
		A × C × D
		B × C × D
Four-way interaction		A × B × C × D

vided or partitioned into eight parts in a three-way analysis: one part for each main and interaction effect as listed in Table 13.1 and the within-cells sum of squares. Each sum of squares has its respective degrees of freedom, which is divided into the sum of squares to obtain the variance estimate or mean square for each of the main and interaction effects.

A three-way design can be visualized as a geometric cube, as displayed in Figure 13.1. In such a display, the three factors could be thought of as rows, columns, and layers. However, the three factors are usually represented by letters such as A, B, and C. Use of letters for each factor allows for expansion to four-factor or five-factor designs. The design displayed in Figure 13.1 is for a 2 × 2 × 2 design and is displayed in matrix form in Figure 13.2. Both types of displays are useful in conceptualizing and interpreting higher order factorial designs. As displayed in Figure 13.2, the main effects for each factor and the two-way interactions are computed and interpreted in the same way as in a two-way analysis. The three-way interactions are usually interpreted as two-way interactions at each level of the third factor. For example, one would compare the A × B interaction at each of the levels of factor C, the A × C interaction at each of the levels of factor B, and the B × C interaction at each of the levels of factor A.

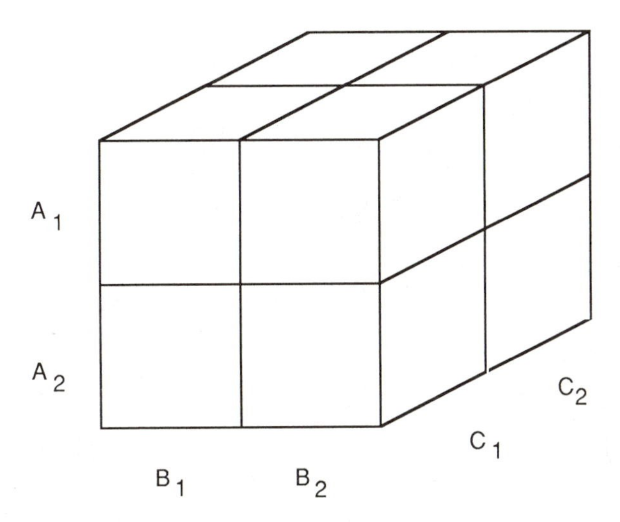

Figure 13.1. Geometric display of a 2 × 2 factorial design.

All factorial designs are interpreted by starting with the highest order of interaction, which for the three-way design would be the three-way interaction. If this interaction is significant, then the investigator would concentrate on the three-way interactions displayed in Figure 13.2. If the three-way interaction is not significant, then the attention should be focused on the two-way interactions. If the two-way interactions are not significant, then the main effects are considered. The reason for this approach is the same as for the two-way factorial design. A nonsignificant interaction indicates that the main effects are consistent at all levels of the second factor and thus can be interpreted as in a one-way analysis. However, a significant interaction indicates that the main effect for one factor is going to vary at different levels of the second factor; thus, one would look at the simple effects by comparing the differences between or among the levels of the first factor at each level of the second factor. In the higher order designs, some variables interact and some do not. For example, if there is a significant main effect for factors A and C and a significant A × B interaction, then the main effect for factor C can be interpreted without qualification. However, the main effect for factor A would have to be interpreted by considering the significant A × B interaction.

The same reasoning applies to the two-way interactions. A significant three-way interaction means that the two-way interactions may be different for different levels of the third factor. The results should be plotted as in Figure 13.2 to understand the meanings of

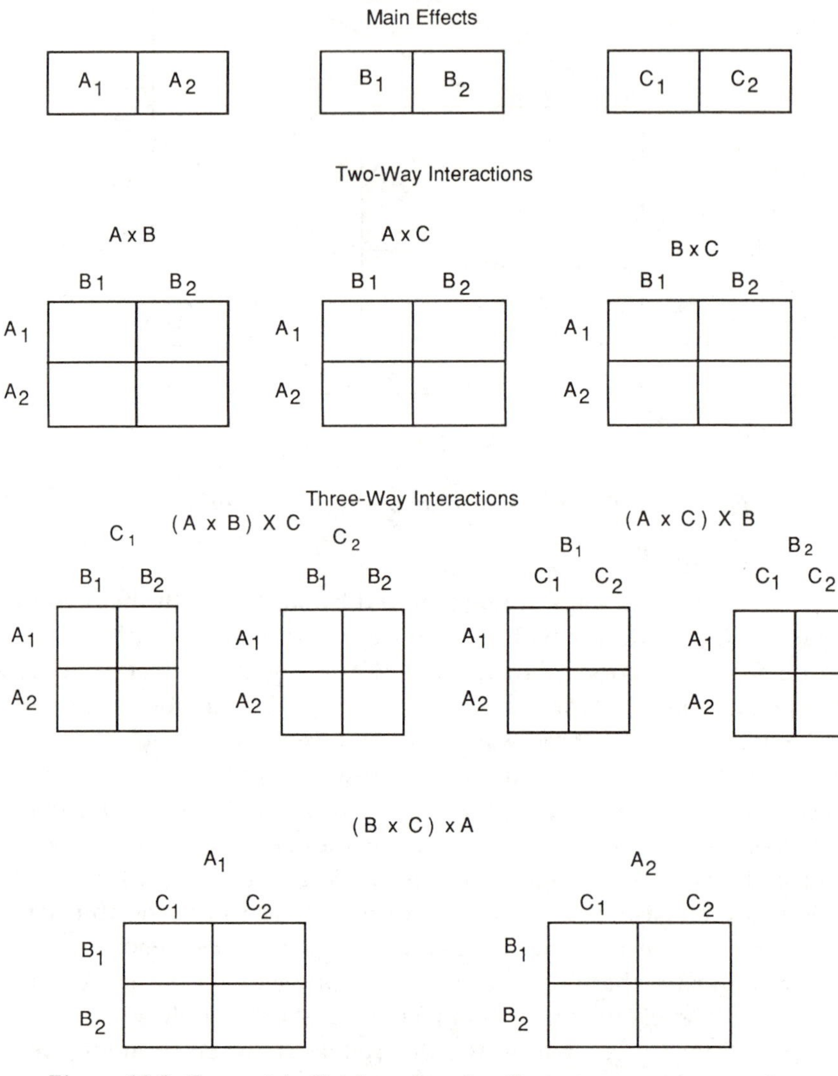

Figure 13.2. Geometric display of main effects, two-way interactions, and three-way interactions in a three-way design.

these interactions. Thus, a significant three-way or triple interaction indicates that the interaction of any two factors may differ for each level of the third factor. Thus, the A × B interaction may be different for level C_1 compared to level C_2, or the A × C interaction may be different at each level of the B factor, or the B × C interaction may be different at each level of the A factor. As with the two-factor design, the interaction for the two variables is often plotted as in Figures

12.4 and 12.5 to determine whether the interaction is ordinal or disordinal. That is, one display can be made by using cell means for the A factor plotted over the levels of the B factor, and the other display can be made by using cell means for the B factor plotted over the levels of the A factor.

The assumptions underlying the three-way analysis are the same as those for the one-way and two-way analyses of variance. As with the two-way design, subjects or cases are randomly assigned to each of the A × B × C treatment conditions, and an equal number of subjects in each of the treatment conditions is usually best. However, procedures, such as those provided by Winer (*1*), Keppel (*2*), and Dayton (*3*), can handle situations having unequal sample sizes. Also, more than one observation or case per treatment condition is best. Otherwise, the proper error term for all of the interaction effects cannot be computed. However, situations may arise in which a large number of treatment conditions are required, such as in a 4 × 4 × 5 design, which has 80 treatment conditions. For this design, to compute a within-cells error term, a minimum of 160 cases or observations would be required. If this number of subjects is not available, Winer (*1*, p. 476) and Kirk (*4*, p. 227) indicated that a factorial design can be computed by using only one subject for each treatment condition. However, the higher order interactions could not be estimated, and this design should be used only when the higher order interactions are thought to not be significant. In fact, when there is one case per cell, the higher order interaction sum of squares is used as the error term to test for lower order interactions and main effects. Also, as will be described in Section 13.6, a fractional factorial design can be used when dealing with designs having many factors and levels of each factor.

Computer programs are available for all of the types of analysis of variance designs, and in practice, these programs are usually used rather than doing the analyses with hand or desk calculators. Most of these programs provide for most of the conditions encountered in any investigation, and the person using these programs can specify the proper error term for a specific analysis, different options with regard to output, and different ways to plot the data.

13.2 Example of a Three-Way Analysis

As an example of the sources of the error estimates for the main and interaction effects for the three-way analysis of variance, the two-way problem used in Chapter 12, where the factors considered were temperature and concentration, will be expanded to include a third factor, which might be two types of catalysts. Thus, the original 2 × 3 design is expanded to take into consideration two types of catalysts. The main reason for adding the third factor is to determine the way in which three factors operate to affect yield separately and in combination. The results of the two-way analysis showed a significant interaction between temperature and pressure with regard to yield for one catalyst. The question answered by expanding the design to include two types of catalyst is whether the type of catalyst makes a difference with regard to yield either separately or in combination with the different levels of temperature and concentration.

A total of seven hypotheses would be tested for the three-way design. These hypotheses are as follows:

1. There is no difference in yield between the two types of catalysts.
2. There is no difference in yield between the two temperature levels.
3. There are no differences in yield among the three concentration levels.
4. There is no A × B interaction.
5. There is no A × C interaction.
6. There is no B × C interaction.
7. There is no A × B × C interaction.

As with the two-way design, the model could be set up by using fixed levels of all factors, which would result in the fixed-effects model. Or, the levels of all factors could be selected at random from a range of all possible levels, which would result in the random-effects model. Or, the levels of some factors could be fixed and for others random, which would result in the mixed-effects model. As for the two-way analysis, most analyses are done by using the fixed-effects model because the levels chosen are usually those deemed most important or representative of those commonly used. However, if random levels are chosen, the main advantage is greater generalization to all possible levels of that factor. But in practice, because of the limitations of how many levels may be practical or possible to set up, the fixed-effects model is usually chosen, with levels chosen to represent those commonly used. The only difference in running the different analyses is in the selection of the proper error term for testing the significance of some effects. Otherwise, the computations and the way the design is conceptualized are the same.

The raw data for this three-way analysis are presented in Figure 13.3. There are two levels of the catalyst (factor A), three levels of concentration (factor C), and two temperature levels (factor B) all selected at fixed points; thus, a fixed-effects model results. As shown in Figure 13.3, there are a total of 2 × 3 × 2 = 12 cells, five observations per cell, and thus 60 observations needed for the analysis. The cell totals, sum of the squared values in each cell, the mean, and the standard deviation of the values in each cell are presented in Figure 13.4. Although the means and standard deviations are not needed for the computations required for the ANOVA, both are needed for a complete interpretation of the results and are easy to compute at this stage because the sums and sums of squares for each cell have to be obtained for the ANOVA. Also, these statistics help to get an idea as to what the outcome might be for the ANOVA.

The results for catalyst 1 are identical to the results presented in Chapter 12 because the same data are used and the study has been expanded to include new data for catalyst 2. Plotting the means provides a visual display as to how the factors have affected yield, as shown in Figures 13.5, 13.6, and 13.7. Such a plot would show a different pattern of yield for catalyst 1 compared to catalyst 2. That

	(A_1) Catalyst 1			(A_2) Catalyst 2		
	C_1	C_2	C_3	C_1	C_2	C_3
B_1	3 6 4 2 5	6 4 7 5 6	8 6 9 7 8	7 8 6 7 6	5 7 6 5 5	4 3 5 3 4
B_2	4 2 5 3 4	4 3 1 4 2	3 1 4 1 2	5 3 4 4 5	6 4 5 4 5	5 3 6 4 6

Figure 13.3. Hypothetical data for a three-way analysis of variance.

		A_1			A_2		
		C_1	C_2	C_3	C_1	C_2	C_3
B_1	ΣX	20	28	38	34	28	19
	ΣX^2	90	162	294	234	160	75
	$\bar{X}$	4.0	5.6	7.6	6.8	5.6	3.8
	s	1.58	1.14	1.14	0.84	0.89	0.84
B_2	ΣX	18	14	11	21	24	24
	ΣX^2	70	46	31	91	118	122
	$\bar{X}$	3.6	2.8	2.2	4.2	4.8	4.8
	s	1.14	1.30	1.30	0.84	0.84	1.30

Figure 13.4. Cell sums, means, and standard deviations for the data in Figure 13.3.

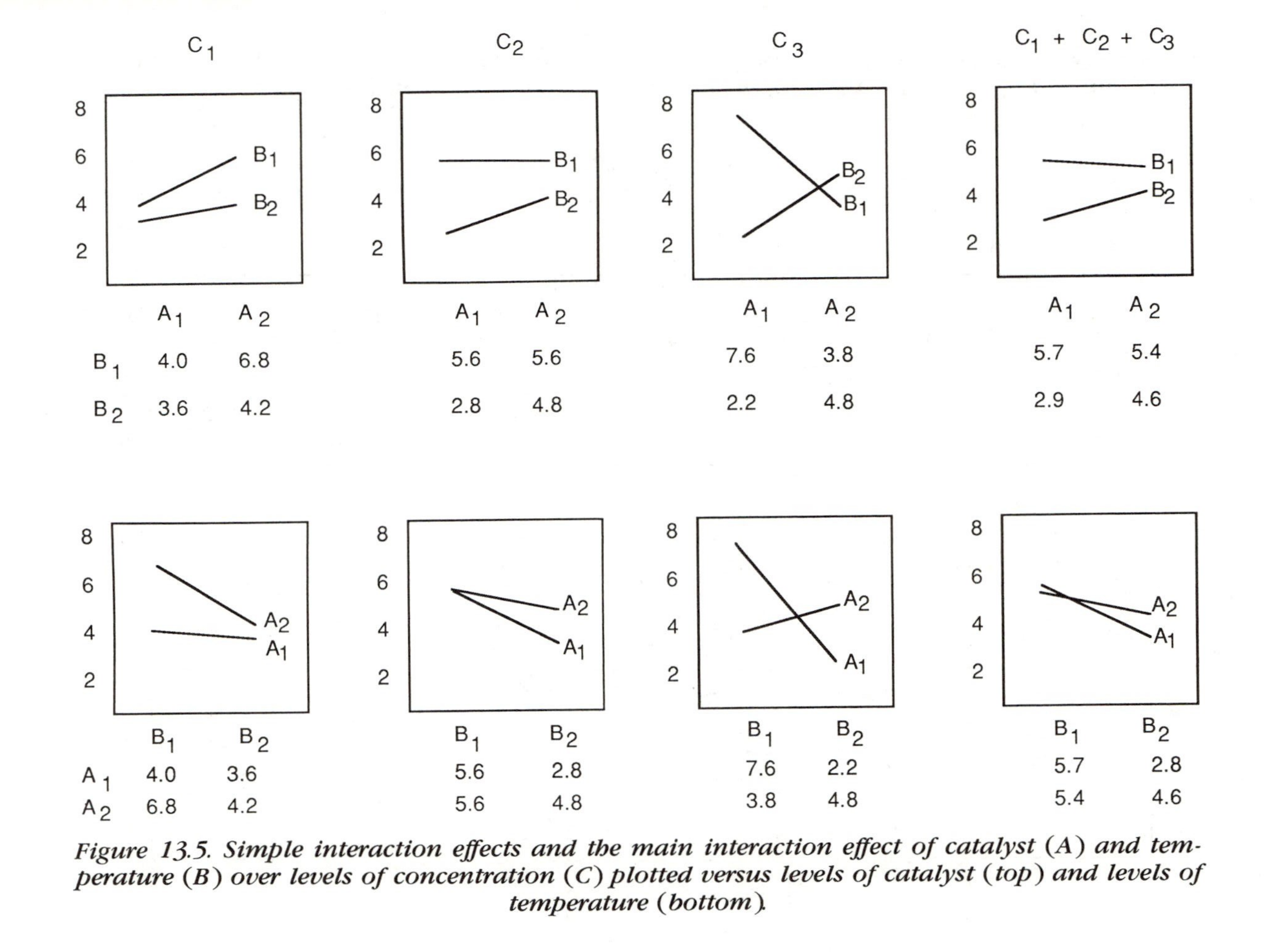

Figure 13.5. Simple interaction effects and the main interaction effect of catalyst (A) and temperature (B) over levels of concentration (C) plotted versus levels of catalyst (top) and levels of temperature (bottom).

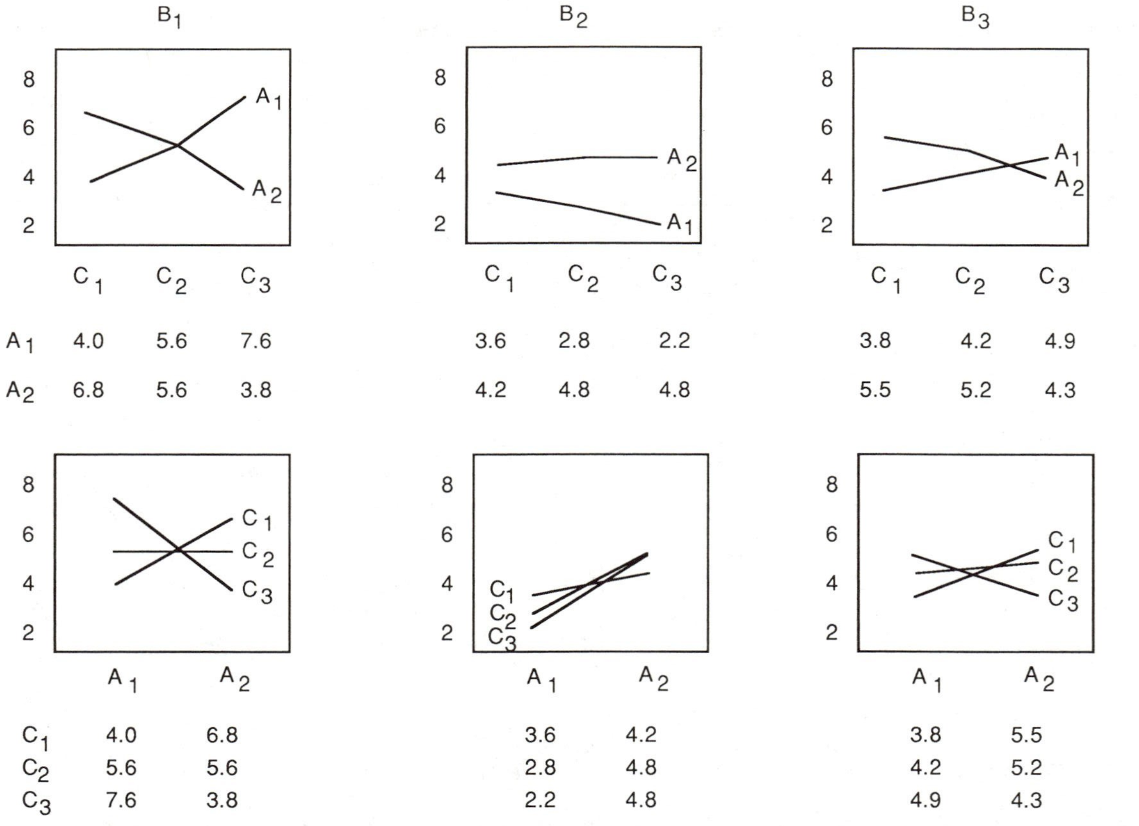

Figure 13.6. Simple interaction effects and the main interaction effect of catalyst (A) and concentration (C) over levels of temperature (B) plotted versus levels of concentration (top) and levels of catalyst (bottom).

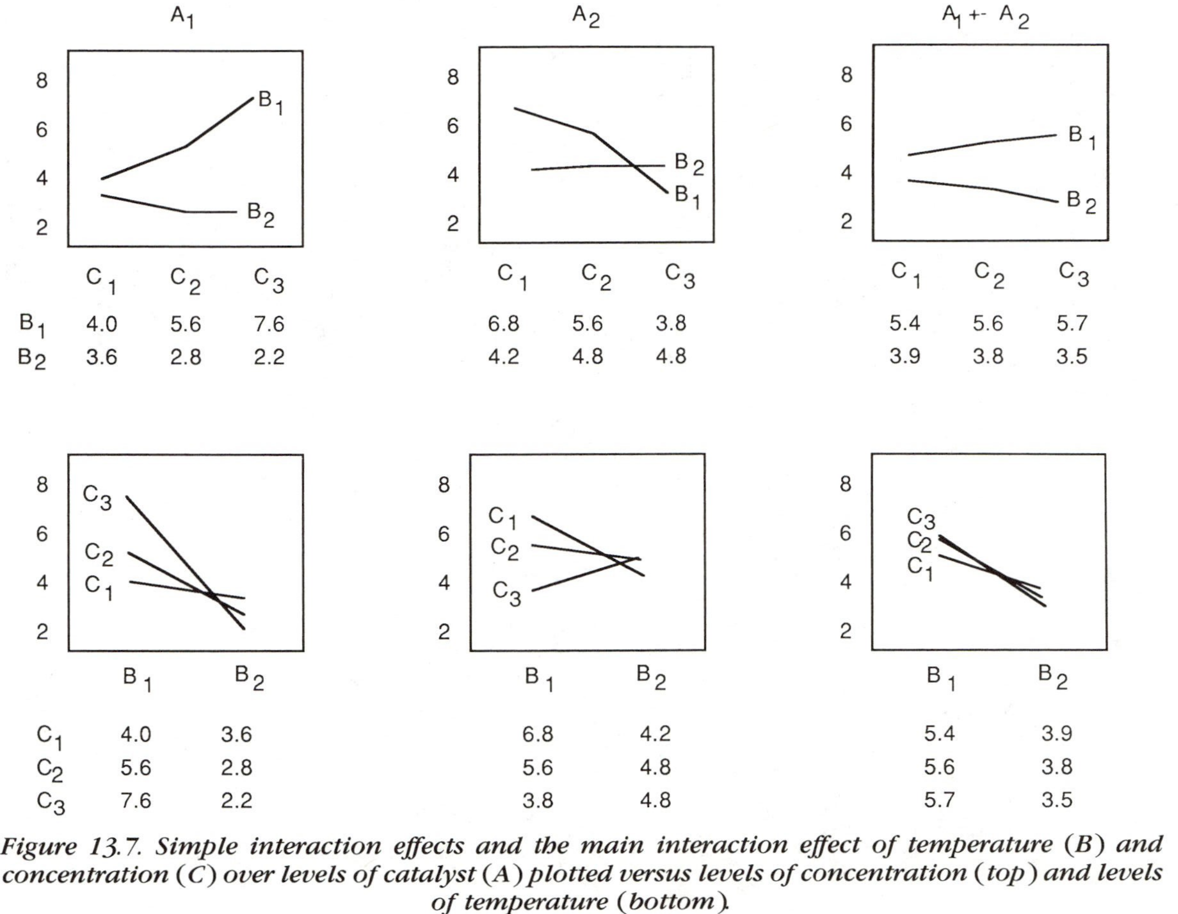

Figure 13.7. Simple interaction effects and the main interaction effect of temperature (B) and concentration (C) over levels of catalyst (A) plotted versus levels of concentration (top) and levels of temperature (bottom).

is, the two-way interaction of temperature and pressure appears to be different for one catalyst compared to the other; thus, a possible significant three-way interaction is indicated. Whether or not this is a significant interaction will be determined by the ANOVA.

Computation of the necessary sums of squares is aided by calculating and recording the totals of the values in each cell and then summing for marginal totals. This summation can be done by creating two-way matrices in which the totals for the A × B factors are summed over the levels of the C factor, the totals for the A × C factors summed over the levels of the B factor, and the totals for the B × C factors summed over the levels of the A factor. Thus, for the A × B matrix, the data in Figure 13.4 yield the cell totals as presented in Figure 13.8. The cell totals are obtained by summing the sums of values in the cells C_1, C_2, and C_3 for A_1 at level B_1, or from Figure 13.4, 20 + 28 + 38 = 86. For A_2, 34 + 28 + 19 = 81. For level B_2 for A_1, 18 + 14 + 11 = 43, and for A_2, 21 + 24 + 24 = 69. Each of these cell sums are based upon 5 × 3 = 15 observations.

Once the cell sums are computed, then the marginal sums are obtained (e.g., 86 + 81 = 167, and 86 + 43 = 129). Each of these marginal sums is based upon 2 × 15 = 30 observations. For the A × C matrix in Figure 13.9, the totals for each cell would be summed over the two levels of the B factor at each level of A and C. For the first sum, 20 + 18 = 38, then 28 + 14 = 42, 34 + 11 = 49, 38 + 21 = 55, 28 + 24 = 52, and 19 + 24 = 43. Each of these cell sums is based upon 2 × 5 = 10 observations. The marginal sums are then computed. For each level of A, these marginal sums are based upon 3 × 10 = 30 observations, and for each level of C, these sums are based upon 2 × 10 = 20 observations. For the B × C

	A_1	A_2	Sum
B_1	86	81	167
B_2	43	69	112
Sum	129	150	279

Each cell sum is based on 15 observations

Figure 13.8. An A × B table of cell values summed over the levels of factor C. Each cell sum is based on 15 observations.

matrix, each cell total is summed over the two levels of the A factor. For these totals, the first sum is 20 + 34 = 54, then 28 + 28 = 56, 38 + 19 = 57, 18 + 21 = 39, 14 + 24 = 38, and 11 + 24 = 35. The marginal totals for each level of B are based upon 3 × 10 = 30 observations, and for each level of C, the marginal totals are based upon 2 × 10 = 20 observations.

The total sum of all values for each matrix equals 279, and this total is always based upon the same number of observations, which for this example is 60. As a check, the row and marginal sums are based upon a total of *n* observations and should also equal 279 when summed.

Once the cell and marginal sums are calculated and tabulated as in Figures 13.8–13.10, then the necessary sums of squares can be calculated. Although the total sum of squares does not have to be calculated for the method presented, calculation of this value will serve as a computational check, and because the sums of the values and the sums of the squared values have been obtained for the cal-

	C_1	C_2	C_3	Sum
A_1	38	42	49	129
A_2	55	52	43	150
Sum	93	94	92	279

Each Cell sum is based on 10 observations

Figure 13.9. An A × C table of cell values summed over the levels of factor B. Each cell sum is based on 10 observations.

	C_1	C_2	C_3	Sum
B_1	54	56	57	167
B_2	39	38	35	112
Sum	93	94	92	279

Each cell sum is based on 10 observations.

Figure 13.10. A B × C table of cell values summed over the levels of factor A. Each cell sum is based on 10 observations.

culation of the means and standard deviations for each cell, then summing these sums for the computation of the total sum of squares is simple. For the example,

$$SS_T = 1493 - \frac{279^2}{60} = 1493 - 1297.35 = 195.65$$

The value 1297.35 is a correction term that enters in the computation for the sum of squares for each main effect. Thus, this value should be calculated and be referenced rather than computing it for each equation.

There are two general ways to proceed with the analysis. One approach is to calculate the treatment sum of squares and then obtain the within sum of squares by subtraction of the treatment sum of squares from the total sum of squares. The second method is to calculate the within sum of squares for each cell and then obtain the treatment sum of squares by subtracting the sum of all of these computed within sum of squares from the total sum of squares. Or both methods could be done as a computational check. From the data in Figure 13.4, the within sum of squares (SS_w) for each cell is computed by using the general equation

$$SS_w = \sum X^2 - \frac{(\Sigma X)^2}{n} \qquad (13.1)$$

For the data in Figure 13.4, the total within sum of squares for all of the cells is equal to 60.4. The treatment sum of squares (SS_b) can be computed by the general equation

$$SS_b = \sum^{R}\sum^{C}\left(\frac{(\Sigma X_c)^2}{n_c}\right) - \frac{(\Sigma X_T)^2}{n_T} \qquad (13.2)$$

where R is the number of rows, C is the number of columns, n_c is the number of cases in each cell, n_T is the total number of cases, X_c is the sum of the values in each cell, and ΣX_T is the sum of the values for all cells. For the data in Figure 13.4,

$$SS_b = \left(\frac{20^2}{5} + \frac{28^2}{5} + \ldots + \frac{24^2}{5} + \frac{24^2}{5}\right) - \frac{279^2}{60} = 135.25$$

This sum of squares is the one partitioned into main and interaction effects. Also, as in the one-way analysis of variance, the treatment and the within sum of squares are equal to the total sum of squares. For the example, 135.25 + 60.40 = 195.65. As with all analysis of variance procedures, a summary analysis of variance table such as Table 13.2 is usually set up, and the sums of squares are recorded as they are computed.

The sums of squares for the main effects and the interaction effects are computed from the marginal totals for the three matrices in Figures 13.8–10. For the main effect due to factor A, which is the type of catalyst, the sum of squares (SS_A) is

$$SS_A = \left(\frac{129^2}{30} + \frac{150^2}{30}\right) - 1297.35 = 7.35$$

Likewise, the sum of squares for the B factor (SS_B), which is the temperature, is

$$SS_B = \left(\frac{167^2}{30} + \frac{112^2}{30}\right) - 1297.35 = 50.41$$

All of the sums are based upon 2 × 15 = 30 observations. The A × B interaction sum of squares can be computed from the data in Figure 13.8 by first computing the sum of squares for cells (SS_{cells}) within that matrix. The sum of squares is computed by the general equation

$$SS_{cells} = \sum^{R}\sum^{C}\left(\frac{(\Sigma X)^2}{n}\right) - \frac{(\Sigma X_T)^2}{n_T} \qquad (13.3)$$

Table 13.2. Analysis of Variance for the Three-Way Example

Source	SS	df	ms	*F*
A (Catalyst)	7.35	1	7.35	5.83*
B (Temperature)	50.41	1	50.41	40.01*
C (Concentration)	0.10	2	0.05	0.04
A × B	16.03	1	16.03	12.72*
A × C	13.90	2	6.95	5.52*
B × C	1.24	2	0.62	0.49
A × B × C	46.22	2	23.11	18.34*
Within	60.40	48	1.26	
Total	195.65	59		

*This difference is significant at the 0.05 level.

where ΣX is the sum of the values in each of the cells for each A × B combination, n is the number of values for each of those cells, and X_T and n_T are for all cells in the matrix. For the data in Figure 13.8,

$$\mathrm{SS_{cells}} = \left(\frac{86^2}{15} + \frac{81^2}{15} + \frac{43^2}{15} + \frac{69^2}{15}\right) - 1297.35 = 73.79$$

As with the computation of other sums of squares, each cell total is divided by the number of observations on which it is based, which in this case is 15 observations. Once the sum of squares for cells is obtained, then for a two-way matrix, the interaction sum of squares can be computed by the general equation

$$\mathrm{SS}_{r \times c} = \mathrm{SS_{cells}} - \mathrm{SS}_r - \mathrm{SS}_c \qquad (13.4)$$

where $\mathrm{SS}_{r\times c}$ is the sum of squares for the interaction of the row and column variables.

For the data in Figure 13.8,

$$\mathrm{SS_{A \times B}} = 73.79 - 50.41 - 7.35 = 16.03$$

These three sums of squares would then be entered in Table 13.2.

This same process would be repeated for the matrix in Figure 13.9 for the sum of squares for the C factor ($\mathrm{SS_C}$), which is concentration. The sum of squares for the A factor has already been computed. So,

$$\mathrm{SS_C} = \left(\frac{93^2}{20} + \frac{94^2}{20} + \frac{92^2}{20}\right) - 1297.35 = 0.10$$

Each of these sums is based upon 20 observations. The interaction sum of squares is computed in the same way as for the A × B interaction (by first computing $\mathrm{SS_{cells}}$). For the A × C matrix,

$$\mathrm{SS_{cells}} = \left(\frac{38^2}{10} + \frac{42^2}{10} + \frac{49^2}{10} + \frac{55^2}{10} + \frac{52^2}{10} + \frac{43^2}{10}\right) - 1297.35 = 21.35$$

Then the sum of squares for interaction would be obtained by

$$\mathrm{SS_{A \times C}} = 21.35 - 7.35 - 0.10 = 13.90$$

The sum of squares for the B × C interaction can be computed if the sum of squares for cells for the matrix in Figure 13.10 is known. The value of SS_{cells} would be

$$SS_{cells} = \left(\frac{54^2}{10} + \frac{56^2}{10} + \frac{57^2}{10} + \frac{39^2}{10} + \frac{38^2}{10} + \frac{35^2}{10}\right) - 1297.35 = 51.7$$

Then,

$$SS_{B \times C} = 51.75 - 50.41 - 0.10 = 1.24$$

The sum of squares for the three-way interaction ($SS_{A \times B \times C}$) can be computed by subtracting the sums of squares computed for the main and two-way interactions from the treatment sum of squares because the treatment sum of squares is partitioned into the main and interaction effects. For the example,

$$SS_{A \times B \times C} = 135.25 - 7.35 - 50.41 - 0.10 - 16.03 - 13.9 - 1.24$$
$$= 46.22$$

The degrees of freedom for each source of variance for the three-way analysis of variance are as follows: For the main effects,

Main Effect	**df**
A	$A - 1 = 2 - 1 = 1$
B	$B - 1 = 2 - 1 = 1$
C	$C - 1 = 3 - 1 = 2$

and for the interaction effects,

Interaction Effect	**df**
A × B	$(A - 1)(B - 1) = (2 - 1)(2 - 1) = 1$
A × C	$(A - 1)(C - 1) = (2 - 1)(3 - 1) = 2$
B × C	$(B - 1)(C - 1) = (2 - 1)(3 - 1) = 2$
A × B × C	$(A - 1)(B - 1)(C - 1) = (2 - 1)(2 - 1)(3 - 1) = 2$
within	$(A \times B \times C)(n - 1) = (2 \times 2 \times 3)(4) = 48$
total	$n_T - 1 = 60 - 1 = 59$

Furthermore, as for analysis of variance procedures, the degrees of freedom for the total sum of squares is equal to the sum of the degrees of freedom for all components, or

$$59 = 1 + 1 + 2 + 1 + 2 + 2 + 2 + 48$$
$$59 = 59$$

The mean squares are then computed by dividing each sum of squares by the appropriate degrees of freedom. Then each F ratio is computed by dividing the mean square for each source by the mean square within, assuming all levels of each factor were fixed. For this example, at the 0.05 level of significance, an $F = 4.04$ would be needed for 1 and 48 degrees of freedom to test the A and B main effects and the A × B interaction. As can be noted from Table 13.2, all of these effects are significant. For the other effects, with 2 and 48 degrees of freedom, an $F = 3.19$ would be needed for significance. Thus, the A × C and A × B × C interactions are significant.

This example was assumed to be a fixed-effects model, for which the correct error term for all main and interaction effects would be the mean square within. If all of the levels of each factor had been chosen at random, the error term for the triple or three-way interaction would have been the mean square within. The appropriate error term for the three two-way interactions would have been the mean square for the three-way or A × B × C interaction, but there are no exact tests of the three main effects. However, Winer (*1*), Kirk (*4*), and Scheffe' (*5*) suggested that approximate F tests or quasi F ratios can be used for determining these main effects. The quasi F ratio (F') as given by Winer (*1*, p. 376) and Kirk (*4*, p. 213) is approximately distributed as the regular F ratio and is computed for each of the three main effects as follows:

$$F'_A = \frac{ms_A}{ms_{A \times B} + ms_{A \times C} - ms_{A \times B \times C}}$$

$$F'_B = \frac{ms_B}{ms_{A \times B} + ms_{B \times C} - ms_{A \times B \times C}}$$

$$F'_C = \frac{ms_C}{ms_{A \times B} + ms_{B \times C} - ms_{A \times B \times C}}$$

If a mixed-effects factorial design is created, the proper error term depends upon which factors are fixed and which are random.

If factor A is fixed and factors B and C are random, then the proper error term for the triple or three-way interaction is the mean square within groups. For the A × B and A × C interaction effects, the proper error term is the mean square for the three-way interaction mean square. For the B × C interaction effect, the proper error term is the mean square within groups. For the main effects for B and C, which are the two random factors, the proper error term is the B × C interaction mean square. There is no exact test for the fixed factor A. If the model has factors A and B as fixed levels and factor C as a random-levels factor, then the mean square for within groups is the proper error term for all effects involving the random factor, which is factor C: the main effect for C; the A × C and B × C two-way interactions; and the three-way, or the A × B × C interaction. The main effect for the fixed-factor A is tested by using the A × C interaction mean square, and the main effect for the fixed factor B is tested by using the B × C interaction mean square. In interpreting the resulting *F* ratios, the proper degrees of freedom associated with the appropriate mean squares would be used in determining the significant *F* ratios.

When random variables are used in the analysis, the *F* ratio is computed by using the mean square for the interaction rather than the mean square within. In such situations, a loss of power occurs because the degrees of freedom associated with the mean square for interaction is less than the degrees of freedom associated with the mean square within. Thus, a larger *F* value is required for significance. The implication of this requirement is that a consideration of this problem should be made before using a random or mixed model. The fixed-effects model probably has the greatest utility for chemistry studies because the variables are usually fixed.

13.3 Interpretation of a Three-Way ANOVA

The results of the three-way analysis of variance are interpreted by first determining if the three-way interaction is significant or not. A significant triple or three-way interaction indicates that the two-way interactions and the main effects will be influenced by this three-way interaction. In other words, the main effects in a significant two-way interaction cannot be interpreted directly because the means for one factor may vary at the different levels of the second factor, and if this interaction is disordinal, the effects would cancel. The same type of situation happens with a three-way interaction. The two-way interactions may be different at different levels of the third factor. For example, the A × B interaction may be different for each level of factor C, the A × C interaction may be different for each level of factor B, or the B × C interaction may be different for each level of factor A. With a significant three-way interaction, any of these two-way interactions might be different, or all of them may be different at the various levels of the third factor. The reason for this is that when a three-way interaction is significant, the variance among the treatment means cannot be accounted for entirely by the main effects and the two-way interactions. If the three-way interaction is not significant, then one would proceed to look at the two-way interactions. If the two-way interactions are significant, then the main effects are interpreted as simple effects as in the two-way analysis of variance. If the two-way interactions are not significant, then the main effects can be interpreted directly.

One way to conceptualize the interaction effects for a significant three-way interaction is to plot the means for each cell in two-way plots such as those in Figures 13.5, 13.6, and 13.7. As with the two-way analyses, the data can be plotted in various ways, depending upon the main interest or concern. For example, in Figure 13.5, the B factor, which is the level of temperature, can be plotted over the two types of catalyst, which is the A factor, or the type of catalyst can be plotted over the two levels of temperature. In these three figures, the data are plotted as the simple interaction effects for each level of the third factor. The data are then plotted over the average values for the levels of the third factor. This plot is analyzed for the two-way interaction in the three-way analysis, and this interaction is referred to as the mean interaction in contrast to the simple interaction effect at each level of the third factor. In each of these figures, the mean for each cell is given and plotted for the simple interaction effect, and the averaged mean over the levels of the third factor are then given and plotted for the main interaction effect. These means are taken directly from Figure 13.4.

The results of the analysis presented in Table 13.2 indicate a significant three-way interaction. Thus, the main two-way interactions and main effects for each factor have to be considered with regard to this effect, and one should concentrate on the simple effects rather than the main effects. From Table 13.2, the A × B two-way interaction is also significant. This significance can be noted from inspection of the plots for the A × B interaction summed over the three levels of the C factor as shown in Figure 13.5. The B factor plotted over the A factor shows an ordinal interaction in which the yield for the first level of the B factor is higher than the second level of the B factor. When type of catalyst is plotted over the two temperature levels, a disordinal interaction results. However, the three-way interaction would indicate different patterns for each of the three levels of the C factor. As can be noted from the other three plots, for each way of plotting these data at each level of the C factor, the interactions are inconsistent from one level to another. Thus, at level C_1, the difference in yield due to temperature (B) for catalyst A_1 is small, whereas for catalyst A_2, the difference in yield is fairly large. At the second concentration level, C_2, the yield for temperature B_1 is the same for both types of catalysts, but for B_2 or the second temperature level, the yield is greater for catalyst A_2 than for A_1. Level C_3 has a disordinal interaction. If the difference in yield for the two temperature levels was of interest, the data would be plotted over temperature levels as in the lower part of Figure 13.5.

The same types of comparisons can be made for the A × C interactions plotted in Figure 13.6. The A × C main interaction is also significant, as can be noted from Table 13.2 and the A × C plots averaged over the levels of the B factor. However, as with the A × B interaction, because the three-way interaction is significant, the A × C interactions will probably be different at each level of the B factor. As can be seen in Figure 13.6, this statement is true because the plots reveal different patterns at each level of C.

From Table 13.2, the main interaction effect for the B × C interaction is not significant. This situation implies no differential effects on yield when temperature (B) and level of concentration (C) are considered. As can be noted from the B × C main interaction when summed over the two levels of the A factor, a consistent effect at all levels of the C factor is shown in Figure 13.7. However, the three-way interaction would indicate different simple interaction effects for each level of C. From inspection of the simple interaction effects, the B × C interaction is much different for catalyst A_1 than catalyst A_2. Thus, for a three-way interaction, even though the mean two-way interaction may not be significant, the simple two-way interaction should be inspected because this is likely to reveal differential patterns.

The plots in Figures 13.5, 13.6, and 13.7 provide one way of giving meaning to the results of the three-way analysis as well as helping to determine what further analyses should be computed. Like the two-way analysis, the simple effects must be examined rather than the main effects when there are higher order interaction effects.

13.4 Individual Comparisons

As with the two-way analysis of variance, the procedures for making comparisons among either marginal or cell means are an extension of those used with the one-way analysis of variance. These procedures can be made a priori on the basis of a theoretical or empirical rationale regardless of the outcome of the overall *F* tests. Or, a posteriori comparisons can be made by using procedures such as the Tukey, Scheffe', Newman–Keuls, or any other preferred method. As with the one-way analysis of variance, such tests should be made only when the overall *F* test is significant for the main effect to which the post hoc analysis is applied. Such tests would be made following a systematic procedure and usually involve comparing means in rows or columns (i.e., by comparing means for or within each factor), which would represent a logical grouping of the means to be compared.

Comparing means that involve two or more factors would be difficult to interpret. For example, a significant difference between the mean in the upper left cell in Figure 13.4 and the mean in the lower right cell would be difficult to interpret. This difference could be attributable to factor A, B, or C because these two means involve different levels of all of these factors. A comparison involving only one factor such as factor C would represent a logical comparison because any differences found would be attributable to differences in level of factor C either over levels of the other factors or within one level of one of the other factors.

If the three-way interaction is significant, then the simple two-way interaction effects would be examined. If these are significant, then possibly the simple effects would be examined at each level of the factor being considered. If the three-way interaction is not significant, then the two-way interactions would be examined. If these are significant, then simple comparisons would be made at each level of the factors involved. If the two-way interactions are not significant, then the main effects would be examined (i.e., comparisons would be made of the marginal means for each cell). For all such post hoc comparisons and for nonorthogonal comparisons, experimentwise error rates must be considered, as discussed in Section 13.5.2. Thus, the best procedure is to restrict the number of comparisons made to only those that form logical groupings that can be meaningfully interpreted. Otherwise, if all A × B × C means in a three-way factorial are compared, the statistical adjustments for the post hoc tests will result in an overcorrection and a too conservative adjustment for all of these comparisons, most of which will probably not be done, and if conducted, will probably not be interpretable. The magnitude of the correction for these comparisons depends upon the number of means compared and the number of observations upon which each mean is based.

13.4.1 Testing for Simple Interaction Effects

If the three-way interaction is significant, then the two-way interactions may be different for the various levels of the third factor. As plotted in Figures 13.5, 13.6, and 13.7, the simple interaction effects may be different from the main two-way interactions. When the three-way interaction is significant, one can test for the simple interaction effects or test for differences in one factor (e.g., factor A) at all possible combinations of the other two factors (e.g., B and C).

Simple interaction effects can be tested for each two-way combination (i.e., A × B interactions for all levels of C, the A × C interactions for all levels of B, and the B × C interactions for all levels of A). The particular simple interaction effect chosen for analysis would depend upon what would be most meaningful to the researcher.

The first step in making these simple interaction tests would be to refer to the two-way matrices as given in Figures 13.8–13.10. However, because these matrices are summed over the level of the third factor, a new set of matrices can be created from the sums in Figure 13.4 as in Figure 13.11, which is used for computing the B × C simple interaction effects. The cell entries are exactly as in Figure 13.4. However, the row and column totals are provided for

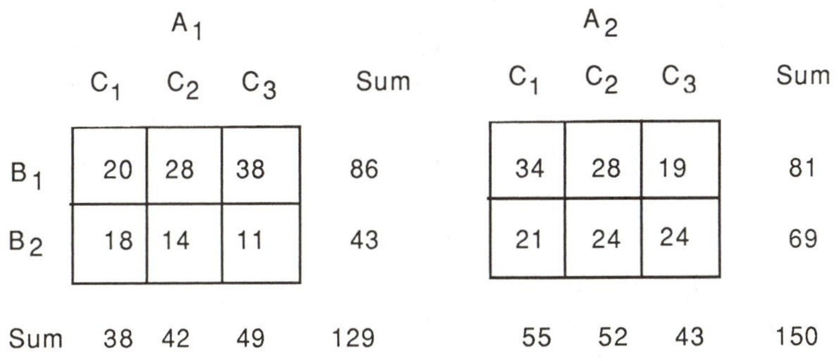

	A_1				A_2			
	C_1	C_2	C_3	Sum	C_1	C_2	C_3	Sum
B_1	20	28	38	86	34	28	19	81
B_2	18	14	11	43	21	24	24	69
Sum	38	42	49	129	55	52	43	150

Figure 13.11. Cell sums for computing B × C simple interaction effects.

the computation of the sums of squares needed for the simple interaction analysis. The two-way sums of squares necessary for the analysis are computed by the general equation

$$SS_{B \times C} \text{ at A} = \frac{\sum (BC)^2}{n} - \frac{\sum (AB)^2}{C(n)} - \frac{\sum (AC)^2}{B(n)} + \frac{\sum (T)^2}{B(C)(n)} \qquad (13.5)$$

where BC represents each cell in the matrix, AB represents each row total, AC represents each column total, T represents the total sums in the matrix, n is the number of observations upon which each cell sum is based, and B and C represent the number of levels for each of these two factors.

For the data in Figure 13.11, the sums of squares for each level of factor A are

$$SS_{B \times C} \text{ at } A_1 = \left(\frac{20^2 + 28^2 + 38^2 + 18^2 + 14^2 + 11^2}{5}\right) - \left(\frac{38^2 + 42^2 + 49^2}{10}\right) - \left(\frac{86^2 + 43^2}{15}\right) + \frac{129^2}{30}$$

$$= 653.8 - 560.9 - 616.33 + 554.7 = 31.27$$

$$SS_{B \times C} \text{ at } A_2 = \left(\frac{34^2 + 28^2 + 19^2 + 21^2 + 24^2 + 24^2}{5}\right) - \left(\frac{55^2 + 52^2 + 43^2}{10}\right) - \left(\frac{81^2 + 69^2}{15}\right) + \frac{150^2}{30}$$

$$= 778.8 - 757.8 - 754.8 + 750 = 16.2$$

As a computational check, the sum of squares for the simple interaction effects is equal to the sum of squares for the B $\times$ C interaction plus the sum of squares for the A $\times$ B $\times$ C interaction. From Table 13.2, $SS_{B\times C} = 1.24$ and $SS_{A\times B\times C} = 46.22$. Thus,

$$1.24 + 46.22 = 31.27 + 16.2$$

$$47.46 = 47.47$$

which are equal within rounding errors.

Once the sums of squares for the simple interactions are computed, they are entered into an analysis of variance table such as in Table 13.3. The degrees of freedom for each interaction ($df_{B\times C}$ for

Table 13.3. Summary Analysis for the Simple A × B Interaction

Source	SS	df	ms	*F*
B × C at A_1	31.27	2	15.64	12.41*
B × C at A_2	16.20	2	8.10	6.43*
Within	60.40	48	1.26	

*This difference is significant at the 0.01 level.

the interaction of B and C) are computed as

$$df_{B \times C} = (B - 1)(C - 1)$$

For each simple effect, the number of degrees of freedom is (2 − 1) (3 − 1) = 2. The *F* ratios are calculated in the usual way: by dividing each mean square by the within mean square from the three-way summary table. For the analysis in Table 13.3, both the *F* tests are significant at the 0.05 level because with 2 and 48 degrees of freedom, an $F = 3.19$ is needed for significance. Thus, a differential effect on yield occurs when considering the interaction of temperature with level of concentration, and as can be noted from Figure 13.7, this interaction is different for catalyst A_1 compared to catalyst A_2. Thus, the simple main effects would be examined for either B or C, or B and C, depending on the interests of the researcher. This procedure would be done for both levels of A.

The same general procedure can be used to test the simple A × C interaction at each level of B, or the simple A × B interaction at each level of C.

13.4.2 Testing for Simple Main Effects

If the three-way interaction is not significant, then the two-way analysis can be tested. As with the two-way analysis, if the two-way interaction is not significant, then the marginal means or the main effects would be analyzed. However, if the two-way analysis is also significant, then one could determine the simple main effects of the interaction by comparing levels of one factor at each level of the second factor. That is, in a three-way analysis, the A × B interaction can be analyzed by testing the variation due to factor A at each level of factor B, or by testing the variation due to factor B at each level of factor A. Likewise, the variation due to factor A at each level of factor C could be tested for the A × C interaction, or the variation due to factor C at each level of factor A could be tested. The same procedure could apply to the B × C interaction.

The computational procedures are the same as for the two-way analysis of simple main effects in Chapter 12. The only adjustment needed is to adjust the equations for the number of observations upon which each total is based. The equation is

$$SS_A \text{ at } B_1 = \frac{\sum (AB)^2}{C(n)} - \frac{(T_B)^2}{A(C)(n)} \qquad (13.6)$$

where SS_A at B_1 represents the sum of squares for factor A at the first level of factor B, AB represents the cell sum, T_B represents the sum for the particular row or column, n represents the number of observations upon which each cell sum is based, A represents the number of levels of the factor A, and C represents the number of levels of factor C. Applying this equation to the B × C interaction as displayed in Figures 13.7 and 13.10 yields

$$SS_C \text{ at } B_1 = \left(\frac{54^2 + 56^2 + 57^2}{10}\right) - \left(\frac{167^2}{30}\right) = 0.47$$

$$SS_C \text{ at } B_2 = \left(\frac{39^2 + 38^2 + 35^2}{10}\right) - \left(\frac{112^2}{30}\right) = 0.87$$

As a computational check, because the simple main effect is based on the sum of squares for the main effect plus the interaction effect, then

$$SS_C \text{ at } B = SS_{B \times C} + SS_C$$

From Table 13.2, $SS_{B \times C} = 1.24$ and $SS_C = 0.10$. Thus,

$$0.47 + 0.87 = 1.24 + 0.10$$

$$1.34 = 1.34$$

The sums of squares are entered into a summary analysis of variance table as in Table 13.4. The number of degrees of freedom for each comparison is equal to $C - 1$ or $3 - 1 = 2$. With 2 and 48 degrees of freedom, an $F = 3.19$ is needed for significance at the 0.05 level. Thus, neither of these F ratios indicate a significant difference.

Table 13.4. Analysis of Variance for the Simple Effects of Factor C for the B × C Interaction

Source	SS	df	ms	*F*
C at B_1	0.47	2	0.24	0.19
C at B_2	0.87	2	0.44	0.35
Within	60.40	48	1.26	

This analysis was done merely to show the computational procedures. If this were an actual problem, this analysis would not be run because of the significant three-way interaction and because the B × C two-way interaction was not significant. However, this example was selected to point out how a significant three-way interaction can influence additional analyses. Because the three-way interaction was significant for this problem, the simple interaction effects would be analyzed as in Section 13.4.1. Because both the simple interaction effects were also significant, one would proceed to analyze the simple effects at each level of the A factor. This type of analysis is sometimes referred to as a simple, simple main effect because the effects are two steps apart from the main effects. That is, the *simple main effect* is an analysis of the sums across two variables (i.e., the different levels of factor A collapsed across one other factor). The *simple, simple main effects* for factor A are based upon sums collapsed across both of the other factors, so in essence across all factors. Actually, this situation is analogous to making such comparisons in the one-way analysis of variance. The basic equation for computing the sum of squares for each simple, simple main effect is the same as equation 13.6 with an adjustment for the number of observations upon which each sum is based. The general equation is

$$SS_A \text{ at } B_iC_i = \frac{\sum (ABC)^2}{n} - \frac{(AB)^2}{A(n)} \qquad (13.8)$$

where SS_A at B_iC_i is the sum of squares for factor A at level *i* of factors B and C, ABC is the cell sum at the particular level of each factor being considered, AB is the sum for that particular row, *n* is the number of observations upon which each cell sum is based, and A is the number of levels for the particular factor being considered.

For the simple, simple main effects for factor C at level A_1 and B_1, from Figure 13.11,

$$SS_C \text{ at } A_1B_1 = \frac{20^2 + 28^2 + 38^2}{5} - \frac{86^2}{3(5)} = 32.53$$

For the other levels of factors A and B,

$$SS_C \text{ at } A_1B_2 = \frac{18^2 + 14^2 + 11^2}{5} - \frac{43^2}{3(5)} = 4.93$$

$$SS_C \text{ at } A_2B_1 = \frac{34^2 + 28^2 + 19^2}{5} - \frac{81^2}{3(5)} = 22.80$$

$$SS_C \text{ at } A_2B_2 = \frac{21^2 + 24^2 + 24^2}{5} - \frac{69^2}{3(5)} = 1.20$$

As a computational check, the sums of squares for the simple, simple main effects of factor C are made up of the sums of squares for the A × C, B × C, and A × B × C interactions plus the sum of squares for C. Thus,

$$32.53 + 4.93 + 22.80 + 1.20 = 0.10 + 13.90 + 1.24 + 46.22$$
$$61.46 = 61.46$$

After the sums of squares are computed, they are entered into an analysis of variance table as in Table 13.5. The degrees of freedom for each of these sources is the number of levels minus 1, or for the example, $C - 1 = 2$. The mean square for each source is then divided by the mean square within. With 2 and 48 degrees of freedom, an $F = 3.19$ is needed for significance at the 0.05 level. Thus, there is a significant difference among the levels of factor C for B_1 but not

Table 13.5. Analysis of the Simple, Simple Main Effects for Factor C

Source	SS	df	ms	*F*
C at A_1B_1	32.53	2	16.27	12.91*
C at A_1B_2	4.93	2	2.47	1.96
C at A_2B_1	22.80	2	11.40	9.05*
C at A_2B_2	1.20	2	0.60	0.48
Within	60.40	48	1.26	

*This difference is significant at the 0.01 level.

for B_2, and this result is consistent for both levels of factor A. From Figure 13.7, this result is what would be expected from the plot of the means.

13.5 Testing for Differences between Means

The procedures presented in Section 13.4.2 dealt with determining whether or not there were significant simple interactions, and if these interactions were significant, determining if the simple effects were significant. In this section, procedures are presented for making individual and multiple comparisons among means, which can be planned (a priori) or unplanned (post hoc). The procedures for making such comparisons, which were presented for the one-way analysis of variance, can be extended directly to factorial designs. A priori comparisons can be made regardless of the outcome of the omnibus F test, but post hoc comparisons should not be made unless the overall F test is significant.

13.5.1 A Priori Tests

A priori tests can be made whenever a factor is set up at more than two levels by creating a set of orthogonal comparisons separately for the levels of each factor, and for which each comparison would have one degree of freedom. For instance, in the example given in Section 13.3, the following a priori comparisons for factor C could be set up by using the following coefficients for the two comparisons possible.

Comparison	C_1	C_2	C_3
Comparison 1	2	−1	−1
Comparison 2	0	1	−1
Marginal sums	93	94	92

The general equation for computing the sum of squares for orthogonal comparisons ($SS_{A_{comp}}$) is

$$SS_{A_{comp}} = \frac{[\sum (C_0)(A)]^2}{(B)(C)(n)[\Sigma(C_0^2)]} \qquad (13.9)$$

where C_0 represents the orthogonal coefficients, A is the marginal sum for each level of factor A, and B and C represent the levels of the factors B and C over which the marginal sums of levels A are obtained.

The only adjustment to the equation used with the one-way analysis of variance is for the number of observations. Because the levels of factor C have been summed over the levels of the other two factors, the actual number of observations upon which each sum is based must be adjusted. For the example, the marginal sums are based upon combining sums over factors A and B. Thus, the denominator will equal $2 \times 2 \times 5 = 20$ rather than simply the number of observations to be used in the denominator for the one-way analysis of variance. For these comparisons, the sums of squares would be

$$SS_{comp1} = \frac{(186 - 94 - 92)^2}{2(2)(5)(2^2 + -1^2 + -1^2)} = \frac{0}{120} = 0$$

$$SS_{comp2} = \frac{(94 - 92)^2}{2(2)(5)(1^2 + -1^2)} = \frac{4}{40} = 0.10$$

The sums of squares should add up to the sum of squares for factor C from the three-way analysis of variance table. From Table 13.2, the sum of squares was 0.10, so these two sums of squares are equal for the computations. The sums of squares could be entered into an analysis of variance table such as in Table 13.6. The degrees of freedom for these comparisons should also equal the number of degrees of freedom for the factor C. Because the mean squares are less than the within mean square, neither *F* ratio will be significant. Because the triple interaction was significant, this main-effect comparison would not be made because the three-way interaction would indicate differential effects. However, because this comparison could have been planned, then one could go ahead and make it realizing

Table 13.6. Analysis of Variance for Main Effects for Factor C

Source	SS	df	ms	*F*
Factor C	0.10	2	0.05	0.04
Comparison 1	0.00	1	0.00	0.00
Comparison 2	0.10	1	0.10	0.08
Within	60.40	48	1.26	

that the result would be suspect due to the three-way interaction. However, an a priori comparison could have been made by comparing the three levels of C only at level 1 of factor A and at level 1 for factor B. By using the same coefficients and the cell totals from Figure 13.4, the following sums of squares would be obtained.

$$SS_{comp1} = \frac{(40 - 28 - 38)^2}{5(2^2 + -1^2 + -1^2)} = 22.53$$

$$SS_{comp2} = \frac{(28 - 38)^2}{5(1^2 + -1^2)} = 10.00$$

For these comparisons, the sum of squares should equal the sum of squares for the simple, simple main effects as calculated in Table 13.5. The two sums of 32.53 are equal; thus, the computations are correct. The sum of squares can be entered into an analysis of variance table such as Table 13.7.

With 1 and 48 degrees of freedom, an $F = 4.04$ is required for significance at the 0.05 level. Thus, as with the outcome of the simple, simple main effects presented in Table 13.5, both differences are significant, indicating that level C_1 is significantly different than the combined levels of C_2 and C_3, and that C_2 is significantly different than C_3.

If any of the factors are based upon quantitative independent variables, then comparisons for trend can be made by using orthogonal polynomials similar to the procedures used for the one-way analysis of variance. The only adjustment for factorial designs is to modify the equation to take into account the number of observations upon which each sum used in the analysis is based. This adjustment is made in exactly the same way as in equation 13.9.

Table 13.7. Analysis of Variance for Simple, Simple Effects for Factor C

Source	SS	df	ms	*F*
C at A_1B_1	32.53	2		
Comparison 1	22.53	1	22.53	17.88*
Comparison 2	10.00	1	10.00	7.94*
Within	60.40	48	1.26	

*This difference is significant at the 0.01 level.

13.5.2 Post Hoc Tests

For post hoc comparisons between all possible pairs of means within a logical grouping of means, any of the post hoc procedures such as Tukey or Scheffe' can be modified for factorial designs. As indicated, usually such comparisons are made within a single row or column rather than comparing cell means that overlap two or more factors. The reason for this is that interpreting such differences is difficult because any difference might be due to any or all of the factors involved.

The modification necessary for the post hoc tests is to adjust for the number of possible pairs of means being compared and the number of observations upon which each mean is based. For post hoc comparisons (e.g., Tukey, Scheffe', Newman–Keuls, least significance difference (LSD), and Duncan), the error term is the within mean square if the assumption of homogeneity of within-group variances is met. For example, the basic equation for the Tukey A test for the one-way analysis of variance for comparing treatment differences by using sums of observations is

$$\mathrm{CR}_T = q_{(k, \mathrm{df}_w)}\sqrt{n(\mathrm{ms}_w)} \qquad (13.10)$$

where CR_T is the critical range for the Tukey test, q is Student's t statistic for the number of treatment groups (k) and degrees of freedom for the within mean square (df_w), n is the number of observations in one subgroup, and ms_w is the mean square within taken from the analysis of variance table. For the three-way analysis of variance, this basic equation is modified to take into consideration the number of means being compared and the number of observations upon which each mean is based. For determining the critical range between treatment sums for factor A, the Tukey A equation would be

$$\mathrm{CR}_T = q_{(\mathrm{A}, \mathrm{df}_w)}\sqrt{\mathrm{B}(\mathrm{C})(n)(\mathrm{ms}_w)} \qquad (13.11)$$

where q is Student's t statistic for the number of levels of factor A and the within degress of freedom (df_w), B and C represent the number of levels of these two factors, n is the number of observations in one subgroup, and ms_w is the mean square within taken from the analysis of variance table. This equation could be used for testing the main effects for factor A, and by changing the A, B, and C positions, for testing the main effects for factors B and C.

If the simple main effects are tested for one factor at a particular level of another factor (e.g., differences between levels of factor A for a particular level of B), the equation becomes

$$CR_T = q_{(A, df_w)}\sqrt{C(n)(ms_w)} \qquad (13.12)$$

where the number of observations is adjusted to account for summing over the levels of factor C.

For determining the critical range between treatment sums for simple, simple main effects, equation 13.10 would be used because this procedure would be similar to the one-way analysis of variance.

The Scheffe' test is adjusted in a similar manner. For determining the critical range for the marginal sums for testing main effects for any factor, the Scheffe' equation is modified in terms of factor A as

$$CR_S = \sqrt{(A - 1)F_{(df_A df_w)}}\ \sqrt{2(C)(B)(n)(ms_w)} \qquad (13.13)$$

where the terms have been identified as for the Tukey test. For determining the critical range for simple main effects such as for differences in levels of factor A for a particular level of factor B, the equation becomes

$$CR_S = \sqrt{(A - 1)F_{(df_A df_w)}}\ \sqrt{2(C)(n)(ms_w)} \qquad (13.14)$$

For simple, simple main effects, the Scheffe' test would be run as for the one-way analysis of variance.

Applying the Tukey A test to the marginal sums for factor C from Figure 13.10 yields

$$CR_T = 3.44\sqrt{2(2)(5)(1.26)} = 17.27$$

where 3.44 is the value of q for $r = 3$ and $df = 40$ because the table does not include $df = 48$, the two 2's denote the two levels of factors A and B, 5 denotes the number of observations, and 1.26 denotes the mean square within. Thus, any difference in the marginal sums greater than 17.27 would be significant at the 0.05 level. The critical range for the simple main effects due to factor C at each level of factor B is determined by

$$CR_T = 3.44\sqrt{2(5)(1.26)} = 12.21$$

Thus, any differences in sums greater than 12.21 would be considered significant. The result is the same as the analysis in Section 13.4.2 (i.e., none of the differences are significant). By applying to the differences among the levels of factor C for each level of A as in Figure 13.9, the same CR would be obtained because the equation involves the same ingredients, and again no significant differences occur among the levels of C for both levels A_1 and A_2.

For determining the critical range for the simple, simple main effects,

$$CR_T = 3.44\sqrt{5(1.26)} = 8.63$$

Thus, any differences in the cell sums equal to or greater than 8.63 would be significantly different at the 0.05 level of significance. For determining the critical range for comparing the simple, simple effects for factors A and B, the critical range would be

$$CR_T = 2.86\sqrt{5(1.26)} = 7.18$$

Thus, in Figure 13.4, the difference between B_1 and B_2 for catalyst A_1 at level of concentration C_1 is not significantly different because the difference between the sums of 20 and 18 is only 2, whereas for levels C_2 and C_3, the differences are significant. Comparing catalyst A_1 with A_2 at concentration level C_1 shows a significant difference at level B_1 because $34 - 20 > 7.18$, but at level B_2, the difference is not significant because $21 - 18 < 7.18$.

By applying the Scheffe' test to the same comparison, (i.e., comparisons of main effects for factor C),

$$CR_S = \sqrt{2(3.19)}\sqrt{2(2)(2)(5)(1.26)} = 17.96$$

which is slightly more conservative than the Tukey test.

For determining the simple main effects due to factor C at each level of factor B,

$$CR_S = \sqrt{2(3.19)}\sqrt{2(2)(5)(1.26)} = 12.70$$

which again is slightly more conservative than the 12.21 critical range for the Tukey test.

For the simple, simple main effects,

$$CR_S = \sqrt{2(3.19)}\sqrt{2(5)(1.26)} = 8.98$$

compared to 8.63 for the Tukey test.

The post hoc tests control for experimentwise error rates only for the particular row or column to which they are applied. If such post hoc tests are made for all rows, columns, and layers of a three-way analysis of variance, then the error rate increases, (i.e., the probability of making a Type I statistical error increases). Keppel (*2*) and Winer (*1*) pointed out that if the experimentwise error rate has been set at, for example, 0.05 for comparisons involving only one row or column in a factorial design, then if comparisons are also made for other rows or columns, the actual error rate is 0.05 times the number of rows or columns for which comparisons are made. Thus, if comparisons are made over three levels of a second factor, then the probability of making a Type I statistical error is $0.05 \times 3 = 0.15$ rather than 0.05. One procedure to adjust for multiple comparisons is to divide the experimentwise error rate, which is the alpha level set for the study, by the number of comparisons made on a post hoc basis. This procedure could be modified to divide the alpha level set for the study by the number of rows or columns being compared. The resulting quotient could then be used as the probability level for each comparison. Thus, if simple, simple main effects are to be tested at the 0.05 level and three levels of a second factor are being considered, then the alpha level should be set at $0.05/3 = 0.016$ for each comparison.

The procedures presented for comparing means are those most frequently used in research. Winer (*1*), Keppel (*2*), Dayton (*3*), and Kirk (*4*) provided additional comparisons and more thorough presentations of these methods.

13.6 Fractional Factorial Designs

The two-way and three-way factorial designs are usually feasible to conduct because the number of treatment combinations and the number of observations necessary for the complete factorial are within reasonable limits. However, the number of treatment combinations becomes quite large when the design includes many levels of the factors being considered or is expanded to include four or more factors, especially if a number of levels is involved for each factor. For example, a $6 \times 5 \times 4$ three-way design would require 120 treatment combinations, a $4 \times 3 \times 3 \times 2$ complete four-way factorial design would require 72 treatment combinations, and a $2 \times 2 \times 2 \times 2 \times 2$ five-way factorial design would require 32 treatment combinations.

If the number of treatment combinations is considered to be too excessive or the costs associated with the number of observations required is prohibitive or exceeds the available budget, one solution is to use a *fractional factorial design*. Such designs, also known as incomplete factorial designs, can be set up to include only a fraction (e.g., one-half, one-third, or one-fourth) or the total treatment combinations required for a complete factorial design. For example, a 3 × 3 × 3 × 3 complete factorial design would require 81 treatment combinations, whereas a one-third fractional design would only require 27 of those 81 treatment combinations.

However, some special assumptions are required for these savings in treatment groups, and some limitations in the results affect how the results should be interpreted. These limitations are due to the fact that in fractional factorial designs, every sum of squares can be given two or more designations, usually involving a main and interaction effect. This situation is called *confounding of effects*, which means the same computed sum of squares is used to estimate both a main effect and an interaction effect, and the two or more designations given to the same sum of squares are called *aliases*. For example, in a 2 × 2 × 2 one-half fractional factorial design, the sum of squares for the main effect for factor A is the same as the B × C interaction sum of squares, the B sum of squares is the same as the A × C sum of squares, and the C sum of squares is the same as the A × B interaction sum of squares. Thus, the main effect for a given factor is completely confounded with the interaction effect of the other two factors.

For a four-way design, the main effects are completely confounded with the three-way interactions of the other three factors, and each of the two-way interactions are confounded with another two-way interaction. For a one-half fractional design, all sums of squares have two aliases, and for a one-fourth fractional design, all sums of squares have four aliases. Thus, fractional designs should be used only when higher order interactions are negligible or theoretically unimportant relative to the main effects and lower order interactions.

Fractional factorial designs can be used when a large number of factors are being studied and the goal is to to obtain information on each main effect and some interaction effects. For example, chemists often need to examine the effect of various conditions (e.g., pressure, temperature, time of reaction, or quality of ingredients) on the yield of a product. However, care must be exercised in setting up such designs because of the confounding effects due to aliasing of treatment and interaction sums of squares. In most chemical studies, interactions of the factors such as time of reaction and the temperature of reaction are well known. Thus, the interaction of these two factors would have to be kept free from aliases with main effects.

In general, fractional factorial designs are used when the higher order interactions can be considered to be zero or small relative to the main effects. Fractional factorial designs are most often used when the number of factors is four or more and when the number of treatment levels for each factor is the same. These designs are most useful for exploratory research in which followup experiments are done in a sequential series of experimentation. The computational procedures are more complex than for the complete factorial designs, and the interpretation is often complicated by confounding effects due to the fact that treatment sums of squares are aliased with interaction sums of squares. Fractional designs require the assumption of zero interaction among these confounding effects, which is often unrealistic in chemical studies. Thus, although these designs have great potential for some types of studies, they are limited in their usefulness for many types of chemical studies. Winer (*1*, pp. 676–684), Dayton (*3*, pp. 130–150), Kirk (*4*, pp. 385–421), and Edwards (*6*, pp. 254–257) provided more complete discussions and examples of using fractional factorial designs. Davies (*7*, pp. 440–494) provided a discussion of fractional factorial experiments in chemical studies.

References

1. Winer, B. J. *Statistical Principles in Experimental Design;* McGraw–Hill: New York, 1971.
2. Keppel, G. *Design and Analysis;* Prentice–Hall: Englewood Cliffs, NJ, 1973.
3. Dayton, C. M. *The Design of Educational Experiments;* McGraw–Hill: New York, 1970.
4. Kirk, R. E. *Experimental Design: Procedures for the Behavioral Sciences;* Brooks/Cole: Belmont, CA, 1968.
5. Scheffe', H. *The Analysis of Variance;* Wiley, New York, 1959.
6. Edwards, A. L. *Experimental Design in Psychological Research;* Holt, Rinehart & Winston: New York, 1968.
7. *The Design and Analysis of Industrial Experiments;* Davies, O. L., Ed.; Longman Group: London, 1978.

Chapter 14

Randomized-Blocks and Latin-Square Designs

Reducing experimental error is one of the principles for increasing the precision of any study. One way to accomplish this is to group observations into relatively homogeneous subgroups on the basis of factors related to the dependent variable. In chemical or industrial research, variations in reaction time, temperature, raw materials, machines, and work shifts, or different laboratory analysts can contribute to the error term when different treatment effects are compared. The *randomized-blocks design* permits the researcher evaluating treatment effects to minimize the effects of one of these extraneous factors by grouping cases into blocks so that the variability among the cases is less than the variability among the blocks. The essence of this design is to keep experimental errors within each group to a minimum, which is accomplished by assigning similar cases to the same group or block. The *Latin-square design* permits the researcher to control two or three extraneous factors in a study.

14.1 Randomized-Blocks Design

The randomized-blocks design is one of the most frequently used designs and is applicable to many scientific and industrial studies. For example, some machines used for testing the resistance of fabrics to wear can only test a few samples per run, and considerable variation can occur from one run to another. Thus, if only four samples can be tested during any one run, then four different fabrics can be tested during each run, and the variation from run to run can be accounted for in the analysis. Each run would be considered a *block*

1453–0/88/0333$08.00/1

in which each of the four fabrics would be randomly placed on the machine. If the wearing qualities of two different substances used as shoe soles were tested, then each person could be considered a block, the two materials could be randomly assigned to the left or right shoe, and the differences in wear due to different walking patterns from one individual to another could then be accounted for. This same analogy can be applied to testing tires for wear by randomly rotating four different tires per car and using each car as a block, thus accounting for the differential effects due to differences in cars. Or, if the effects of six treatments are to be compared on the basis of raw material, and the material is known to vary from batch to batch, then this source of variation can be accounted for by treating six lots from each batch differently. In chemical studies, the randomized-blocks design can also be used to reduce the effect of time trends by using the different time periods as blocks. For example, if four production methods are being compared in a manufacturing process that takes place over 7 days, then each month can serve as a block in which the sequence for testing each process is randomly assigned to a week, and the order of testing is changed randomly in each block. Thus, any effect on the manufacturing process due to time trends would be accounted for.

The main advantage of the randomized-blocks design is that the design has greater power than a completely randomized design because the effects of variation among cases can be reduced. The variation within each block is anticipated to be less than the variation expected if cases were randomly assigned to subgroups. This is the same assumption underlying related or matched cases as discussed in Chapter 10 for the related t test. The randomized-blocks design, also called the *treatment-by-levels design*, originated from agricultural experiments where strips of land were divided into neighboring plots or blocks. Contiguous plots of land were more likely to be similar than plots from different locations in the same field with respect to general soil condition, moisture, sunshine, and other factors affecting yield. Thus, by grouping adjacent plots into blocks, differences important to yield were controlled in the experiments.

The randomized-blocks design can be used in many industrial studies in which any number of treatments can be compared with as many blocks as desired. The analysis is similar to that used in factorial designs. The increased precision of this design compared to the completely randomized design results from partitioning the total sum of squares into three components (treatment, blocks, and blocks by treatment or error) compared to only two components in the random design (treatment and within or error). Thus, the error term used in the analysis will be smaller because more of the total variance has been accounted for. The amount of reduction depends upon the amount of among-blocks variability and the relationship of the blocking variable to the criterion variable in the study. Selection of a blocking variable should be made on the basis of its relationship to the criterion variable because, in general, the stronger this relationship, the greater the power of the randomized-blocks design compared to the completely randomized design and the greater the homogeneity of cases within blocks compared to the variability among blocks.

The blocking variable can be either discrete or continuous. If it is continuous, categories must be established to form levels for the blocking variable. The number of levels must be chosen with two considerations in mind: (1) the precision gained from the reduction in variability of cases within each block, and (2) the loss of degrees of freedom for the error term. As will be shown, the number of degrees of freedom for the error term in the randomized-blocks design is $B - 1$ fewer than in the completely randomized design, where B is the number of levels or blocks. Thus, a study comparing four treatments by using five cases per treatment would have $20 - 4 = 16$ degrees of freedom for the error term in the completely random design but only 12 degrees of freedom for the randomized-blocks design if five blocks were set up. An optimal number of blocks should be set on the basis of three factors:

1. the cases available,
2. the number of treatments, and
3. the degree of correlation between the blocking variable and the criterion variable.

Generally, the number of levels can increase as the number of subjects increases, as the correlation between the blocking and criterion variable increases, or for smaller numbers of treatments. Feldt (*1*) provided a table of the optimal number of levels for various combinations of the three factors. Feldt's table indicates that when the correlation between the block variable and the criterion variable is only around $r = 0.20$, then a minimum of five or six subjects per treatment-level combination is needed. When $r = 0.60$ or more, three subjects per treatment-level combination are needed when two treatments are to be compared. If there are four or five treatments to be compared, then one could have a minimum of two subjects per treatment-level condition. For example, with $r = 0.60$ and two treatments to be compared, if 30 subjects were available, then five levels of the blocking variable could be set with three subjects per treatment-level combination.

The major assumption that has to be made when using the randomized-blocks design is that the block and treatment effects are additive (i.e., the cases within each block have the same trend with respect to the treatments). In other words, no significant block-by-treatment interaction occurs. This means that, although the general level of the results may be different for the different blocks, the effects of the treatments are generally similar for all blocks. In practice, the interactions, if they exist, are negligible compared to the treatment effects. Because the interaction mean square is used as the error term, if the interaction effect is large, then the resulting F ratios will be too conservative and the results may be misleading. For many studies, such as testing positions on a machine, the positions on the machine can be assumed to have an effect independent of the treatment because different materials can be randomly assigned to positions. However, for chemical studies, concentration levels, catalysts, temperature, and pressure levels are factors that often interact. Thus, care must be taken when setting up randomized-blocks designs so that an interacting factor is not selected as a blocking variable. The randomized-blocks design is most useful when the various sources of error can be assumed to act independently, such as differences in testing machines, time of day, position of testing, and run-to-run variations.

14.1.1 Example of a Randomized-Blocks Design

The resistance to corrosion of four aluminum alloys in a chemical plant atmosphere is an example of a study where the randomized-blocks design can be effectively used. Five sites in a factory were chosen on the basis of different environmental conditions. At each site, a plate made from each of the four alloys was exposed for 6 months. Thus, each site served as a block, and within each block, the four plates made from different alloys were placed in a random order. Thus, the variation due to different sites was eliminated because each plate was tested at each site. The null hypothesis of major interest is that no significant differences occur among the four alloys with regard to resistance to corrosion. The 0.01 level could be used to test this hypothesis.

Hypothetical results of metallurgical tests to determine the amount of corrosion for each of the 20 plates are given in Table 14.1. The randomized-blocks design can be perceived as a basic two-way factorial design, and the computational procedures are essentially the same. The total sum of squares is partitioned into three sources: treatment, blocks, and blocks by treatment. The equation for the treatment sum of squares (SS_{TR}) is

$$SS_{TR} = \sum^{k} \left(\frac{TR^2}{n_{TR}} \right) - \frac{(\Sigma X)^2}{n_T} \qquad (14.1)$$

where k is number of treatments, TR is the sum for each treatment, ΣX is the sum of all observations, n_{TR} is the number of cases in each treatment group, and n_T is the total number of cases included in the study.

The equation for computing the sum of squares for blocks (SS_B) is

$$SS_B = \sum^{R} \left(\frac{(B^2)}{n_b} \right) - \frac{(\Sigma X)^2}{n_T} \qquad (14.2)$$

where R is the number of blocks, B is the sum for each block, n_b is the number of cases in each block, and ΣX and n_T are as defined for equation 14.1.

The total sum of squares (SS_T) is computed by

$$SS_T = \Sigma X^2 - \frac{(\Sigma X)^2}{n_T} \qquad (14.3)$$

Table 14.1. Corrosion Amount for Four Alloys at Five Sites in a Factory

Site	Alloy A	Alloy B	Alloy C	Alloy D	Sums
1	8	10	11	11	40
2	9	8	11	10	38
3	7	9	9	10	35
4	6	8	10	9	33
5	7	6	9	9	31
Sums	37	41	50	49	177

where X is the observed value for each case or observation, and n_T is the total number of observations.

The sum of squares for blocks by treatment ($SS_{B \times TR}$) can be computed as

$$SS_{B \times TR} = SS_T - SS_{TR} - SS_B \tag{14.4}$$

For the data in Table 14.1, the sums of squares are as follows:

$$SS_T = 8^2 + 9^2 + 7^2 \ldots + 9^2 + 9^2 - \frac{177^2}{20} = 44.55$$

$$SS_{TR} = \left(\frac{37^2}{5} + \frac{41^2}{5} + \frac{50^2}{5} + \frac{49^2}{5}\right) - \frac{177^2}{20} = 23.75$$

$$SS_B = \left(\frac{40^2}{4} + \frac{38^2}{4} + \frac{35^2}{4} + \frac{33^2}{4} + \frac{31^2}{4}\right) - \frac{177^2}{20} = 13.30$$

$$SS_{B \times TR} = 44.55 - 23.75 - 13.30 = 7.50$$

The computed sums of squares are presented in an analysis of variance table as in Table 14.2. The number of degrees of freedom is computed as follows: the total degrees of freedom (df_T) equals the total number of cases minus 1, or $20 - 1 = 19$; the treatment degrees of freedom (df_{TR}) equals the number of treatments minus 1, or $4 - 1 = 3$; the blocks degrees of freedom (df_B) equals the number of blocks minus 1, or $5 - 1 = 4$; and $df_{B \times TR} = (df_{TR})(df_B) = (4 - 1)(5 - 1) = 12$. The sums of squares are divided by their respective degrees of freedom to obtain the mean squares. Because there is only one observation per cell, there is no estimate of an error mean square and no separate within-cell variance to pool. Hence, no need arises to test for homogeneity of within-cell variance as is the case with the completely random analysis. The F ratios for treatment and block effects are computed by dividing the respective mean squares by the treatment-by-levels interaction mean square. For the fixed-effects models (i.e., when both the treatment and blocking variables are fixed), the use of the interaction mean square results in a conservative test of the null hypothesis. Thus, given a true null hypothesis, fewer hypotheses would be rejected. However, if the null hypothesis is rejected, the actual level for the test is less than the level of significance set.

For the random-effects model, the interaction mean square is appropriate, and F is interpreted without qualification. For the mixed-effects model, the test for treatments is valid but not for blocks if the levels for treatments are fixed. Because the main concern in the example presented is to test the hypothesis of no significant treatment effect, no problem occurs unless the block effect was of interest. Fortunately, the fixed-treatment effect is most often used when there is only one case per cell.

For the analysis in Table 14.2, with 3 and 12 degrees of freedom, the *F* ratio for treatments is significant at the 0.01 level of significance. Thus, the null hypothesis is rejected, and the conclusion is that the differential treatment effects are significant. With a significant *F* ratio, post hoc tests such as the Tukey or Scheffe' test could be run to determine which treatments are significantly different from each other. Or a priori tests could be made regardless of the outcome of the overall *F* ratio. In either case, the proper error term to use for these analyses is the interaction mean square. The procedures for both a priori and post hoc tests are the same as those for factorial designs.

The *F* ratio for the block effect is also significant at the 0.05 level, a result indicating that blocking has been successful in reducing the amount of error variance.

If there is more than one case per cell, the randomized-blocks design is set up and computed exactly like a two-way factorial design. Because there will then be within-cell variance, this variance within each cell can be pooled to estimate experimental error. Therefore, all main effects as well as interaction effects can be estimated as in a regular factorial design. For such studies, one can also block on two variables, and a situation similar to a three-way design results.

14.1.2 Relative Efficiency and Strength of Association

As pointed out by Dayton (*2*) and Kirk (*3*), the relative efficiency or power of one design over a completely randomized design can be calculated and used to determine if the more sophisticated design

Table 14.2. Analysis of Variance for the Data in Table 14.1

Source	SS	df	ms	*F*
Treatments	23.75	3	7.92	12.57
Blocks	13.30	4	3.33	5.29
B × TR	7.50	12	0.63	
Total	44.55	19		

is worth the extra effort. This test is simple to do and easy to interpret. Essentially, the test is a ratio of the error variances and is based on Fisher's (*4*) concept of the amount of variance accounted for by the two designs. In general, the equation, corrected for degrees of freedom, is

$$\mathrm{RE} = \frac{(\mathrm{df}_1 + 1)(\mathrm{df}_2 + 3)(\mathrm{ms}_2)}{(\mathrm{df}_2 + 1)(\mathrm{df}_1 + 3)(\mathrm{ms}_1)} \times 100 \qquad (14.5)$$

where RE is the relative efficiency; ms_1 and df_1 are the mean square and degrees of freedom, respectively, for the block design; and ms_2 and df_2 are for the one-way analysis of variance. Thus, for the randomized-blocks design, the randomized-blocks analysis, and the simple one-way analysis can be made by disregarding the blocks, and then the ratio of the mean squares could be compared. Or, the mean square for the completely randomized design (ms_w) can be computed by

$$\mathrm{ms}_w = \frac{\mathrm{SS}_B + \mathrm{SS}_{B\times \mathrm{TR}}}{\mathrm{df}_B + \mathrm{df}_{B\times \mathrm{TR}}} \qquad (14.6)$$

where df_B and $\mathrm{df}_{B\times \mathrm{TR}}$ are the degrees of freedom for blocks and interaction, respectively; and SS_B and $\mathrm{SS}_{B\times \mathrm{TR}}$ are the mean squares for blocks and interaction, respectively. The values of df_B, $\mathrm{df}_{B\times TR}$, SS_B, and $\mathrm{SS}_{B\times \mathrm{TR}}$ are taken from the summary analysis of variance table. For the data in Table 14.2,

$$\mathrm{ms}_w = \frac{13.3 + 7.5}{4 + 12} = 1.3$$

and

$$\mathrm{RE} = \frac{13(19)(1.3)}{17(15)(0.63)} \times 100 = 199.87$$

A relative efficiency greater than 100% indicates that the randomized-blocks design is more efficient than the completely randomized design. Thus, for the example, the randomized-blocks design is 200% more efficient than the completely randomized one-way analysis of variance would have been.

As with the one-way analysis of variance, omega squared (ω^2) provides a measure of the association or relationship between the treatment and criterion variables. For the fixed-effects randomized-blocks design,

$$\omega^2 = \frac{SS_{TR} - (k - 1)ms_{B \times TR}}{SS_T + ms_{B \times TR}} \qquad (14.7)$$

where k is the number of treatments.

For the data in Table 14.2,

$$\omega^2 = \frac{23.75 - (4 - 1)(0.63)}{44.55 + 0.63} = 0.48$$

The results indicate that 48% of the variance in the criterion or dependent variable is accounted for by the independent variable.

14.2 Incomplete-Block Designs

As with factorial designs, if the number of treatment-by-levels combinations is relatively large, the number of cases required for each block may not be obtained. For such situations, an incomplete-block design may be used in which all treatments are not replicated within each block. Several types of incomplete designs can be used, but because the computational procedures are complex, they will not be covered in this book. Dayton (*2*) and Natrella (*5*) discussed such designs.

One type of incomplete-block design can be used when the experimental error is small and when the cost of running cases is high. Natrella (*5*) and Cochran and Cox (*6*) described such a *chain-block design*. For such a design, only a few more observations are needed than the number of treatments. However, in this design, differences in treatment effects should be substantially larger than experimental error. In this design, some treatments are only observed once, whereas others are observed twice.

14.3 Latin Squares

Whereas the randomized-blocks design permits a researcher to minimize the effects of one source of error in a study, the Latin square extends the basic principle to two noninteracting sources of error. The treatments in a Latin-square design are grouped into replicates in two ways, which correspond to rows and columns in the factorial design layout. Thus, the Latin-square design provides the researcher more opportunity than the randomized-blocks design to control for errors in a study. The design requires that each source of error be divided into an equal number of blocks or levels, and the number of treatment levels must be equal to the number of levels in each row and column. In the Latin-square design, the rows represent the blocks for one control variable, the columns represent the levels for the second control variable, and the letters within the matrix represent the different treatments. As with the randomized-blocks design, the major assumption is that the control variables cannot interact with each other or the dependent variable.

The Latin-square concept originated from an ancient puzzle that dealt with the number of different ways that Latin letters could be arranged in a square table so that each letter appeared once and only once in each row and column. If three treatments, three rows, and three columns are used, then there are 12 different ways that the three letters can be placed in the matrix. A square is said to be in standard form if the first row and first column are in numerical or alphabetical order, such as those in Figure 14.1.

In each matrix, each letter appears once and only once in each row and each column. In practice, this situation means that each treatment occurs at each level for both the row and column variables, and that $n_R = n_C = n_T$. The number of different matrices becomes very large as the number of levels increases. For three levels of each variable, one standard square and 12 possible arrangements of this standard form occur. For four levels, 144 arrangements are possible for each of the four standard squares, with a total of 576 possible arrangements for the four standard squares. For a 5×5 matrix, 161,280 arrangements of the 25 standard squares are possible, 812,851,200 arrangements are possible for the 9408 standard squares in a 6×6 matrix, and 16,942,000 standard-square arrangements are possible for a 7×7 matrix. Sets of Latin-square matrices were given by Natrella (*5*), Fisher and Yates (*7*), and Federer (*8*). These sets can also be made up by the researcher with a table of random numbers. Theoretically, the researcher should randomly select a Latin-square from the population of squares available. However, this is not practical because of the large populations of possible squares. As outlined by Fisher and Yates (*7*), the following procedure in selecting a random-order Latin square is usually acceptable for most applications. The general procedure is to randomly select one of the standard squares and then randomize the order of rows and columns. For example, to select a 5 × 5 Latin square, three sets of five random digits (1–5) would be drawn from a table of random numbers. For example, the three sets may be as follows: 2, 1, 4, 3, and 5; 4, 5, 1, 3, and 2; and 3, 4, 1, 5, and 2. If the first set is used, 2 is the first digit, and the second 5 × 5 Latin square in Figure 14.1 can be selected.

A B C D E
B C D E A
C D E A B
D E A B C
E A B C D

If the second set of five random numbers is used, the rows are then arranged in the order 4, 5, 1, 3, and 2 so that the matrix becomes

D E A B C
E A B C D
A B C D E
C D E A B
B C D E A

3 x 3:

ABC
BCA
CAB

4 x 4:

ABCD	ABCD	ABCD	ABCD
BADC	BCDA	BDAC	BADC
CDBA	CDAB	CADB	CDAB
DCAB	DABC	DCBA	DCBA

5 x 5:

ABCDE	ABCDE
BAECD	BCDEA
CDAEB	CDEAB
DEBAC	DEABC
ECDBA	EABCD

6 x 6:

ABCDEF
BFDCAE
CDEFBA
DAFECB
ECABFD
FEBADC

Figure 14.1. Examples of standard Latin squares.

Finally, if the third set of random numbers is used, the columns are arranged as 3, 4, 1, 5, and 2, which yields the final Latin square.

A B D C E
B C E D A
C D A E B
E A C B D
D E B A C

The Latin-square design also had its origins in agricultural research, where the two sources of error to be controlled were the two directions of a field, and the "square" was actually a square of ground. However, the use of the Latin-square design has been expanded to applications where there are two sources of error that might affect the experimental results. For example, some wear tester can test four materials during any run. Thus, each run could be considered a block, and four materials tested during each run could be randomly assigned to positions on the machine. If variations occur from one run to another, these could also be blocked so that the row blocks might identify the different runs, the column blocks might denote the different positions on the machine, and the letters within the matrix might represent the four different materials to be tested. As another example, suppose five fertilizers are to be compared, and the soil and other conditions show from experience that fertility depends on wind direction, direction of sunlight, and rainfall. Thus, two sets of blocks could be set up in both directions, and then the five treatments could be randomly assigned to these blocks through the use of the Latin-square design. Each block would be divided into five units. The Latin-square design might also be used to test the output of six machines that yield differences from one time period to another and from operator to operator. Six time periods could be set up as six blocks, and six operators could be selected for the study. Treatments could then be randomly assigned to cells within this matrix by using the Latin-square design.

14.3.1 Example of a 5 × 5 Latin Square

When investigating batch chemical processes, several factors affect the yield of the chemical. These factors include variation from one batch to another and variations in temperature, raw materials, and agitation rates. Suppose that a chemical engineer wants to determine the effect on yield due to variation in processing temperature. Variation from one batch to the next produces variations in yield as do variations in the raw material due to impurities, texture, and consistency within the material. If five temperature levels are used, one way of setting up the experiment would be to design a three-way factorial study with five treatment levels and as many levels of the other two factors as are thought to exist. However, the variation between batches or between lots of raw material might not be feasible to control and categorize. The main effects of these two factors may not be important to assess, but variation in these two factors may be important to control. Thus, the Latin-square design would be the most efficient design to use.

The study could be set up as the 5 × 5 Latin-square design in Section 14.3. The columns could represent different lots of raw material for which there is sufficient material for five batches of the chemical product. The rows represent the different batches that are analyzeda consecutive time period, and the letters within the square denote the five different temperature levels. Hypothetical data for yield of the chemical product are given in Table 14.3. Within the table are the yields for each of the temperature levels (A, B, C, D, and E). The null hypothesis to be tested is that no differences in yield of the chemical product are due to changes in processing temperature, or $\mu_A = \mu_B = \mu_C = \mu_D = \mu_E$. The significance level could be set up at any level, but for this study, the 0.05 level is selected.

Table 14.3. Yield of Chemical Product for Various Lots of Raw Temperature Levels

Batch	Lot 3 (Temp.)	Lot 4 (Temp.)	Lot 1 (Temp.)	Lot 5 (Temp.)	Lot 2 (Temp.)	Sum
	(A)	(B)	(D)	(C)	(E)	
4	12	10	8	8	7	45
	(B)	(C)	(E)	(D)	(A)	
5	10	9	4	7	12	42
	(C)	(D)	(A)	(E)	(B)	
1	7	7	11	5	9	39
	(E)	(A)	(C)	(B)	(D)	
3	5	11	6	8	6	36
	(D)	(E)	(B)	(A)	(C)	
2	6	6	9	9	7	37
Sum	40	43	38	37	41	199

NOTE: The letters A, B, C, D, and E represent different temperature levels.

The sums of squares necessary for the Latin-square analysis are computed by using the same procedures as the randomized-blocks design, with the additional partitioning of the total sum of squares into four rather than three sources. The total sum of squares is computed by first obtaining the sum of all values and the sum of all squared values, and then using the equation

$$SS_T = \Sigma X^2 - \frac{(\Sigma X)^2}{n_T} \tag{14.8}$$

where X is the observed value for each case, and n_T is the total number of observations. The sum of squares for rows (SS_r) is computed by

$$SS_r = \sum^{R} \left(\frac{(\Sigma X_r)^2}{n_r} \right) - \frac{(\Sigma X_T)^2}{n_T} \tag{14.9}$$

where R is the number of rows, ΣX_r is the sum of the values for each row, ΣX_T is the sum of all values within the matrix, n_r is the number of levels for each row, and n_T is the total number of observations.

The sum of squares for columns (SS_c) is

$$SS_c = \sum^{C} \left(\frac{(\Sigma X_c)^2}{n_c} \right) - \frac{(\Sigma X_T)^2}{n_T} \tag{14.10}$$

where C is the number of columns, ΣX_c is the sum of the values for each column, n_c is the number of levels for each column, and ΣX_T and n_T are defined as in equation 14.9.

The total sum of squares for treatment (SS_{TR}) is obtained by first obtaining the sum of values for each treatment. Then the sum of squares is computed by

$$SS_{TR} = \sum^{TR} \left(\frac{(\Sigma X_{TR})^2}{n_{TR}} \right) - \frac{(\Sigma X_T)^2}{n_T} \tag{14.11}$$

where TR is the number of treatments, ΣX_{TR} is the sum of values for each treatment, n_{TR} is the number of levels of the other factors that each treatment is investigated at, and ΣX_T and n_T are defined as in equation 14.9.

The error or residual sum of squares (SS_{error}) is computed by subtraction:

$$SS_{error} = SS_T - SS_r - SS_c - SS_{TR} \qquad (14.12)$$

For the hypothetical data in Table 14.3, the engineer would first obtain the sums of values for each treatment. These are obtained by adding the cell entries corresponding to each temperature level. For this data,

$$\Sigma X_A = 12 + 12 + 11 + 11 + 9 = 55$$

$$\Sigma X_B = 10 + 10 + 9 + 8 + 9 = 46$$

$$\Sigma X_C = 8 + 9 + 7 + 6 + 7 = 37$$

$$\Sigma X_D = 8 + 7 + 7 + 6 + 6 = 34$$

$$\Sigma X_E = 7 + 4 + 5 + 5 + 6 = 27$$

The necessary sums of squares obtained by using equations 14.8–14.12 are as follows:

$$SS_T = 1701 - \frac{199^2}{25} = 116.96$$

$$SS_r = \frac{45^2}{5} + \frac{42^2}{5} + \frac{39^2}{5} + \frac{36^2}{5} + \frac{37^2}{5} - \frac{199^2}{25} = 10.96$$

$$SS_c = \frac{40^2}{5} + \frac{43^2}{5} + \frac{38^2}{5} + \frac{37^2}{5} + \frac{41^2}{5} - \frac{199^2}{25} = 4.56$$

$$SS_{TR} = \frac{55^2}{5} + \frac{46^2}{5} + \frac{37^2}{5} + \frac{34^2}{5} + \frac{27^2}{5} - \frac{199^2}{25} = 94.96$$

$$SS_{error} = 116.96 - 10.96 - 4.56 - 94.96 = 6.48$$

These sums of squares are then entered into an analysis of variance table as in Table 14.4. The degrees of freedom are computed as follows: $df_T = n_T - 1$; $df_r = n_r - 1$; $df_c = n_c - 1$; $df_{TR} = n_{TR} - 1$; and for the error or residual, $df_{error} = (k - 1)(k - 2)$, where k is the number of levels for each dimension of the table. For the example, $df_{error} = (5 - 1)(5 - 2) = 12$. The mean squares are obtained by dividing each sum of squares by its appropriate degrees of freedom, and then each F ratio is obtained by dividing the three

Table 14.4. Analysis of Variance for the Data in Table 14.3

Source	SS	df	ms	*F*
Treatments	94.96	4	23.74	43.96
Rows	10.96	4	2.74	5.07
Columns	4.56	4	1.14	2.11
Error	6.48	12	0.54	
Total	116.96	24		

mean squares for treatments, rows, and columns by the error mean square. With 4 and 12 degrees of freedom, an $F = 3.26$ is needed for significance at the 0.05 level. The results presented in Table 14.4 indicate a significant treatment and row effect. Thus, the differences from batch to batch are significant, and blocking on this variable has been effective in reducing error variance, but the differences from lots of raw material (columns) have little effect on yield.

Post hoc tests could be run to determine which pairs of treatments are significantly different from each other, or a priori tests could be made. For this example, the Scheffe' procedure can be used to compute the critical range (CR_S) for treatment sums.

$$CR_S = \sqrt{(k - 1)(F_{df_{k-1,error}})(2)(n)(ms_{error})} \qquad (14.13)$$

where k is the number of levels of treatments, $F_{df_{k-1,error}}$ is the table value for the degrees of freedom for treatments and error, n is the number of replications or sample observations, and ms_{error} is obtained from the analysis of variance table. For the given data,

$$CR_S = \sqrt{4(3.26)(2)(5)(0.54)} = 8.39$$

Thus, any difference between treatment sums greater than 8.39 would be significantly different at the 0.05 level. The differences between each pair of treatment sums are presented in Table 14.5. Thus, treatment A is significantly different from all other treatments as is treatment B, treatment C is not significantly different from treatment D but is significantly different from treatment E, and the difference between treatments D and E is not significant.

14.3.2 Relative Efficiency

As with the randomized-blocks design, the relative efficiency or power of the Latin-square design the completely randomized and the randomized-blocks design can be computed. The general equation

Table 14.5. Difference between Treatment Sums

Treatment	A	B	C	D
B	9*			
C	18*	9*		
D	21*	12*	3	
E	28*	19*	10*	7

*These values are significantly different at the 0.05 level.

for comparing the Latin-square design with the simple one-way analysis of variance, with a correction for differences in degrees of freedom, is as follows:

$$\mathrm{RE} = \left(\frac{\mathrm{ms}_w}{\mathrm{ms}_{\mathrm{error}}}\right)\left(\frac{(k-1)(k-2)+1}{(k-1)(k-2)+3}\right)\left(\frac{k(k-1)+3}{k(k-1)+1}\right) \times 100 \tag{14.14}$$

where RE is the relative efficiency, and k is the number of levels of each dimension. The value of ms_w can be computed from the data for the Latin-square analysis by

$$\mathrm{ms}_w = \frac{\mathrm{SS}_r + \mathrm{SS}_c + \mathrm{SS}_{\mathrm{error}}}{\mathrm{df}_r + \mathrm{df}_c + \mathrm{df}_{\mathrm{error}}} \tag{14.15}$$

where SS and df values for rows, columns, and errors are taken from the analysis of variance table. For the data in Table 14.4, ms_w can be estimated as

$$\mathrm{ms}_w = \frac{10.96 + 4.56 + 6.48}{4 + 4 + 12} = 1.10$$

If the treatment values observed for treatments A–E were run as a one-way analysis of variance, then $\mathrm{ms}_w = 1.10$, as presented in Table 14.6. Thus, the relative efficiency could be computed by using estimated ms_w, or by running the analysis as a one-way analysis of variance and using the ms_w from this analysis.

For this problem, the relative efficiency would be

$$\mathrm{RE} = \left(\frac{1.10}{0.54}\right)\left(\frac{4(3)+1}{4(3)+3}\right)\left(\frac{5(4)+3}{5(4)+1}\right) \times 100$$

$$= 2.04 \times 0.87 \times 1.1 \times 100 = 195.23$$

Table 14.6. One-Way Analysis of Variance for the Data in Table 14.3

Source	SS	df	ms	F
Between	94.96	4	23.74	21.58
Within	22.00	20	1.10	
Total	116.96	24		

Therefore, the Latin-square analysis would have been 195% more efficient than the one-way analysis of variance.

The treatment sum of squares in Table 14.4 and the between sum of squares for the one-way analysis of variance in Table 14.6 are equal (i.e., both equal 94.96). However, the within sum of squares for the one-way analysis (22.00) is partitioned into three sources in the Latin-square design. Thus, if the blocking is effective, the error sum of squares for the Latin-square analysis is reduced. However, the degrees of freedom are also smaller for the Latin-square analysis.

The relative efficiency of the additional blocking variable for the Latin-square design can also be compared to the randomized-blocks design that uses only one blocking variable. The general equation for this comparison is

$$\mathrm{RE} = \left(\frac{\mathrm{ms}_{B\times \mathrm{TR}}}{\mathrm{ms}_{\mathrm{error}}}\right)\left(\frac{(k-1)(k-2)+1}{(k-1)(k-2)+3}\right)\left(\frac{(k-1)^2+3}{(k-1)^2+1}\right) \times 100 \qquad (14.16)$$

The value of $\mathrm{ms}_{B\times \mathrm{TR}}$ can be estimated by

$$\mathrm{ms}_{B\times \mathrm{TR}} = \frac{\mathrm{ms}_R + (k-1)\mathrm{ms}_{\mathrm{error}}}{k} \qquad (14.17)$$

where ms_R is the mean square for the added dimension from the Latin-square analysis, either rows or columns; $\mathrm{ms}_{\mathrm{error}}$ is from the Latin-square analysis; and k is the number of levels of each dimension. For the data in Table 14.4, if the added dimension was rows, then

$$\mathrm{ms}_{B\times \mathrm{TR}} = \frac{2.74 + 4(0.54)}{5} = 0.98$$

The relative efficiency would then be

$$\text{RE} = \left(\frac{0.98}{0.54}\right)\left(\frac{4(3) + 1}{4(3) + 3}\right)\left(\frac{(4)^2 + 3}{(4)^2 + 1}\right) \times 100$$

$$= 1.81 \times 0.87 \times 1.12 \times 100 = 176.37$$

Thus, the Latin-square design is 176% more efficient than the randomized-blocks design. If the added dimension was columns, then the relative efficiency would be

$$\text{ms}_{B \times \text{TR}} = \frac{1.14 + 4(0.54)}{5} = 0.66$$

$$\text{RE} = \left(\frac{0.66}{0.54}\right)\left(\frac{4(3) + 1}{4(3) + 3}\right)\left(\frac{4^2 + 3}{4^2 + 1}\right) \times 100$$

$$= 1.22 \times 0.87 \times 1.12 \times 100 = 119$$

Thus, by adding columns as the second blocking variable, the Latin-square design is only slightly more efficient than the randomized-blocks design using only rows as the blocking variable. The relative efficiencies can be estimated from the relative sizes of the mean squares for each source.

14.3.3 Latin Squares with More Than One Observation per Cell

A Latin-square design can be set up with more than one observation per cell. The main reason for using more observations is to increase the precision of the study by reducing the error term. Essentially, the design is a fractional replication of the complete factorial design. For example, consider a $3 \times 3 \times 3$ factorial design, which would require 27 treatment combinations. If one was primarily interested in treatment effects and the influence of two control factors, then the 3×3 Latin square requiring only nine treatment combinations would be efficient to use. However, if only one observation per cell is obtained, then df = 2. To increase the number of degrees of freedom for the error term, more than one observation per cell can be used. In this type of design, the total sum of squares is partitioned into five sums of squares: one for each of the three dimensions, a residual sum of squares, and a within-cells sum of squares. The residual sum of squares contains part of the variation associated with the main effects plus interaction effects if these are present. If there is zero interaction, then the residual mean square should be equivalent to or less than the mean square for within cells. For this analysis, the number of degrees of freedom for each of the main factors is $k - 1$, where k is the number of levels of each factor. The number of degrees of freedom for the residual sum of squares is $(k - 1)(k - 2)$, and for the within sum of squares, $k^2(n - 1)$.

Suppose that three testing procedures are used by a corporation to measure the concentration of a chemical in a liquid with high-performance liquid chromatography (HPLC), and that variations occur in these measures among the three procedures. Some of this variation might be due to different operators as well as day-to-day variation in temperature, batches, and other factors. A study can be set up so that three random samples from each of four batches are analyzed by each testing procedure by different operators on different days. Thus, four samples are analyzed for each procedure by each of the three groups of operators on three different days. A Latin square can be set up as in Table 14.7 so that columns represent the different operators. Thus, three analyses of the same four batches of liquid are analyzed, and differences among procedures, operators, and day-to-day variation can be analyzed. Operator I would analyze four samples by using procedure A on the first day, four more samples by using procedure B on the second day, and four more samples by using procedure C on the third day. Operator II would also analyze four samples on each day by using procedures B, C, and A, in that order. Operator III would use procedures C, A, and B for the first, second, and third days, respectively.

The results of the four analyses could be presented as in Table 14.7, which are fictitious data for computational purposes but would represent the level of concentration of the chemical for each sample in the study. The sums for procedure A would be $23 + 26 + 23 = 72$, the sums for procedure B would be $22 + 27 + 27 = 76$, and the sums for procedure C would be $32 + 33 + 30 = 94$. The sums of squares for total, rows, columns, and treatments are computed from equations 14.8–14.11, respectively, by using the marginal and treatment sums that are based upon the sums of each set of four observations and the number of observations on which each sum is based. Two additional sums of squares are needed for this analysis. The residual sum of squares is computed by subtracting the main

Table 14.7. Dioxamine Levels in a Liquid Analyzed by Three Operators on Three Different Days

Statistic	Operator I	Operator II	Operator III	Sum
		Day 1		
n				
1	5	7	9	
2	7	5	7	
3	6	6	7	
4	5	4	8	
ΣX_{cell}	23	22	31	76
		Day 2		
n				
1	8	9	8	
2	6	7	5	
3	8	8	7	
4	5	9	6	
ΣX_{cell}	27	33	26	86
		Day 3		
n				
1	8	7	8	
2	7	5	5	
3	8	6	7	
4	7	5	7	
ΣX_{cell}	30	23	27	80
Sum	80	78	84	242

effects and correction term from the sum of squares for cells. The residual sum of squares (SS_{res}) is computed by

$$SS_{res} = \sum^{C} \left(\frac{(\Sigma X_{cell})^2}{n_{cell}} \right) - \frac{(\Sigma X_T)^2}{n_T} - SS_r - SS_c - SS_{TR} \quad (14.18)$$

where ΣX_{cell} is the sum of the values in each cell; ΣX_T is the sum of all values in the matrix; n_{cell} is the number of observations in each cell; n_T is the total number of observations; and SS_r, SS_c, and SS_{TR} are the sums of squares for rows, columns, and treatments, respectively. The sum of squares for error (SS_{error}) is the within sum of squares for each cell and can be computed for each cell from equation 14.8, which is modified for the sums and number of observations in each cell, and then summed over all cells. Or, SS_{error} can be computed by subtraction:

$$SS_{error} = SS_T - SS_r - SS_c - SS_{TR} - SS_{res} \tag{14.19}$$

For the data in Table 14.7,

$$SS_T = 1690 - \frac{242^2}{36} = 63.22$$

$$SS_r = \frac{76^2}{12} + \frac{86^2}{12} + \frac{80^2}{12} - \frac{242^2}{36} = 4.21$$

$$SS_c = \frac{80^2}{12} + \frac{78^2}{12} + \frac{84^2}{12} - \frac{242^2}{36} = 1.55$$

$$SS_{TR} = \frac{72^2}{12} + \frac{76^2}{12} + \frac{94^2}{12} - \frac{242^2}{36} = 22.88$$

$$SS_{res} = \frac{23^2}{4} + \frac{22^2}{4} + \frac{31^2}{4} + \frac{27^2}{4} + \frac{33^2}{4} + \frac{26^2}{4} + \frac{30^2}{4}$$

$$+ \frac{23^2}{4} + \frac{27^2}{4} - \frac{242^2}{36} - 4.21 - 1.55 - 22.88 = 1.08$$

$$SS_{error} = 63.22 - 4.21 - 1.55 - 22.88 - 1.08 = 33.50$$

As a computational check, the sum of the nine sums of squares computed for each cell also totals 33.50. These sums of squares are then summarized in an analysis of variance table such as Table 14.8. The number of degrees of freedom for each main effect equals the number of dimensions minus 1, or $(k - 1)$; the total degrees of freedom equals the total number of observations minus 1; the residual degrees of freedom equals $(k - 1)(k - 2)$; and the within or error

Table 14.8. Analysis of Variance for the Data in Table 14.7

Source	SS	df	ms	F
Treatment	22.88	2	11.44	9.23
Rows (days)	4.21	2	2.11	1.70
Columns	1.55	2	0.78	0.63
Residual	1.08	2	0.54	0.44
Error	33.50	27	1.24	
Total	63.22	35		

degrees of freedom equals $k^2(n_{\text{cell}} - 1)$, where k is the number of treatments, number of rows, or number of columns, and n_{cell} is the number of observations in each cell. Mean squares are obtained by dividing each sum of squares by its degrees of freedom, and F ratios are computed by dividing each mean square by the error mean square because each of these represent fixed dimensions. With 2 and 27 degrees of freedom, an $F = 5.49$ is required at the 0.01 level of significance. Thus, only the F ratio for treatment or testing procedure is significant. Thus, the major source of variation is due to differences in testing procedures, not to differences due to operators or day-by-day variations. The F ratio for the residual term is a partial check on the degree of first-order interactions (i.e., $\text{TR} \times R$, $\text{TR} \times C$, and $R \times C$ interactions). If this F ratio is significant, then interactions are present, and because this term is confounded with main effects, the results would be questionable. Because this F ratio is not significant, then the interpretation of the three main effects can be interpreted as valid estimates of these sources of variance.

The analysis can be continued to determine where the significant differences among testing procedures originate. Post hoc tests such as the Scheffe' test can be used to determine which pairs of machines are significantly different. By using the Scheffe' test, the critical difference for treatment sums (CR_S) is computed by

$$\text{CR}_S = \sqrt{(k - 1)F_{\text{df}_{\text{T,error}}}}\ \sqrt{2(n)(\text{ms}_{\text{error}})} \qquad (14.20)$$

$$= \sqrt{(2)(5.49)}\ \sqrt{2(4)(1.24)} = 10.43$$

From the sums for treatments given in Table 14.7, testing procedure C is significantly different from procedures A and B, but these two procedures are not significantly different from each other.

14.3.4 Greco-Latin-Square Designs, Hyper-Greco-Latin-Square Designs, and Youden-Square Analysis

The basic Latin-square design can be expanded to eliminate more than two sources of extraneous error by adding dimensions to the design. As pointed out by Dayton (*2*), a Latin-square design is always a $1/k$ replicate of a k^3 or three-way factorial design, and if one more dimension is added, the design then becomes a $1/k$ replicate of a k^4 or four-way factorial design. To represent a Greco-Latin-squares design, Greek letters or numbers are added to each letter in the basic Latin-square matrix with the restriction that the added letter or number occurs once and only once in each row and column, and once and only once with each original letter. The computation of a Greco-Latin-square can be quite tedious, especially for matrices of 5 × 5 or more, but examples of such matrices are presented in various advanced statistics textbooks (*9*). The design is created by imposing one Latin square, such as

A	B	C	D
B	A	D	C
C	D	A	B
D	C	B	A

on another Latin square, such as

1	2	3	4
4	3	2	1
2	1	4	3
3	4	1	2

The resulting design is

A1	B2	C3	D4
B4	A3	D2	C1
C2	D1	A4	B3
D3	C4	B1	A2

Each letter occurs once and only once in each row, in each column, and with each number, and the same is true for the numbers. For example, the three sources of error to be controlled might be plant sites for columns, machine positions for rows, different operators as numbers, and different treatments as letters.

The computations are the same as for the Latin-square design except one more sum of squares has to be computed, which in this notation would be the numbers within the matrix. The sums of values for computing this sum of squares are obtained in the same way as for the treatments or letters in the Latin-square design. The degrees of freedom for each of the four dimensions would be $k - 1$, and for the error or residual sum of squares, $(k - 1)(k - 3)$. For the

4×4 Greco-Latin square, the degrees of freedom would be shown as:

Sum of Squares	Degrees of freedom
columns	$k - 1 = 4 - 1 = 3$
rows	$k - 1 = 4 - 1 = 3$
numbers	$k - 1 = 4 - 1 = 3$
letters	$k - 1 = 4 - 1 = 3$
error	$(k - 1)(k - 3) = 3 \times 1 = 3$
total	15

Thus, for a 4×4 Greco-Latin square, the error sum of squares would have df = 3, a 5×5 design would have df = 8, a 6×6 design would have df = 15, and a 7×7 design would have df = 24. This means that unless the blocking on the three variables is effective and the measurements and other factors are fairly reliable, the mean square for error will be large relative to treatment effects and will require large treatment differences to be significant. The Greco-Latin design can be expanded by adding a fourth error factor to be controlled. This situation is the hyper-Greco-Latin-square design. In each cell, the added dimension is denoted by a Greek letter in addition to the English letter and number. For example, for the Greco-Latin-square design just presented, a fourth source of error, different times of the day, could be included, which could be represented by Greek letters. The matrix might then be as follows:

$$\begin{matrix} \phi A1 & \chi B2 & \psi C3 & \omega D4 \\ \omega B4 & \psi A3 & \chi D2 & \phi C1 \\ \chi C2 & \phi D1 & \psi A4 & \psi B3 \\ \psi D3 & \omega C4 & \phi B1 & \chi A2 \end{matrix}$$

The degrees of freedom for each of the five dimensions would be $k - 1$ or $4 - 1 = 3$, and for the error term, $(k - 1)(k - 3) = 3 \times 1 = 3$. However, for this 4×4 design, this result would yield a total of 18 degrees of freedom, which is three more than the total degrees of freedom available. The hyper-Greco-Latin-square design can only be used for designs with more than five levels (i.e., where k 5). For this design, the degrees of freedom for each dimension is $k - 1$; for the error term, the degrees of freedom is computed as $(k - 1)(k - 4)$.

A special type of Latin-square design that permits more flexibility in design is the *Youden square*, first presented by Youden (*10*) for use in agricultural research. This design has the same number of rows and treatments, but the number of columns is always one or more columns fewer than the number of treatments or rows. For example, a 4×4 Latin-square design has to be set up so that there are four columns, four rows, and four treatments, for a total of 16 observations. A Youden-square analysis could be set up using only three columns that would require only 12 observations, yet differences among the four treatments would be tested. The computational procedures are similar but more complicated than for the Latin-square design. The reader is referred to Natrella (*5*), Cochran and Cox (*6*), or Kirk (*3*) for examples of Youden-square arrangements and computational procedures.

14.3.4 Advantages and Limitations of Latin-Square Designs

The major advantage of the Latin-square design (as well as the randomized-blocks design) is that fewer observations are needed to compare treatment effects while controlling for two or more extraneous factors (or one factor in the randomized – blocks design). For example, consider a $3 \times 3 \times 3$ factorial design that requires 27 groups of observations, each group receiving one of the 27 treatment combinations. If five cases per group are required for reliability, then a total of 135 observations would be necessary. By setting up the design as an incomplete factorial design or a Latin-square design, only nine observations would be needed to obtain estimates of row, column, and treatment effects. For a $4 \times 4 \times 4$ design, a total of 64 treatment combinations would be needed for the complete factorial design, but only 16 observations as a Latin-square design. If seven treatments are to be compared at each of seven levels of two other factors, a total of 343 treatment combinations would have to be set up for the complete $7 \times 7 \times 7$ factorial design. Estimates of each of these three factors can be obtained from a 7×7 Latin-square design that requires only 49 observations. The Latin-square design is always a $1/k$ replicate of a k^3 series complete factorial design.

Because of this efficiency, the Latin-square design has great appeal to experimenters in chemistry as well as many other areas. Although the time savings achieved can be considerable, especially when observations are expensive, some problem areas must always be considered. The main assumption that has to be made when using the Latin-square design is that the factors must affect results independently of each other. That is, no significant interaction occurs among the factors. In practice, this usually means that the interactions among the factors are small or negligible compared to the effects that are important to estimate, which is usually the treatment effect. The treatment effect is valid only if the row-by-column interaction is negligible, otherwise this interaction is confounded with the treatment effect. As in all incomplete factorial designs, main effects are confounded with interaction effects, and in the Latin-square design, the result is to increase the experimental error and to possibly cause the results to be misleading. Errors due to the factors to be controlled are assumed to be random, normally distributed, and independent of the other factors.

However, in chemical studies, independence is the assumption most likely to be violated. In chemical processes, the factors that usually required controlling such as temperature, concentration levels, processing time, and pressure usually act in an interacting rather than an independent way. Thus, the Latin-square design would not be suitable for such an investigation. If interactions among the factors are thought to exist, then the appropriate procedure would be to use the full factorial design.

With regard to the assumption of negligible interactions, Latin-square designs are most appropriate when applied to testing methods, marketing research, or productivity procedures where the assumption that the factors to be controlled act independently of each other is tenable. If there is concern about the interactions of the factors included in a study, a test for this assumption can be made by using Tukey's test for nonadditivity (*11*), which is described by Kirk (*3*) and Winer (*12*). If this test is significant, then interaction effects are added into main and residual effects, and the tests of main effects may not be valid.

Another limitation of the Latin-square design is that the number of treatment levels, rows, and columns must be the same. For example, problems are caused if five materials are to be tested, and the testing machine can handle only four materials. Or, problems would be created in testing four or five types of materials if three rather than four production shifts are typical. The Youden-square analysis can be used in some situations where the number of columns is fewer than the number of rows and treatments.

One of the more serious limitations of the Latin-square design is that unless there is more than one observation per cell, there is a small number of degrees of freedom for the error term. In the 2 × 2 Latin square, there are no degrees of freedom for the error term, 2 for the 3 × 3 matrix, 6 for the 4 × 4 matrix, 12 for the 5 × 5 matrix, 20 for the 6 × 6 matrix, and 30 for the 7 × 7 matrix. For most studies, usually a 5 × 5 Latin square is the smallest square acceptable because df = 12 for the error term. However, below that size, the results have to show a large treatment effect and the blocking has to be effective for the treatment effect to be significant. To obtain more degrees of freedom for the error term, the study can be replicated with additional Latin squares of the same size that will also increase the treatment mean square. For example, with three 3 × 3 Latin squares, the degrees of freedom increases from 2 to 10, and five 4 × 4 Latin squares have 30 degrees of freedom compared to only 6 degrees of freedom for one 4 × 4 Latin square. Computational procedures for replication of Latin squares were given by Edwards (*13*) and Winer (*12*). As pointed out by Kirk (*3*), when using replication of Latin squares, some interactions among variables can be accounted for in a way similar to how these are accounted for in incomplete factorial designs having $n > 1$ per cell.

Another way to increase the number of degrees of freedom for the error term is to use more than one subject per cell, which is then analyzed as a fractional replication of the complete factorial design. An example of this use of Latin square was given in Section 14.3.3.

In summary, the Latin-square designs effectively and efficiently control for two or more extraneous variables that are thought to be related to the dependent variable. It accomplishes this by using a reduced number of treatment combinations, and thus a smaller number of observations compared to a complete factorial design. However, these designs require zero or negligible interactions among the factors that are set up, and also require the same number of levels for each dimension or factor.

References

1. Feldt, L. S. *Psychometrika* **1955,** *23,* 335–354.
2. Dayton, C. M. *The Design of Educational Experiments;* McGraw-Hill: New York, 1970.
3. Kirk, R. E. *Experimental Design: Procedures for the Behavioral Sciences;* Brooks/Cole: Belmont, CA, 1968.
4. Fisher, R. A. *The Design of Experiments;* Hafner: New York, 1960.
5. Natrella, M. G. *Experimental Statistics;* U. S. Government Printing Office: Washington, 1963.

6. Cochran, W. G.; Cox, G. M. *Experimental Designs;* Wiley: New York, 1957.
7. Fisher, R. A.; Yates, F. *Statistical Tables for Biological, Agricultural and Medical Research;* Oliver & Boyd: Edinburgh, 1963.
8. Federer, W. T. *Experimental Design: Theory and Application;* Macmillan: New York, 1955.
9. Peng, K. C. *The Design and Analysis of Scientific Experiments;* Addison-Wesley: Reading, PA, 1967.
10. Youden, W. J. *Experimental Designs to Increase Accuracy of Greenhouse Studies;* Boyce Thompson Institute: Yonkers, NY, 1940.
11. Tukey, J. W. *Biometrics* **1955,** *11,* 111–113.
12. Winer, B. J. *Statistical Principles in Experimental Design;* McGraw-Hill: New York, 1971.
13. Edwards, A. L. *Experimental Design in Psychological Research;* Holt, Rinehart & Winston: New York, 1968.

Problems

1. Suppose that a drug company wants to test the effect of three drugs on reaction time. The experimenter knows that reaction time will vary from person to person; thus, this variation should be controlled. But a limited number of subjects are available to try these drugs. Therefore, to control for differences in reaction time among the subjects, all subjects will take all drugs over a period of time during which the effects of one drug will not carry over to influence or interact with the other drug. The reaction time, in milliseconds, for each subject for each of the three drugs is as shown.

Subject A	Drug A	Drug B	Drug C
1	14	18	22
2	12	15	19
3	21	28	29
4	16	21	20
5	19	26	28
6	13	15	15
7	18	22	21
8	10	14	13
9	16	17	16
10	12	18	17

The design selected is the randomized-blocks design, for which each randomly selected subject serves as a block and takes the drugs in a random sequence so that the order of administration of the drugs is different for each subject. The design is a mixed-effects model because drugs are a fixed factor and subjects are random. Are there any significant differences in reaction time among the three drugs? Test the null hypothesis at the 0.01 level of significance.

2. If there are significant differences among the three drugs, which drugs are significantly different from each other?
3. Is the randomized-blocks design more efficient in testing the hypothesis in problem 1 than if a completely randomized analysis of variance design would have been used? If yes, how much more effective is the randomized-blocks design?
4. Suppose that for problem 1, five drugs are to be tested and only five subjects are available. The researcher believes that not only would there be differences in reaction times among the five subjects, but also that reaction time would vary from day to day. Thus, both of these factors should be controlled in a study. A Latin-square design was selected, where columns represent subjects, rows represent days, and letters within the matrix represent the five different drugs. The first 5 × 5 Latin square matrix in Figure 14.1 was used. Reaction time, in milliseconds, for each subject under five conditions is as shown.

Days	Subject 1	Subject 2	Subject 3	Subject 4	Subject 5
1	A14	B18	C22	D18	E16
2	B15	A12	E14	C19	D20
3	C29	D17	A21	E18	B28
4	D17	E19	B21	A16	C20
5	E17	C28	D16	B26	A19

Are there differences in reaction times that can be attributed to the different drugs? Test the null hypothesis of no differences among the five drugs at the 0.01 level.

5. If there are significant differences among the five drugs, which drugs are significantly different from each other?
6. How much more efficient is the Latin-square design than the randomized-blocks design after controlling for the day of drug administration?

Chapter 15

Hierarchal Designs

The designs presented in the last four chapters are essentially crossed-treatment designs in which all possible combinations of the levels of two or more treatments occur together in the study. Such designs are preferred because the differential effects of the treatments or factors can be examined as they affect the results either separately or for all combinations of the treatments or factors. However, a study might be designed so that crossed treatments are not possible, or a study might be designed by using a series of subsamples. Such designs are known as nested or hierarchal classifications. In such designs, factors not crossed with each treatment are nested within the treatment levels, and a study with complete nesting is referred to as a *hierarchal design.*

An example of nesting of a factor would be a study to determine the precision of a test method used by all laboratories available to a company. Any variation in the results could be due to differences within each laboratory or differences among technicians. Because laboratory technicians cannot be randomly assigned to different laboratories, they represent a nested factor. The researcher selects three laboratories from those using this testing procedure and sends identical samples to each with instructions that three technicians from each laboratory analyze two of the samples. Thus the design could be set up as in Figure 15.1A. The three laboratories are listed at the top and the technicians along the side so that columns represent laboratories and rows represent technicians. Technicians A, B, and C perform tests only in their own laboratory and not in laboratories II and III. Likewise, Technicians D, E, and F perform tests only in laboratory II, and Technicians G, H, and I perform tests only in laboratory III. Thus, technicians are nested under laboratories. Com-

1453–0/88/0363$06.00/1

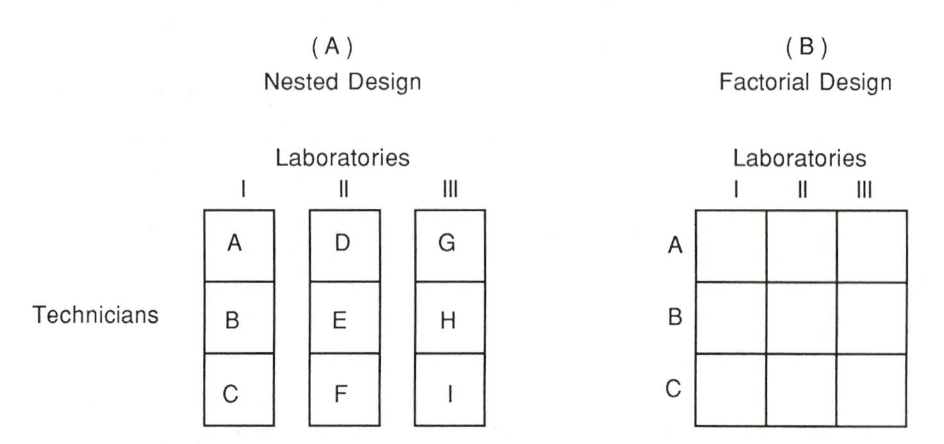

Figure 15.1. Example of a nested design (A) compared to a factorial design (B) with both factors crossed.

pare this design to the factorial design in Figure 15.1B, in which each technician performs tests in each laboratory. Here, interaction effects can be calculated because the two factors are completely crossed. To highlight the differences between these two designs, columns are usually used for displaying the nested design, whereas the rectangle is used to display the factorial design such as in Figure 15.1.

Another way of depicting the hierarchal nature of the nested design is to use a diagram such as in Figure 15.2. This figure shows more clearly the sequence of setting up a nested design and the hierarchal structure. That is, random samples from the same material are sent to three laboratories to be analyzed by three different technicians in each laboratory. The design layout is similar to that of the one-way analysis of variance. However, the one important difference is that technicians in the nested classification usually cannot be randomly assigned to different laboratories. Thus, the effects due to technicians cannot be controlled by randomization but are confined

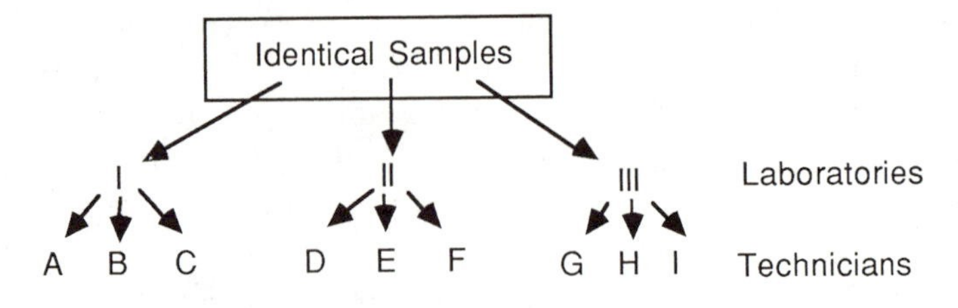

Figure 15.2. A nested, or hierachal, design.

to the laboratory in which they are working. Effects due to differences in technicians are restricted to only one laboratory or level of that factor (laboratories) and are said to be nested under that factor. The interaction effect between these two factors cannot be evaluated because a given group of technicians appears only under one of the laboratories. In order to evaluate the interaction effect, technicians from each group would have to be assigned to each laboratory. As with the randomized-blocks and the Latin-square designs, the nested design is limited to situations in which there are no interactions of the factors being studied. If there are interactions, these are negligible relative to the treatment factor. In most situations in which the nested design might be used, not being able to evaluate interaction effects would not present a serious problem because such effects are rarely of intrinsic value. If they were important, then the factorial design should be used. The nested design is used most often when an important factor cannot be controlled through randomization or other methods.

The unique effects associated with the nested factor can be estimated by partitioning the total variance into three parts:

- the variance due to treatment effects,
- the variance associated with differences in the means of the nested factor, and
- the variance within groups.

The variance due to treatment effects is the variance due to differences in laboratories for the example displayed in Figures 15.1 and 15.2. The variance associated with the differences in means of the nested factor corresponds to the technicians in the example. This variance can be estimated separately for each level of the treatment factor and then summed or pooled over all levels of the treatment factor. The variance within groups is computed as in the one-way analysis of variance and then the results are summed over all groups.

15.1 Computational Procedures

The sums of squares for the nested design are computed as follows. The total sum of squares (SS_T) is calculated by using the equation

$$SS_T = \Sigma X_T^2 - \frac{(\Sigma X_T)^2}{n_T} \qquad (15.1)$$

where ΣX_T is the observed value for all of the observations in the study, X^2 is the squared value of each observed value, and n_T is the total number of observations in the study.

The sum of squares for treatment (SS_{TR}) is computed by using the equation

$$SS_{TR} = \frac{\Sigma(\Sigma X_a)^2}{n_a} - \frac{(\Sigma X_T)^2}{n_T} \qquad (15.2)$$

where ΣX_a is the sum of values for each treatment, n_a is the number of values within each treatment, and ΣX_T and n_T are as defined in equation 15.1.

The sum of squares for the nested factor can be obtained by computing the sum of squares for each treatment level and then summing over the levels of treatment. At each treatment level, the sum of squares for the nested factor ($SS_{b(a)}$) is

$$SS_{b(a)} = \frac{\Sigma(\Sigma X_b)^2}{n_b} - \frac{(\Sigma X_{b(a)})^2}{n_{b(a)}} \qquad (15.3)$$

where ΣX_b is the sum of the values for each subgroup of the nested factor, n_b is the number of observations for each of these subgroups, $\Sigma X_{b(a)}$ is the sum of all values at that treatment level, and $n_{b(a)}$ is the number of observations at that treatment level.

The within-group sum of squares (SS_w) is computed for each subgroup by using the equation

$$SS_w = \Sigma X^2 - \frac{(\Sigma X)^2}{n} \qquad (15.4)$$

and then these sums of squares are totaled for all subgroups. As a computational check,

$$SS_T = SS_{TR} + SS_{b(a)} + SS_w \qquad (15.5)$$

The number of degrees of freedom for each sum of squares is as follows. For the total sum of squares, the number of degrees of freedom is equal to the total number of observations minus 1, or $n_T - 1$. For the treatment sum of squares, number of degrees of freedom would equal the number of treatments minus 1, or $A - 1$. For the nested factor, the number of degrees of freedom is $A(B -$

1), where A is the number of treatments, and B is the number of subgroups within each treatment. The within degrees of freedom can be computed by $AB(n - 1)$, or simply the number of degrees of freedom in each subgroup summed all subgroups.

For the data in Table 15.1, which represents hypothetical data for the hierarchal design depicted in Figure 15.2, the sums of squares would be

$$SS_T = 1265 - \frac{147^2}{18} = 64.50$$

$$SS_{TR} = \frac{40^2 + 46^2 + 61^2}{6} - \frac{147^2}{18} = 39.00$$

The nested sum of squares equals the sum of squares for all treatment levels:

$$SS_{b(a1)} = \frac{13^2 + 14^2 + 13^2}{2} - \frac{40^2}{6} = 0.33$$

$$SS_{b(a2)} = \frac{16^2 + 15^2 + 15^2}{2} - \frac{46^2}{6} = 0.33$$

$$SS_{b(a3)} = \frac{19^2 + 21^2 + 21^2}{2} - \frac{61^2}{6} = 1.34$$

The nested sum of squares would thus be 2.0.

The within-group sum of squares is found by computing each subgroup first. The first two subgroups would be computed as follows:

$$SS_{w_{a1b1}} = 89 - \frac{13^2}{2} = 4.5$$

$$SS_{w_{a1b2}} = 100 - \frac{14^2}{2} = 2.0$$

The last subgroup would be

$$SS_{w_{a3b3}} = 221 - \frac{21^2}{2} = 0.5$$

Adding the sum of squares for each of these nine subgroups yields the within group sum of squares, which for this example is 23.5.

Table 15.1. Hypothetical Test Results for a Nested Design

Statistic	Laboratory I			Laboratory II			Laboratory III		
	A	B	C	D	E	F	G	H	I
n									
1	8	6	7	9	6	7	11	9	11
2	5	8	6	7	9	8	8	12	10
$\sum X$	13	14	13	16	15	15	19	21	21
$\sum X^2$	89	100	85	130	117	113	185	225	221
$\sum x^2$	4.5	2.0	0.5	2.0	4.5	0.5	4.5	4.5	0.5
$\overline{X}$	6.5	7.0	6.5	8.0	7.5	7.5	9.5	10.5	10.5
Treatment Sum		40			46			61	
Treatment Mean		6.67			7.67			10.17	

NOTE: The letters A–I represent the different technicians.

As a computational check,

$$SS_T = SS_{TR} + SS_{b(a)} + SS_w$$

$$64.5 = 39.0 + 2.0 + 23.5$$

$$64.5 = 64.5$$

The sums of squares are then entered into an analysis of variance table such as Table 15.2. Computing *F* ratios depends upon whether each level of the nested factor represents a fixed or random variable. If all factors are at fixed levels, then the appropriate error term for computing all *F* ratios is the mean square within. If the nested variable is random, then the appropriate error term to use for testing the hypothesis of no significant difference in treatments is the mean square for the nested factor. If the nested factor were at fixed levels, then the appropriate error term would have been the mean square within.

For the example described in the beginning of the chapter, laboratories would represent a fixed dimension, and technicians would represent a random factor. Thus, to test for laboratory or treatment differences, the *F* ratios would be computed as $19.5 \div 0.33 = 59.09$, which is significant at the 0.01 level. The degrees of freedom for this *F* ratio are 2 and 6, and the table value required for significance at the 0.01 level is 10.92. To determine if the differences among the nested factor are significant, an *F* ratio is formed by dividing the mean square for the nested factor by the mean square within. For the results summarized in Table 15.2, this ratio is 0.13 and, with 6 and 9 degrees of freedom, is not significant because an $F = 5.80$ is required at the 0.01 level of significance. The results of this analysis indicate significant differences among the three laboratories but not among the technicians within each laboratory. Thus, the treatment effect, which is the difference among laboratories, is highly significant, but the

Table 15.2. Analysis of Variance for the Nested Design

Source	SS	df	ms	*F*
A (treatment)	39.0	2	19.50	59.09
B (nested factor)	2.0	6	0.33	0.13
C (within)	23.5	9	2.61	
Total	64.5	17		

nested factor is not significant and contributes very little to the total variance.

In most uses of the nested design, the treatment is usually at fixed levels, and the nested factor is random. The number of observations for each subgroup should be the same, otherwise the numerator and denominator for the *F* ratio are based on sources of variance that are based on different numbers of observations. As pointed out by Snedecor and Cochran (*1*), such *F* ratios will not be distributed in the same way as the theoretical distribution. The analysis for the nested design has the same assumptions as the one-way analysis of variance (randomization or independent errors, normally distributed variables, and homogeneity of variances among the subgroups). However, Havlicek and Peterson (*2*) showed that if the sample sizes are the same, violations of normality and homogeneity of variance have little effect on the *F* ratio.

As pointed out by Kirk (*3*) and Dayton (*4*), one of the main problems in using a hierarchal design is the typically small number of degrees of freedom associated with the mean square for the nested factor when that factor is a random variable, which is usually the case. Thus, the power of the nested design is reduced. When the nested factor is random, the power of the test can be increased only by increasing the number of levels of the nested factor. If the nested factor is fixed, then the within mean square is used as the error term for all *F* ratios, and power of the test can be increased by increasing the sample size of each group. If the number of degrees of freedom is relatively small (e.g., less than 25), Kirk (*3*, p. 233) suggested pooling of the nested mean square ($ms_{b(a)}$) with the mean square within (ms_w) as a way of providing a more adequate error term. However, pooling introduces a positive bias that is difficult to interpret statistically. Thus, caution must be taken when pooling sources of variance. Paull (*5*) and Green and Tukey (*6*) discussed pooling in more detail.

15.2 Comparisons among Treatment Means

As with the other analysis of variance designs, comparisons among treatment means can be made on an a priori or post hoc basis. For a priori tests, orthogonal contrasts can be set up to absorb the degrees of freedom in the design. For the example in Table 15.1, the nine subgroups of technicians would yield eight degrees of freedom. These eight degrees of freedom are divided so that two are available for treatment comparisons and six for the nested factor (Table 15.2). Contrasts that could be set up are presented in Table 15.3. The first contrast for the treatment effect involves comparing laboratory I with laboratories II and III. The second contrast is for comparing laboratory II with laboratory III. The six contrasts for the nested factor are set up so that comparisons of technicians are made within each laboratory. The analysis is completed exactly like that for the one-way analysis described in Section 11.3.2. For the nested factor, contrasts are always made within one level of the treatment factor in which the nesting takes place, and the degrees of freedom associated with the nested factor can be completely accounted for by using this procedure.

Post hoc tests can be made by using any of the procedures for the one-way analysis of variance (e.g., the Tukey or Scheffe' test when the overall F ratio is significant). In making such tests, the proper number of observations upon which each sum is based and the proper error term must be used. The proper error term is the same error term used to compute the F ratio for that effect. Thus, in the mixed design as in the example used for Table 15.2, the appropriate error term for treatment comparisons should be ms_{nested} and, for nested comparisons, ms_w. If both the treatment and the nested factors were fixed, then the appropriate error term would have been ms_w for both. For comparing treatment effects or differences between laboratories

Table 15.3. Orthogonal Contrasts for the Nested Design

	Laboratory I			Laboratory II			Laboratory III		
Statistic	**A**	**B**	**C**	**D**	**E**	**F**	**G**	**H**	**I**
Means	6.5	7.0	6.5	8.0	7.5	7.5	9.5	10.5	10.5
Contrasts									
A_1	2	2	2	−1	−1	−1	−1	−1	−1
A_2	0	0	0	1	1	1	−1	−1	−1
B_1	1	1	−2	0	0	0	0	0	0
B_2	1	−1	0	0	0	0	0	0	0
B_3	0	0	0	1	1	−2	0	0	0
B_4	0	0	0	1	−1	0	0	0	0
B_5	0	0	0	0	0	0	1	1	−2
B_6	0	0	0	0	0	0	1	−1	0

NOTE: The letters A−I in the column headings represent the different technicians.

at the 0.01 level by using the Scheffe' procedure, the critical range (CR_S) would be

$$\begin{aligned} CR_S &= \sqrt{(a - 1)F_{df_{a\,nested}}}\sqrt{2(n)(ms_{nested})} \qquad (15.6) \\ &= \sqrt{2(10.92)}\sqrt{2(6)(0.33)} \\ &= 4.67(1.99) = 9.29 \end{aligned}$$

Thus, any treatment sums greater than 9.29 would be significantly different at the 0.01 level of significance. For the data in Table 15.1, laboratory III is significantly different from the other two laboratories, but laboratories I and II are not significantly different from each other.

15.3 Nesting in Factorial Designs

The design considered in this chapter is a one-way classification with one nested factor. The nested factor consists of either intact subgroups that cannot be crossed with the other factor representing the one-way classification or control factors that may increase the precision of the study. This basic design can be expanded to include additional factors, and these factors can be completely nested with each factor or can be nested within one factor and crossed with others. As with other factorial designs, these designs can be very complex, depending upon the number and types of factors that are included in a study.

If each factor is nested within each of the other factors, the result is a completely nested, or hierarchal design. For example, consider the design presented in Figure 15.3, which is a general layout for comparing the yield of an organic chemical from two production processes, which, because of technical factors, cannot be put in operation at all plants. Process I can only be considered for plants A, B, and C, whereas process II can only be considered for plants D, E, and F. Thus, plants are nested under process. Because operators cannot be transferred to the different plants, operators are nested under plants. Data that would be used for the analysis are shown as "X" values under each operator. In this design, operators (factor C) are completely nested under plants (factor B), and plants are completely nested under the first factor, which is the type of process. Because all factors are completely nested, no interactions can be computed among the factors. However, the analysis would provide a test of the main effects due to factor A (process), factor B within A, and factor C within B within A. Computational procedures

	PROCESS (A)					
	I			II		
Plants (B)	A	B	C	D	E	F
Operators (C)	1 2 3	4 5 6	7 8 9	10 11 12	13 14 15	16 17 18
Data	X X X	X X X	X X X	X X X	X X X	X X X
	X X X	X X X	X X X	X X X	X X X	X X X
	X X X	X X X	X X X	X X X	X X X	X X X
	X X X	X X X	X X X	X X X	X X X	X X X

Figure 15.3. A complete nested design.

for completing the analyses can be found in advanced statistics textbooks (*3, 4, 7–8*).

In some studies, a factor could be nested under another factor but crossed with other factors. For the example just given, if each process in each plant could be tested, then the design would be a 2 × 3 factorial design as shown in Figure 15.4. However, if operators were thought to be an important factor but could not be transferred to each plant, then operators would be nested under plants but crossed with the process. Thus, the analysis would provide tests of the main effects for factors A and B, the interaction effects of factors A and B, the effect of factor C operating within factor B, and an AC interaction effect within factor B. Computational equations, how to deal with unequal sample sizes, and additional designs for various

(B) Plants		(C) Operators	Process (A) I	Process (A) II
	A	1	X X X X	X X X X
	A	2	X X X X	X X X X
	A	3	X X X X	X X X X
	B	4	X X X X	X X X X
	B	5	X X X X	X X X X
	B	6	X X X X	X X X X
	C	7	X X X X	X X X X
	C	8	X X X X	X X X X
	C	9	X X X X	X X X X

Figure 15.4. A partial nested design.

combinations of nested and crossed factors are given in the references cited for the completely nested design (*3, 4, 7, 8*).

Kirk (*3*) and Keppel (*8*) pointed out that the final design selected for a given study must be based on many factors. A design using nested factors is usually simpler to set up and conduct, and often may be the only design that is feasible because of the given conditions. However, the factorial design with crossed factors provides more information and should be used where the interaction of those factors is important. If the differential effects of the factors are important (i.e., how the effects of one factor vary at different levels of another factor), then the major concern of the study demands an interaction model and thus the factorial design. If the researcher is concerned mostly with the main effects and can perceive a study as a series of separate studies, then the nested design would provide the necessary information and might be easier to conduct. Also, in the completely crossed design, because the effects of a factor can be estimated at each level of all other factors, any differences due to each factor are equally spread over those factors. However, when the nested design is used, any differences within each factor have to be controlled through random selection within the nested factor. For the design shown in Figure 15.3, operators in each plant should be randomly selected so that the results can be generalized to all operators within each plant. The design would then be a random-factor design as far as the nested factor is concerned, and consequently, the degrees of freedom for the denominator for testing main effects will probably be low. Thus, the power of the statistical test will also be low.

The statistical model selected for any study must be based on many factors, the most important of which are the nature of the questions the researcher is asking and the specific information that is necessary to be obtained from a given study. The references cited in this section provide detailed discussions of the advantages and disadvantages of these two basic designs as well as the post hoc tests that can be used with each design (*3,4,7,8*).

References

1. Snedecor, G. W.; Cochran, W. G. *Statistical Methods*; Iowa State University Press: Ames, IA, 1971.
2. Havlicek, L. L.; Peterson, N. L. *Psych. Reports* **1974,** *34,* 1074–1114.
3. Kirk, R. E. *Experimental Design: Procedures for Behavioral Sciences*; Brooks/Cole: Belmont, CA, 1968.

4. Dayton, C. M. *The Design of Educational Experiments*; McGraw-Hill: New York, 1970.
5. Paull, A. E. *Annals of Math. Stat., 21,* 539–556.
6. Green, B. F.; Tukey, J. *Psychometrica 25,* 127–152.
7. Winer, B. J. *Statistical Principles in Experimental Design*; McGraw-Hill: New York, 1971.
8. Keppel, G. *Design and Analysis: A Researcher's Handbook*; Prentice Hall: Englewood Cliffs, NJ, 1973.

Problem

1. Suppose that two production processes (process I and II) in the manufacturing of a certain chemical are to be compared. Because of the plant facilities, process I can be implemented only in plants with a certain type of equipment, and process II can be implemented only in plants with a different type of equipment. Also, plant differences could affect yield of the manufactured chemical. Thus, a hierarchal design is needed because plants have to be nested under the type of production process, as the processes cannot be implemented in all types of plants. For this study, a random sample of plants is selected to try each process. Four production samples were taken from each of the four plants randomly selected for each process. Hypothetical data indicating percentage yield of the chemical for each sample are shown in the following tables. Test the hypothesis of no difference in yield for process I compared to process II and assess the effect due to the different plants. Test the hypothesis at the 0.01 level of significance.

Percentage Yield for Process I

Plant A	Plant B	Plant C	Plant D
28	24	25	22
27	28	30	28
24	29	27	26
26	26	27	24

Percentage Yield for Process II

Plant E	Plant F	Plant G	Plant H
21	27	21	24
26	25	24	23
22	26	26	22
25	25	22	24

Chapter 16

Statistical Selection Guide

Researchers often try to find an appropriate statistical analysis after the data are obtained from routine operational procedures. Sometimes, the data have been collected in a way such that no statistical analysis is appropriate. Sometimes, some statistical analyses are not appropriate because the basic assumptions underlying these analyses have not been met. Lack of planning may also cause a researcher to select a statistical analysis that is not appropriate or correct for the proposed study. At the other extreme, the researcher may resort to running many statistical analyses in order to find a significant result. This approach not only violates the basic probability theory underlying inferential statistics, but also may not yield the desired information or be appropriate for the complexity of the research study. Lack of planning may also result in using only one statistical procedure, whereas additional analyses or a different type of analysis could have provided much more information or information that might have been more appropriate for the research question.

Statistical analyses should be planned concurrently with a research study to ensure that the desired procedures can be implemented. Research studies should be conceived so that all aspects of data collection fit a predetermined statistical analysis. In that way, statistical analyses will not only be feasible to run but also will provide a valid analysis of the data.

Statistical analyses should be selected to fit the needs of the specific research. As pointed out by Lindquist (*1*, p. vi), a researcher should develop a research procedure and select statistical analyses that will provide the answers required for a specific research study rather than trying to fit a research study into a readily available model, which often results in incorrect inferences from the results. One

1453–0/88/0377$06.00/1

sequence in developing a useful statistical analyses is to list the analyses that will provide information with regard to description, significance statements, confidence limits, and evaluation of the results. Essentially, this procedure means that definitive answers to some basic questions should be provided before a study begins. These questions pertain to the main purpose of the study, the kinds of decisions that must be made from the results, the kinds of data taken (i.e., the measures or observations used), and the conditions or circumstances of data collection.

16.1 Answering Questions with Statistics

The basic purposes of statistical analyses were provided in Chapter 1. Essentially, that chapter discussed the types of information statistical analyses could provide. In this section, a guide to statistical analyses used to answer specific questions will be provided. References will be made to specific sections in this book for each question or, for those analyses not covered in this book, to specific sources.

16.1.1 How Can the Findings of a Study Be Described and Presented?

For all studies, the findings must be presented. What values were found? How are these values distributed? How can the results be presented so that another person can readily see what the results were?

Many ways were presented in Chapter 3 to organize and present findings. A number of descriptive statistical procedures can be used to summarize and present results, such as basic frequency distributions, percentages, graphic presentations, and various types of tables and charts. Selecting which methods should be used is determined by considering the target audience and which procedures that audience is accustomed to.

16.1.2 What Is the Average or Most Typical Value?

A number of methods can be used to summarize findings with regard to central tendency. Depending upon the type of data, the mode, median, or mean would be appropriate for nominal, ordinal, or interval/ratio data, respectively. The geometric mean could be used for data in the form of geometric series, especially when the data are in the form of ratios. The harmonic mean is useful when working with data given as averages of rates of work, time, cost, or distance. These measures were presented in Chapter 4.

16.1.3 How Much Variation Is in the Data?

As presented in Chapter 5, a number of statistical procedures can be used to describe the variability of data. For nominal data, the range of items or values can be presented. For ordinal data, the range as well as the interquartile range can be used. For interval/ratio data, the variance and standard deviation can be used as well as various types of ranges.

16.1.4 What Is the Shape of the Distribution of Values?

For interval/ratio data, measures of skewness can be used to indicate whether the data are normally distributed or tend toward the high or low values. How peaked the distribution is can be communicated by the measure of kurtosis.

16.1.5 What Is the Relative Position of a Value within a Distribution?

Various standard measures, such as percentile rank and standard score values, can be used to communicate relative position. As presented in Chapter 7, the normal curve is the basis for determining the relative position for most studies.

16.1.6 How Are Variables Related?

Several measures of correlation can answer this type of question. For nominal data, the phi (ϕ) or contingency coefficient can be used, depending upon whether the study contains only two variables or more than two variables. For ordinal data, the rank-difference correlation can be used. For interval data, the Pearson correlation coefficient can be used for linear relationships, and eta (η) can be used for curvilinear relationships. If one variable is to be correlated with a number of other variables, then multiple correlation can be used. These procedures were presented in Chapter 6. If a set of multiple variables is to be correlated with another set of multiple variables, then the canonical correlation can be used (*13,16,17,22–24*).

16.1.7 How Can Future Performance Be Estimated or Predicted?

For interval/ratio data, statistical regression can be used to predict one variable from another as described in Section 6.3. If more than one predictor variable is available, then multiple regression can be used as described in Section 6.5. Section 9.6 discusses ways to determine if the relationship is significantly different from zero correlation.

16.1.8 How Does the Quality or Performance of a Given Product Compare to a Standard?

A number of one-sample tests are designed to compare data from one sample to a known or hypothetical standard. If the data are categorical, then the standard error of a proportion can be used to determine if the obtained proportion is within chance limits of the standard (Section 9.4). The chi-square (χ^2) goodness-of-fit test can be used to determine whether an obtained distribution of values is similar or different from the standard normal distribution (Section 10.5).

If the data are in the form of ordered ranks or ordinal data, then the Kolmorgorov–Smirnov one-sample test can be used to determine the agreement between the obtained distribution and the known or theoretical distribution (*2–8*).

If the data are interval or ratio, then a number of procedures can be used. Confidence limits for the mean can be used (Section 9.3), or a z ratio for large samples or a t ratio for small samples can be used to test the hypothesis of no significant difference between the mean for the sample compared to the mean for the standard or hypothetical value (Section 9.3). A two-tailed test can be used if one wants to know if there is a difference either higher or lower, than the standard, and a one-tailed test can be used if one wants to know whether the given mean for the product is only higher or only lower than the standard.

16.1.9 How Do Two Products or Groups of Items as Individuals Differ with Regard to Average Quality or Performance?

For comparing two products, the first consideration is whether the two sets of data for the two products are independent or matched in some way. If random samples are taken for each product, then the data are usually considered independent. If items or cases from each product are matched on a one-to-one basis on some variable related to the criterion or measure of concern, then the data are considered paired, matched, or related. The second consideration is the type of data being analyzed.

For nominal data, the chi-square (χ^2) test of independence (Section 10.5) can be used to compare the distributions from the two products when the data are independent. For matched data, the McNemar test for significant changes (Section 10.4.1) can be used for bivariate data or when two groups are to be compared.

For ordinal data, the Mann–Whitney test or the Kolmorgorov–Smirnov two-sample test can be used to test whether two independent groups have been drawn from the same population. For matched data, the Wilcoxon matched-pairs T test can be used. Nonparametric statistics are covered in several textbooks (*2–8*).

For interval data, the z ratio for large samples or the t test for small samples can be used (Section 10.1). Procedures are available for both independent and related samples.

For all of these tests, a two-tailed test can be used to test the null hypothesis of no difference regardless of direction, or a one-tailed test can be used to test a directional hypothesis that one group is higher than the other.

16.1.10 Are There Differences in the Average Quality or Performance of Several Products or Groups of Items or Individuals?

As with comparing two products or groups, one has to consider whether the data are independent or matched, and the type of data that are available.

For nominal data, the χ^2 test of independence can be used for independent data (Section 10.5). For matched data, the Cochran Q test (Section 10.4.2) can be used. For ordinal data from independent samples, the nonparametric Kruskal–Wallis one-way analysis of variance can be used, or if pairs of products or groups are to be compared, the Wilcoxon–Wilcox multiple comparisons test can be used. For related or matched data, the Friedman multisample test for related groups can be used (*2–8*).

For interval data, the analysis of variance can be used for independent groups, followed by post hoc tests to determine which pairs of means are significantly different from each other (Section 11.3.1). A priori tests can be made to compare specific pairs of groups or combinations of groups (Section 11.3.2), and trend analysis can be used when the independent variable is on a quantitative scale (Section 11.4). For related samples or data for which several measures are taken, the repeated-measures analysis of variance can be used. Winer (*9*) and Keppel (*10*) provided information on the repeated-measures analysis of variance.

16.1.11 How Does the Variability of a Given Product Compare to the Variability of a Standard?

The standard error of a standard deviation can be used to determine whether the variability of the sample is significantly different from a known or hypothetical standard (Section 9.5). Confidence limits can be set up on the basis of this standard error, and such limits are useful for quality control work.

16.1.12 How Do Two Products Differ with Regard to Variability?

For comparing the variability of two products, one has to consider the type of data and whether the data are considered independent or matched.

For independent samples and nominal data, the χ^2 goodness-of-fit test can be used to determine whether the two distributions are the same or significantly different (Section 10.5.1). For ordinal data, the nonparametric Kolmogorov–Smirnov two-sample test can be used. For interval data, the F test comparing two variances (Section 10.2) can be used for independent data, and the t test (Section 10.2) can be used for matched data.

16.1.13 What Sample Size Is Necessary and How Can a Sample Be Taken?

As pointed out in Section 8.6, the sample size needed for a study depends upon several factors, the most important of which are (1) the variability of the variables in the population to which inferences are to be made, (2) the degree of precision desired, (3) the number of treatment groups or levels, (4) the probability of making Type I and Type II statistical errors, and (5) the minimum treatment effect that an investigator is interested in detecting. Several procedures for estimating sample sizes were presented in Section 8.6 as well as two tables for determining sample sizes in the Appendix (Tables L and M).

A number of sampling procedures that can be used to select cases were presented in Section 8.7. The most often used method is simple random selection using the table of random numbers, which is Table A in the Appendix. Other procedures are systematic sampling, stratified random sampling, and cluster sampling.

16.1.14 What Are the Basic Experimental Designs and What Are Their Advantages?

The majority of studies in the physical sciences are designed to determine the effects of different procedures or treatments upon the mean value of a specified criterion for a specified target population. The selection of the experimental or statistical design is based primarily on the types of questions to be answered by the investigation; how many factors are to be considered or controlled; the number of levels for each of these factors; and the complexity of the factors, procedures, and observations that have to be considered. The following is a brief description of the basic designs.

The one-way analysis is the simplest of the analysis of variance designs in that there is only one treatment or classification variable with two or more levels. The analysis involves groupings of subjects only on one dimension and is restricted to the treatment levels selected. Several types of one-way designs based on how subjects are assigned to treatments or levels of the same treatment are available. The first type is a completely random design in which each individual in a random sample of subjects is randomly assigned to treatment groups. The second type, known as a repeated-measures design, is where one group of subjects receives all treatments. The third type is where subjects are matched on a variable or variables that are known to be related to the criterion variable. The completely random design was discussed in Chapter 11. The repeated-measures design is discussed in advanced textbooks such as those by Winer (*9*) and Keppel (*10*).

Factorial designs are those in which the effects of two or more independent variables can be simultaneously studied. The main advantage of the factorial design is that the effects of the independent variables can be ascertained separately as well in combination. In a complete factorial design, all combinations of the different levels of all factors can be compared. This design provides more control of the variables than the simple one-way design and usually results in wider generalizations of the results because the results are studied at all levels of two or three factors rather than over one factor. Factorial designs were discussed in Chapters 12 and 13.

The randomized-blocks design and the various types of Latin-square designs can be used to reduce the amount of error in a study by reducing the variation within groups by matching subjects or cases into blocks. If the blocking is done on the basis of a variable that is highly correlated with the criterion variable, then the error term in the analysis of variance is substantially reduced; thus, the likelihood of obtaining a significant result increases. When the blocking is done on two or more variables, one of the more complex Latin-square designs can be used. These designs were covered in Chapter 14.

Hierarchal or nested designs are used in situations in which levels of one factor cannot be studied at any of the levels of another factor. For example, laboratory technicians usually cannot be randomly assigned to different laboratories and thus would represent a nested factor in a study of different laboratory testing procedures. Nested designs were discussed in Chapter 15.

16.2 Common Statistical Packages

A large number of packaged statistical programs are available for most of the mainframe computers and for personal computers. Most of these packages contain programs from very simple descriptive statistics to complex multivariate analyses. In addition to these packages, many universities offer statistical packages or computer programs for specific analyses. The following list is not comprehensive, and other computer programs can be found through literature reviews, especially the journals devoted to computers and computed software; through personal contacts; and through computer representatives.

BMDP
BMDP Statistical
Software, Inc.
1964 Westwood Blvd.
Suite 202
Los Angeles, CA 90025

MIDAS
Statistical Research Laboratory
University of Michigan
Ann Arbor, MI 48109

SAS
SAS Institute, Inc.
P.O. Box 8000
SAS Circle
Gary, NC 27511

SPSS
SPSS, Inc.
444 N. Michigan Ave.
Chicago, IL 60611

STSC
STSC, Inc.
2115 E. Jefferson St.
Rockville, MD 20852

Systat
Systat, Inc.
2902 Central St.
Evanston, IL 60201

OSIRIS
Institute for Social Research
University of Michigan
Ann Arbor, MI 48109

COFAMM
(Confirmatory Factor Analysis
with Model Modification)
National Educational Resources
Chicago, IL

ARTHUR
Laboratory of Chemometrics
Department of Chemistry
University of Georgia
Athens, GA 30602

TROLL
Computer Research Center for
Economics and Management
Science
Massachusetts Institute of
Technology
Cambridge, MA 02141

Universities That Offer Statistical Packages

University of Chicago Computation Center
1155 East 60th St.
Chicago, IL 60637

Biometric Laboratory
University of Miami
P.O. Box 248011
Coral Gables, FL 33124

Princeton University Computation Center
87 Prospect Ave.
Princeton, NJ 08544

Harvard University Computation Center
1730 Cambridge Street
Cambridge, MA 02138

University of California Computation Center
209 Evans Hall
Berkeley, CA 94720

Variance is the basic concept that the statistician works with. The statistical expression "analysis of variance" is very appropriate because most statistical analyses partition, isolate, and identify the variation in the dependent variable that is due to different independent variables and possibly confounding or error factors. The statistical methods presented in this book give a basic foundation of this concept for univariate analyses, and these concepts and principles can be expanded to multivariate situations where there are two or more independent and dependent variables. Some of the more useful multivariate methods and where they are used are given in this section so that the reader can identify the appropriate statistical method for a specific research problem.

If one is dealing with two sets of variables (e.g., a number of predictor variables such as intelligence, aptitudes, and experience) and a number of criterion variables (e.g., productivity, absenteeism, and morale), and these sets of variables are hypothesized to be related, then *canonical correlation* should be used to test the hypothesis (*11–18*). If measurements for a number of variables are obtained for a set of objectives or individuals, and the researcher wants to determine if they could be divided into relatively homogeneous subgroups that can be distinguished from other subgroups, then *cluster analysis*, also referred to as segmentation analysis or taxonomy analysis, would be used (*19–21*). If subgroups of objects or individuals are dichotomous in nature, such as users versus nonusers or users of several different brands of a product, and the researcher wants to determine if these subgroups can be classified and distinguished on the basis of how they are measured on a set of variables, then *discriminate analysis*, sometimes referred to as discriminant analysis, can be used (*11–18, 122*). Another type of problem often encountered in research is where a large number of measures are taken for a group of objects or individuals (e.g., 30 aptitude and personality measures for a group of workers), and the researcher wants to determine if these measures can be reduced to a smaller number of dimensions or factors (e.g., ability, cooperativeness, and motivation). *Factor analysis* should be used for such problems (*13–18, 21–24*). Finally, if one is interested in determining any significant differences among several dependent variables (e.g., percent yield, purity, viscosity, and composition) due to several independent variables (e.g., type of mixture, temperature, pressure, time of reaction, purity of ingredients, and manufacturing process), then the *multivariate analysis of variance* (MANOVA) should be used (*11–18, 23,24*).

These multivariate analyses involve complicated and extensive computations that are beyond the scope of this book. However, most of the computer programs listed in Section 16.2 include multivariate analyses. The references given for each analysis should provide enough information for those who wish to apply these analyses and interpret the results.

References

1. Lindquist, E. F. *Design and Analysis of Experiments in Psychology and Education;* Houghton Mifflin: New York, 1953.
2. Bradley, J. V. *Distribution-Free Statisticals Tests;* Wiley: New York, 1968.
3. Conover, W. J. *Practical Nonparametric Statistics;* Wiley: New York, 1971.
4. Daniel, W. W. *Applied Nonparametric Statistics;* Houghton Mifflin: Boston, 1978.
5. Hollander, M.; Wolf, D. A. *Nonparametric Statistical Methods;* Wiley: New York, 1973.
6. Siegel, S. *Nonparametric Statistics for the Behavioral Sciences;* McdGraw–Hill: New York, 1956.
7. Sprent, P. *Quick Statistics;* Penguin Books: New York, 1981.
8. Walsh, J. E. *Handbook of Nonparametric Statistics;* Van Nostrand: Princeton, NJ, 1965.
9. Winer, B. J. *Statistical Principles in Experimental Design;* McGraw–Hill: New York, 1971.
10. Keppel, G. *Design and Analysis;* Prentice–Hall: Englewood Cliffs, NJ, 1973.
11. Anderson, T. W. *An Introduction to Multivariate Statistical Analysis;* Wiley: New York, 1958.
12. Kachigan, S. K.; *Statistical Analysis;* New York, 1986.
13. Cooley, W. W.; Lohnes, P. R. *Multivariate Data Analysis;* Wiley: New York, 1971.
14. Harris, R. J. *A Primer of Multivariate Statistics;* Academic: New York, 1975.

15. Hope, K. *Methods of Multivariate Analysis;* Gordon and Breach Science: New York, 1969.
16. Morrison, D. F. *Multivariate Statistical Methods;* McGraw–Hill: New York, 1967.
17. Overall, J. E.; Klett, C. J. *Applied Multivariate Analysis;* McGraw–Hill: New York, 1972.
18. Press, S. J. *Applied Multivariate Analysis;* Holt, Rinehart & Winston: New York, 1972.
19. Hartigan, J. A. *Clustering Algorithms;* Wiley: New York, 1975.
20. Jardine, N.; Sibson, R. *Mathematical Taxonomy;* Wiley: New York, 1971.
21. Massart, D. L.; Kaufman, L. *The Interpretation of Analytical Chemical Datad by the Use of Cluster Analysis;* Wiley: New York, 1983.
22. Lachenbruch, P. A. *Discriminant Analysis;* Hafner: New York, 1975.
23. Harman, H. H. *Modern Factor Analysis;* University of Chicago Press: Chicago, 1960.
24. McDonald, R. P. *Factor Analysis and Related Methods;* Lawrence Erlbaum Associates: Hillsdale, NJ, 1985.

Answers to Problems

Chapter 4

1. The three measures of central tendency for each distribution are as follows:

Statistic	A	E	G
Mean	5.00	6.67	7.96
Median	5.00	6.94	8.67
Mode	5.00	7.50	9.00

   Distribution A is a normal distribution; therefore, the mean would be the most appropriate measure of central tendency to report because this could be used in further computations. Distributions E and G are skewed. The means are influenced by the relatively few low values, and the mode appears to be too high, especially for distribution G. Thus, for these two distributions, the median probably best represents the majority of the values.

2. Because each mean is computed on a different number of test tubes purchased, the weighted mean obtained by using equation 4.4 should be used. The weighted mean is 0.780. If the four means were averaged without a consideration of the different number of tubes in each sample, the average would have been 0.8125, which would not have been correct.

3. Because the average rate of change over the 6-year period is of interest, the geometric mean (GM) should be used. The first step would be to compute the percentage change from one year to the next as shown. Then, the geometric mean would be computed. This computed mean indicates an average yearly increase of 16.3%. Then, starting with the year 1980, multiplying each year's production cost by 1.163 would give an estimate of the yearly

1453–0/88/0389$06.00/1

cost for 1986 ($111.35). Or, the 1985 data would result in an estimated cost of $96 × 1.163 = $111.65.

Yearly cost ($)	45	54	60	74	82	96
Change (%)		120	111	123	111	117

From this data,

$$\log \text{GM} = \frac{2.0792 + 2.0453 + 2.0899 + 2.0453 + 2.0682}{5}$$

$$= \frac{10.3279}{5} = 2.06558$$

$$\text{GM} = \text{antilog } 2.06558 = 116.3$$

4. The engineer is interested in the lower and upper values that would cut off the middle 90% of the distribution. One way to do this would be to compute the percentile points at P_{05} and P_{95}. By using equation 4.1, the lower limit would be computed as 3.781, and the upper limit would be 4.645. Thus, 90% of the samples should contain percentages of carbon between these two values.

Chapter 5

1. The mean is 6.0, the median is 7.5, the mode is 8, and the standard deviation is 3.02. To compute skewness and kurtosis, $m_2 = 8.2$, $m_3 = -13.8$, and $m_4 = 117.4$. Skewness is therefore −0.588, indicating a slightly negative skew. Kurtosis is −1.25, indicating a slightly platykurtic or flat distribution. The interquartile range is $8 - 3 = 5$. Thus, 50% of the values would fall between the values of 8 and 3, whereas the mean and standard deviation indicate that 68% of the values fall between the values of 9 and 3.
2. Because the sample sizes vary from plant to plant, both the weighted mean, obtained by using equation 4.4, and the weighted standard deviation procedure, described by equation 5.13, have to be used. For this data, the weighted mean is 78.91, and the weighted standard deviation is 13.69.

3. The mean is 4.184, and the standard deviation is 0.257.

Chapter 6

1. From equation 6.4, $r = 0.957$. Thus, the measure of lead content for the samples analyzed has a high degree of consistency. The standard error of measurement (SE_{meas}) is computed from equation 6.31 as

$$SE_{meas} = 10.10\sqrt{1 - 0.957} = 2.09$$

   where 10.10 is the standard deviation of the measures for vial A. Thus, 68% of the time, the measured lead content will vary ± 2.09 from the measure taken. That is, for the first sample of vial A, the actual measure would probably fall between 9.91 and 14.09.

2. From equation 6.4, $r = 0.535$. To compute the standard error of estimation (SE_{est}), the standard deviation of the predicted variable must be calculated. For this problem, the standard deviation for the measures of the amount of tin extracted is 1.99. Thus,

$$SE_{est} = 1.99\sqrt{1 - 0.535^2} = 1.68$$

   Thus, predicted values will be within 1.68 of the actual value 68% of the time. This statement will hold for the limits of the problem (i.e., for refluxing time that varies from 30 to 75 min).

3. By assuming a linear relationship between these two variables, a regression equation can be set by using equation 6.6. The first step is to compute the regression coefficient b_{yx}. By using equation 6.7, the regression coefficient for predicting yield (Y) from amount of ammonium chloride (X) is computed as

$$b_{yx} = (0.80)\frac{8}{1.5} = 4.27$$

   The a value is computed by using equation 6.11 as follows:

$$a_{yx} = 160 - (4.27)(14.5) = 98.09$$

   The complete regression equation is then

$$\hat{Y} = (4.27)(X) + 98.09$$

and for $X = 16$,

$$\hat{Y} = (4.27)(16) + 98.09 = 166.41$$

The standard error of estimation can be computed by using equation 6.13. For this problem,

$$SE_{est} = 8\sqrt{1 - 0.80^2} = 4.8$$

Thus, the predicted yield of the organic chemical would be 166.41 lb if the percentage of ammonium chloride was set at 16%, and 68% of the time the yield would vary from 161.61 to 171.21 lb/batch.

Chapter 9

1. One solution would be to test the null hypothesis of no significant difference between the sample mean of 0.162% and the mean of 0.15% stated by the manufacturer. The hypothesis could be tested by using the z ratio in equation 9.8. The 0.05 level of significance could be selected for this situation. From equation 9.2,

$$SE_M = \frac{S}{\sqrt{n}} = \frac{0.04}{\sqrt{40}} = 0.0063$$

$$z = \frac{0.162 - 0.15}{0.0063} = 1.905$$

Because a $z = 1.96$ is needed for significance at the 0.05 level, the random sample does not deviate significantly from the manufacturer's claim of 0.15% sulfur content. The confidence limits would be computed by using equation 9.7 because the sample size is less than 100. A t value of 2.021 would be used when df = 40.

$$CL = 0.162 \pm (2.021)(0.0063) = 0.162 \pm 0.013$$

or from 0.149% to 0.175%. Thus, this answer shows that the manufacturer's claim is within the 95% confidence limits for the mean of the sample. Also, the claim seems to be correct on the basis of this one sample.

2. The standard error of the mean would give the confidence limits

where the population mean would likely be, and if critical, the 99% confidence limits would be the most appropriate to use because this would include where the mean is likely to be 99% of the time when repeated samples are taken. Using equation 9.2,

$$SE_M = \frac{1.1}{\sqrt{120}} = 0.10$$

Because $n > 100$, equation 9.6 can be used to compute the confidence limits at the 99% level.

$$CL = 20.5 \pm (2.58)(0.10) = 20.5 \pm 0.258$$

Thus, the strength of filaments is likely to vary from 20.24 to 20.76 oz for 99% of the filaments produced.

3. Because the problem deals with estimating the number of customers who might buy this product, equation 9.12 could be used.

$$SE_{\%} = \sqrt{\frac{(45)(550)}{100}} = 4.97$$

$$CL = 45 \pm (1.96)(4.97) = 45 \pm 9.74$$

Thus, from 35.26% to 54.74% of the doctors are likely to buy this product, and given the 400,000 estimated population, between 141,040 and 218,960 products are likely to be purchased by these doctors.

4. From Table F, with df = 8, an $r = 0.765$ is needed for significance at the 0.01 level. The obtained $r = 0.535$ is not equal to or greater than the table value for r at the 0.01 level; thus, no evidence is found to reject the null hypothesis of $r = 0$, and the conclusion is that there is no significant correlation between refluxing time and the amount of tin extracted. Thus, the amount of tin extracted from the product does not appear to relate to refluxing time for this data.

Chapter 10

1. Because the sample sizes are large, the z ratio described by equation 10.2 could be used. By using equations 9.2 and 10.1, the

standard error of the difference between the two means (SE_{D_M}) would be

$$SE_{M_1} = \frac{0.10}{\sqrt{400}} = 0.005$$

$$SE_{M_2} = \frac{0.22}{\sqrt{300}} = 0.0127$$

$$SE_{D_M} = \sqrt{0.005^2 + 0.0127^2} = 0.0136$$

$$z = \frac{0.80 - 0.92}{0.0136} = -8.82$$

where SE_{M_1} and SE_{M_2} are the standard errors for the first and second means, respectively. At the 0.001 level of significance, a $z = 3.30$ is needed for a two-tailed test. Thus, the obtained z is significant, and the conclusion is that there is a significant difference in the mean per-unit production costs between plant A compared to plant B. Plant A has significantly lower per-unit production costs.

2. To determine if there are significant differences in the two variances, equation 10.9 can be used because the data are from independent samples.

$$F = \frac{0.22^2}{0.10^2} = 4.84$$

For df = 299 for the larger variance and df = 399 for the smaller variance, Table D shows that the closest F ratio for these degrees of freedom needed for significance at the 0.01 level is 1.39, if a conservative approach is taken and the fewer degrees of freedom in the table are used (i.e., df = 200 and 200). Thus, the obtained F is significant, and the conclusion is that the variance in per-unit costs are less in plant A.

3. The null hypothesis is that no reduction in the amount of analysis time results from using the new equipment compared to the old equipment. Because the data are related (the same technicians are used to eliminate variation in speed of technicians), the related t test described by equation 10.8 should be used. The mean difference is 0.50, and the standard deviation of those differences is 1.90. The related t test is then computed as follows: if $n = 10$,

$D = 5, D^2 = 35, \Sigma x^2_D = 35 - [(5^2)/10] = 32.5$, then

$$t = \frac{0.5}{\sqrt{\frac{32.5}{(10)(9)}}} = 0.83$$

With df = 10 − 1 = 9, a t = 2.821 is needed for a one-tailed test at the 0.01 level. On this basis, no significant reduction in analysis time is found between the new and old equipment.

4. The directional hypothesis would be that no reduction in variance in time occurs when using the old equipment compared to the new equipment. Because the data are related, equation 10.10 has to be used. The r value between the two analysis times is 0.847. The two standard deviations are $S_1 = 3.23$ and $S_2 = 1.93$. Then t value is then computed as

$$t = \frac{(3.23^2 - 1.93^2)(\sqrt{8})}{2(3.23)(1.93)(1 - 0.847^2)} = 2.87$$

With df = 8, the t required for significance at the 0.05 level is 1.86 for a one-tailed test. Thus, the result indicates a significant difference, and the variance in the amount of analysis time appears to be significantly less when the new equipment is used.

5. With df = 8, an r = 0.765 is required as shown in Table F. Thus, the obtained r (0.847) is significant at the 0.01 level.

6. The value of interest is the frequency of patients who have improved taking the new compared to the old medicine. Therefore, the χ^2 analysis can be used because the two samples are independent. The frequencies can be set up as follows:

	Improved		**Not Improved**		
Medicine	**O**	**E**	**O**	**E**	**Row Sums**
New	160	145	40	55	200
Old	130	145	70	55	200
Column sums	290		110		

The expected frequencies for each cell are computed as follows: for the improved patients, 290 × 200 divided by 400 = 145. Chi-square (χ_2) is then

$$\chi^2 = \frac{(160 - 145)^2}{145} + \frac{(130 - 145)^2}{145} + \frac{(40 - 55)^2}{55} + \frac{(70 - 55)^2}{55}$$

$$= 11.28$$

With df = 1, a χ_2 = 6.635 is needed at the 0.01 level. Because the obtained χ_2 exceeds this value, the conclusion is that a significant difference occurs in the number of patients who have improved with the new medicine compared to the old medicine. Thus, the new medicine appears to be more effective than the old medicine in bringing about an improvement in more patients. However, because "improvement" is not defined, caution must be exercised with specific statements of effectiveness.

Chapter 11

1. Because five groups are to be compared, the one-way analysis of variance would be the appropriate statistic to use. The necessary sums are provided as shown, along with the means and standard deviations necessary for describing the results.

Statistic	A	B	C	D	E	Sums
n	8	8	8	8	8	40
ΣX	51	55	59	63	62	290
ΣX^2	333	383	443	505	484	2148
$\bar{X}$	6.38	6.88	7.38	7.88	7.75	7.25
S	1.06	0.83	1.06	1.13	0.71	1.08

$$SS_T = 2148 - \frac{290^2}{40} = 2148 - 2102.5 = 45.50$$

$$SS_w = 2148 - \frac{51^2}{8} + \frac{55^2}{8} + \frac{59^2}{8} + \frac{63^2}{8} + \frac{62^2}{8}$$

$$= 2148 - 2115 = 33.00$$

$$SS_b = 21115 - 2102.50 = 12.50$$

As a computational check, 45.5 = 33.0 + 12.5.

The analysis of variance would be

Source	SS	df	ms	F
Between	12.5	4	3.125	3.314
Within	33.0	35	0.943	
Total	45.5	39		

With df = 4 and 35, at the 0.05 level, an F = 2.64 is needed. Thus, the null hypothesis of no difference among the means is rejected, and the conclusion is that the coatings are not equally effective in their resistance to corrosion for this type of tin plate.

2. Because planned comparisons were not made, a posteriori tests have to be made by using the Tukey A or the Scheffe' test. Using the Tukey A test,

$$CR_T = 4.10 \sqrt{\frac{0.943}{8}} = 1.41$$

where 4.10 is from Table I with k = 5 and df_w = 30. Thus, any differences among means greater than 1.41 would be considered significantly different at the 0.05 level of significance. For the means listed in the answer to problem 1, the differences are

	A	B	C	D
B	0.50			
C	1.00	0.50		
D	1.50*	1.00	0.50	
E	1.37	0.87	0.37	0.13

Thus, only the difference between means A and D are significant at the 0.05 level.

3. Because several sets of data are to be compared to one standard, the Dunnett test (equation 11.15) can be used by using set A as the standard. For this data, d_k = 2.56 for five means to be com-

pared (Table J) and df = 35 for the mean square within for a two-tailed test. Then,

$$C_{\text{diff}} = 2.56\sqrt{\frac{2(0.943)}{8}} = 1.24$$

Any difference between the mean for set A and the other means that is greater than 1.24 would be significantly different at the 0.05 level. Thus, the means for sets D and E are significantly different from the mean for set A, and from inspection of the means in the answer to problem 1, are significantly higher.

4. Because an a priori test can be made for this data, orthogonal contrasts can be used for this comparison. The weights would be as shown, because the sum of these weights have to equal zero. The sum of squares for this comparison is computed from equation 11.17 by using the following procedure:

	A	B	C	D	E
Sums	51	55	59	63	62
C weights	3	3	−2	−2	−2
Sums × C weights	153	165	−118	−126	−124

$$SS_{\text{comp}} = \frac{[(3)(51) + (3)(55) + (-2)(59) + (-2)(63) + (-2)(62)]^2}{8(3^2 + 3^2 + -2^2 + -2^2 + -2^2)}$$

$$= \frac{2500}{240} = 10.42$$

The F ratio for this comparison would be computed by using the mean square from problem 1:

$$F = \frac{10.42}{0.943} = 11.05$$

which is significant at the 0.05 level for df = 1 and 35. Thus, a significant difference exists in these two types of coatings.

Chapter 12

1. The two-way analysis is a mixed design because there are two fixed levels of catalyst and six randomly selected levels of concentration. The sums of values, sums of squared values, means, and standard deviations for each cell are as shown. The sums of squares are then computed and put in the analysis of variance table. The F ratio for rows is computed by dividing the mean square for rows by the interaction mean square because this is the fixed factor. The F ratio for columns is computed by dividing the column mean square by the within mean square because this is the random factor. The F ratio of 4.05 for rows is not significant because with 1 and 5 degrees of freedom, an $F = 6.61$ is needed for significance at the 0.05 level. The F ratio for columns is significant because an $F = 2.62$ is needed for significance at the 0.05 level for df = 5 and 24. The interaction is not significant. The sums needed for the analysis are as shown.

		Concentration Levels						
Group	**Statistic**	**13**	**16**	**17**	**22**	**26**	**27**	**Sums**
A	ΣX	22	24	27	31	35	36	175
	ΣX^2	166	194	245	325	409	434	1773
B	ΣX	20	25	26	31	32	33	167
	ΣX^2	138	211	230	323	344	365	1611
A and B	ΣX	42	49	53	62	67	69	342
								3384

$$SS_T = 3384 - \frac{342^2}{36} = 135$$

$$SS_r = \frac{175^2}{18} + \frac{167^2}{18} - \frac{342^2}{36} = 1.78$$

$$SS_c = \frac{42^2}{6} + \frac{49^2}{6} + \frac{53^2}{6} + \frac{62^2}{6} + \frac{67^2}{6} + \frac{69^2}{6} - \frac{342^2}{36} = 95.68$$

$$SS_w = 3384 - \frac{22^2}{3} + \frac{24^2}{3} + \ldots \frac{32^2}{3} + \frac{33^2}{3} = 35.36$$

$$SS_{r \times c} = 135 - 1.78 - 95.68 - 35.36 = 2.18$$

Source	SS	df	ms	F
Rows	1.78	1	1.78	4.05
Columns	95.68	5	19.14	13.02
Interaction	2.18	5	0.44	0.30
Within	35.36	24	1.47	
Total	135.00	35		

The F ratio for comparing the two catalysts is not significant at the 0.05 level because an $F = 6.61$ is needed with df = 1 and 5. However, the F ratio for columns, which represents different levels of concentration, is significant and indicates a significant difference among the yields for the various levels of concentration. To determine which pairs of means are significantly different, the Scheffe' test can be used. By using the column sums, the critical range is equal to 26.32 points at the 0.05 level of significance. Computation of the Scheffe' test is as follows:

$$CR_S = \sqrt{(6 - 1)(2.62)}\ \sqrt{(2)(3)(6)(1.47)} = 26.32$$

The only difference that is significant is for concentration level 13 compared to concentration level 27 because the sums for these two columns differ by more than 26.32 points. If the levels of concentration were at equal distances, trend analysis could have been used to determine the trend of yield with increasing levels of concentration.

Thus, no significant difference in yield is found between the two catalysts. There is a significant difference in yield for concentration level 13 compared to level 27, but no significant interaction. This result indicates the yields for the two catalysts are similar at each level of concentration.

Chapter 14

1. To compute the randomized-blocks design, the first step is to obtain the marginal sums as shown. Then, the sums of squares for total, columns, blocks, and columns by blocks can be computed.

	Marginal Sums			Row
	A	B	C	Sum
1	14	18	22	54
2	12	15	19	46
3	21	28	29	78
4	16	21	20	57
5	19	26	28	73
6	13	15	15	43
7	18	22	21	61
8	10	14	13	37
9	16	17	17	49
10	12	18	17	47
	151	194	200	545

$$SS_T = 14^2 + 12^2 \ldots + 16^2 + 17^2 - \frac{545^2}{30}$$

$$= 10{,}609 - 9900.83 = 708.17$$

$$SS_c = \frac{151^2}{10} + \frac{194^2}{10} + \frac{200^2}{10} - 9900.83 = 142.87$$

$$SS_B = \frac{54^2}{3} + \frac{46^2}{3} \ldots + \frac{47^2}{3} - 9900.83 = 513.48$$

$$SS_{c \times B} = 708.17 - 142.87 - 513.48 = 51.82$$

These sums of squares are then entered into an analysis of variance table.

Source	SS	df	ms	F
Drugs	142.87	2	71.44	24.81
Subjects	513.48	9	57.05	19.81
Interaction	51.82	18	2.88	
Total	708.17	29		

With 2 and 18 degrees of freedom, an $F = 6.01$ is significant at the 0.01 level, and because the columns or drug effect is a fixed variable, the F ratio is valid and significant. Thus, the drugs are significantly different in their effectiveness with regard to reaction time.

2. Any of the post hoc tests can be used to determine which drugs are significantly different from each other. For this problem, the Tukey HSD test described by equation 11.10 can be used. For this problem,

$$\mathrm{CR}_T = q_{(df=3,18)} \sqrt{\frac{\mathrm{ms}_w}{n}}$$

$$= 4.70 \sqrt{\frac{2.88}{10}} = 2.52$$

where 4.70 is from Table I. Thus, any difference between means equal to or greater than 2.52 will be significant at the 0.01 level. Thus, the mean for drug A (15.10) is significantly different from the means for drugs B (19.40) and C (20.00), but drugs B and C are not significantly different from each other.

3. To determine the relative efficiency (RE) of the randomized-blocks design over a completely randomized ANOVA, the RE can be computed by first computing what the ms_w would be for the one-way ANOVA. Using equation 14.6,

$$\mathrm{ms}_w = \frac{513.48 + 51.82}{9 + 18} = 20.94$$

Then,

$$\mathrm{RE} = \frac{(\mathrm{df}_1 + 1)(\mathrm{df}_2 + 3)(\mathrm{ms}_2)}{(\mathrm{df}_2 + 1)(\mathrm{df}_1 + 3)(\mathrm{ms}_1)} \times 100$$

$$= \frac{(19)(30)(20.94)}{(28)(21)(2.88)} \times 100 = 704.8$$

Thus, the randomized-blocks design is about 705% more effective than the completely randomized design in that the experimental error is greatly reduced, thereby making the test statistically more powerful.

4. The sums of values for rows, columns, and drugs are as follows:

	Row	Column	Drug Description	Drug Value
1	88	92	A	82
2	80	94	B	108
3	113	94	C	118
4	93	97	D	88
5	106	103	E	84

By using equations 14.8–14.11, the sums of squares for the analysis would be computed as follows:

$$SS_T = 9702 - \frac{480^2}{25} = 9720 - 9216 = 486.00$$

$$SS_r = \frac{88^2}{5} + \frac{80^2}{5} \ldots\ldots + \frac{106^2}{5} - 9216 = 143.60$$

$$SS_c = \frac{92^2}{5} + \frac{94^2}{5} \ldots + \frac{103^2}{5} - 9216 = 14.80$$

$$SS_{TR} = \frac{82^2}{5} + \frac{108^2}{5} \ldots + \frac{84^2}{5} - 9216 = 206.40$$

$$SS_{error} = 486 - 143.6 - 14.8 - 206.4 = 121.20$$

Source	SS	df	ms	F
Drugs	206.4	4	51.6	5.11
Rows	143.6	4	35.9	3.55
Columns	14.8	4	3.7	0.37
Residual	121.2	12	10.1	
Total	486.0	24		

With df = 4 and 12, an $F = 5.41$ is needed for significance at the 0.01 level. Thus, none of the calculated F ratios are significant, and the conclusion is that there is no significant difference in the effectiveness of the five drugs.

5. Because the overall F ratio is not significant, post hoc tests should not be computed. If the F ratio was significant, then the Tukey or Scheffe' test could have been applied.

6. The relative efficiency of the Latin-square design compared to the randomized-blocks design can be estimated by using equation 14.16. The value of $ms_{B \times TR}$ can be estimated by using equation 14.17. For this problem,

$$ms_{B \times TR} = \frac{35.9 + (4)(10.1)}{5} = 15.26$$

$$RE = \left(\frac{15.26}{10.1}\right)\left(\frac{(4)(3) + 1}{(4)(3) + 3}\right)\left(\frac{4^2 + 3}{4^2 + 1}\right) \times 100 = 147.13$$

Thus, the Latin-square design is 147% more efficient than the randomized-blocks design without blocking for days.

Chapter 15

From equations 15.1–15.4, the sums of squares can be computed for the treatment effect, the nested factor, within groups, and for the total combined group by using the sums, sums of squares, and means for each plant as shown:

Plant	ΣX	ΣX^2	$\bar{X}$
	Process I		
A	105	2765	26.25
B	107	2877	26.75
C	109	2983	27.25
D	100	2520	25.00
	Process II		
E	94	2226	23.50
F	103	2655	25.75
G	93	2177	23.25
H	93	2165	23.25
Totals	804	20,368	25.125

For process I, the treatment sums total 421, and the treatment means total 26.31. For process II, the treatment sums total 383, and the treatment means total 23.94. Also,

$$SS_T = 20{,}368 - \frac{804^2}{32} = 167.50$$

$$SS_{TR} = \frac{421^2 + 383^2}{16} - \frac{804^2}{32} = 45.125$$

$$SS_{b(a1)} = \frac{105^2 + 107^2 + 109^2 + 100^2}{4} - \frac{421^2}{16} = 11.19$$

$$SS_{b(a2)} = \frac{94^2 + 103^2 + 93^2 + 93^2}{4} - \frac{383^2}{16} = 17.69$$

The total sum of squares for blocks is thus 11.19 + 17.69 = 28.88. For the within sum of squares,

$$SS_{w_{a1b1}} = 2765 - \frac{105^2}{4} = 8.75$$

$$SS_{w_{a1b2}} = 2877 - \frac{107^2}{4} = 14.75$$

The last within sum of squares would be

$$SS_{w_{a2b4}} = 2165 - \frac{93^2}{4} = 2.75$$

Thus, the total within sum of squares is 93.50.

The sums of squares would then be entered into the analysis of variance table:

Source	**SS**	**df**	**ms**	**F**
(A) Treatment	45.125	1	45.13	9.38
(B) Nested	28.880	6	4.81	1.23
(C) Within	93.500	24	3.90	
Total	167.505	31		

Because this is a mixed-effects design with process as a fixed factor and plants as a random factor, the proper error term for the treatment factor is the nested mean square. The computed F ratio is 9.38, and with df = 1 and 6, this value is not significant at the 0.01 level because an F = 13.74 is required. Thus, no significant difference exists between the two production processes. The differences among the different plants under each process is also not significant because with 6 and 24 degrees of freedom, an F = 3.67 is needed. Thus, variation among plants does not seem to be a significant factor in production yield of the chemical. If the treatment effect had been significant, then one would have to be cautious in making conclusions because these differences could have been due to differential effects of the process or due to differences in the type of equipment or other factors associated with each subgroup of plants.

Abbreviations and Symbols

Greek

α	*1.* Probability of making a Type I statistical error. *2.* Level of significance set by the researcher.
α_{ew}	Probability of making an experimentwise error.
β	*1.* Probability of making a Type II statistical error. *2.* Symbol used to designate the standard regression coefficients in multiple regression.
$1 - \beta$	Power of a statistical test.
η	Correlation ratio for nonlinear relationships.
μ	Mean of a population.
μ_0	Population mean specified by the null hypothesis.
μ_1	(1) Population mean specified by the alternative hypothesis. (2) Population mean of group 1.
μ_2	Population mean of group 2.
ρ	*1.* Population Pearson product-moment correlation. *2.* Spearman rank difference correlation coefficient.
Σ	Summation symbol.
$\Sigma\Sigma X$	Double summation obtained by first summing the values of X in each group and then summing over all groups.
σ	Population standard deviation.

1453–0/88/0407$06.00/1

σ^2 Population variance.

σ_M Standard error of a population mean.

ϕ Correlation between two dichotomous variables.

χ^2 Pearson's chi-square statistic used to determine the significance of difference between two frequency distributions or dependence in a cross-tabulation.

ω^2 Measure of the strength of association between independent and dependent variables in the analysis of variance.

English

a Intercept on the *x* or *y* axis in regression analysis.

$A \times B$ Interaction of two variables.

b Slope of the regression line in regression analysis.

cf Cumulative frequency.

C *1*. Contingency coefficient. *2*. Number of columns.

CI Confidence interval for a statistic.

CL Confidence limits for a statistic.

CR Critical range used in analysis of variance post hoc methods.

CV Coefficient of variation, also referred to as the relative standard deviation (RSD).

d Effect size, which is calculated as $(\mu_1 - \mu_0)/\sigma$.

df Degrees of freedom.

$D_{z'}$ Difference between respective Fisher z′ values.

E	Index of forecasting efficiency.
f	Frequency of values in a distribution or cell.
f_1	Frequency of event 1.
f_2	*1.* Frequency of values below a given value in a frequency distribution. *2.* Frequency of event 2.
f_e	Expected frequency in chi-square analysis.
f_o	Observed frequency in chi-square analysis.
F	Ratio of between variance divided by within variance in the analysis of variance or in comparing two independent variances.
H_0	Null hypothesis.
H_1	Alternative hypothesis.
HM	Harmonic mean.
i	*1.* Width of an interval. *2.* Index of summation.
k	*1.* Number of subgroups, categories, or levels in analysis. *2.* Coefficient of alienation.
k^2	Coefficient of nondetermination.
LL	Lower limits of an interval.
M	Moment about the arithmetic mean used to obtain measures of skewness or kurtosis.
ms	Mean square or variance estimate in the analysis of variance.
M	Mean of the values in a distribution or group of values.
M_D	Mean difference between two measures.

Mdn — Median value, or P_{50}.

n — Number of cases in a sample; size of sample.

N — Total number of cases in a population.

p — Proportion.

$\overline{p}_w$ — Weighted mean for two proportions.

P — Probability of an event P.

P_{50} — Pecentile point in a distribution above and below which 50% of the cases lie.

$\overline{q}_w$ — The weighted mean for two proportions of nonfavored events or nonoccurence of an event P in a dichotomy, or $1 - \overline{p}_w$.

Q — *1.* Probability of an event Q. In a dichotomy, $Q = 1 - P$, which is the probability of nonoccurrence of an event P. *2.* Cochran's Q for extending McNemar's test to three or more groups.

$r_{\max xy}$ — The maximum correlation possible between two unreliable variables X and Y.

r_{Txy} — Hypothetical true correlation between two variables X and Y corrected for attenuation.

r_{xx}^2 — Reliability coefficient for the measure of variable x.

r_{xy} — Pearson product-moment correlation coefficient between two variables x and y.

r_{xy}^2 — Coefficient of determination, which indicates the percentage of variance common between two variables.

R — *1.* Range of values in a distribution.
2. Coefficient of multiple correlation.
3. Number of rows.

RE	Relative efficiency of one analysis compared to another analysis.
RSD	Relative standard deviation.
S	Standard deviation of a sample of values.
SE_{xy}	Standard error for predicting X from Y.
S^2	Variance of a sample of values.
SD	Standard deviation of a sample of values.
SE	Standard error of a statistic.
SE_{D_M}	Standard error of the difference between means.
SE_{η}	Standard error of the correlation ratio (η).
SE_M	Standard error of a mean.
SE_{meas}	Standard error of measurement.
SS	Sum of squares, which is calculated as Σx^2.
SS_b	Sum of squares between groups.
SS_c	Sum of squares for columns.
$SS_{r \times c}$	Sum of squares for interaction.
SS_r	Sum of squares for rows.
SS_T	Total sum of squares.
SS_w	Sum of squares within groups.
t	Fisher's or Student's test of difference between two means.
T	(1) Total of rows and columns in a table; used to com-

	pute Cochran's Q. (2) Wilcoxon matched-pairs signed-ranks test.
UL	Upper limit for an interval.
x	Deviation value, which is calculated as $X - \overline{X}$.
x^2	Squared deviation value, which is calculated as $(X - \overline{X})^2$.
Σx^2	Sum of squares, which is ccalculated as $\Sigma(X - \overline{X})^2$.
X	A variable, or an individual value for that variable.
$\overline{X}$	Mean of a sample of values of variable X.
$\overline{X}_w$	Weighted mean of several groups for variable X.
Y	A variable (*see* X).
$\hat{Y}$	Predicted value in regression analysis.
z	Normal curve unit, which is the ratio of the difference between a sample and a population mean divided by the standard error of the mean difference.
z'	Fisher's r to z transformed values.
Z	Deviation score based on $M = 50$ and $S = 10$.

Appendix

Statistical Tables

Appendix

Table A. Random Numbers

28 89 65 87 08	13 50 63 04 23	25 47 57 91 13	52 62 24 19 94	91 67 48 57 10
30 29 43 65 42	78 66 28 55 80	47 46 41 90 08	55 98 78 10 70	49 92 05 12 07
95 74 62 60 53	51 57 32 22 27	12 72 72 27 77	44 67 32 23 13	67 95 07 76 30
01 85 54 96 72	66 86 65 64 60	56 59 75 36 75	46 44 33 63 71	54 50 06 44 75
10 91 46 96 86	19 83 52 47 53	65 00 51 93 51	30 80 05 19 29	56 23 27 19 03
05 33 18 08 51	51 78 57 26 17	34 87 96 23 95	89 99 93 39 79	11 28 94 15 52
04 43 13 37 00	79 68 96 26 60	70 39 83 66 56	62 03 55 86 57	77 55 33 62 02
05 85 40 25 24	73 52 93 70 50	48 21 47 74 63	17 27 27 51 26	35 96 29 00 45
84 90 90 65 77	63 99 25 69 02	09 04 03 35 78	19 79 95 07 21	02 84 48 51 97
28 55 53 09 48	86 28 30 02 35	71 30 32 06 47	93 74 21 86 33	49 90 21 69 74
89 83 40 69 80	97 96 47 59 97	56 33 24 87 36	17 18 16 90 46	75 27 28 52 13
73 20 96 05 68	93 41 69 96 07	97 50 81 79 59	42 37 13 81 83	92 42 85 04 31
10 89 07 76 21	40 24 74 36 42	40 33 04 46 24	35 63 02 31 61	34 59 43 36 96
91 50 27 78 37	06 06 16 25 98	17 78 80 36 85	26 41 77 63 37	71 63 94 94 33
03 45 44 66 88	97 81 26 03 89	39 46 67 21 17	98 10 39 33 15	61 63 00 25 92
89 41 58 91 63	65 99 59 97 84	90 14 79 61 55	56 16 88 87 60	32 15 99 67 43
13 43 00 97 26	16 91 21 32 41	60 22 66 72 17	31 85 33 69 07	68 49 20 43 29
71 71 00 51 72	62 03 89 26 32	35 27 99 18 25	78 12 03 09 70	50 93 19 35 56
19 28 15 00 41	92 27 73 40 38	37 11 05 75 16	98 81 99 37 29	92 20 32 39 67
56 38 30 92 30	45 51 94 69 04	00 84 14 36 37	95 66 39 01 09	21 68 40 95 79
39 27 52 89 11	00 81 06 28 48	12 08 05 75 26	03 35 63 05 77	13 81 20 67 58
73 13 28 58 01	05 06 42 24 07	60 60 29 99 93	72 93 78 04 36	25 76 01 54 03
81 60 84 51 57	12 68 46 55 89	60 09 71 87 89	70 81 10 95 91	83 79 68 20 66
05 62 98 07 85	07 79 26 69 61	67 85 72 37 41	85 79 76 48 23	61 58 87 08 05
62 97 16 29 18	52 16 16 23 56	62 95 80 97 63	32 25 34 03 36	48 84 60 37 65
31 13 63 21 08	16 01 92 58 21	48 79 74 74 72	08 64 80 91 38	07 28 66 61 59
97 38 35 34 19	89 84 05 34 47	88 09 31 54 88	97 96 86 01 69	46 13 95 65 96

Continued on next page.

1453–0/88/0413$10.25/1

Table A.—Continued

32 11 78 33 82	51 99 98 44 39	12 75 10 60 36	80 66 39 94 97	42 36 31 16 59
81 99 13 37 05	08 12 60 39 23	61 73 84 89 18	26 02 04 37 95	96 18 69 06 30
45 74 00 03 05	69 99 47 26 52	48 06 30 00 18	03 30 28 55 59	66 10 71 44 05
11 84 13 69 01	88 91 28 79 50	71 42 14 96 55	98 59 96 01 36	88 77 90 45 59
14 66 12 87 22	59 45 27 08 51	85 64 23 85 41	64 72 08 59 44	67 98 36 65 56
40 25 67 87 82	84 27 17 30 37	48 69 49 02 58	98 02 50 58 11	95 39 06 35 63
44 48 97 49 43	65 45 53 41 07	14 83 46 74 11	76 66 63 60 08	90 54 33 65 84
41 94 54 06 57	48 28 01 83 84	09 11 21 91 73	97 28 44 74 06	22 30 95 69 72
07 12 15 58 84	93 18 31 83 45	54 52 62 29 91	53 58 54 66 05	47 19 63 92 75
64 27 90 43 52	18 26 32 96 83	50 58 45 27 57	14 96 39 64 85	73 87 96 76 23
80 71 86 41 03	45 62 63 40 88	35 69 34 10 94	32 22 52 04 74	69 63 21 83 41
27 06 08 09 92	26 22 59 28 27	38 58 22 14 79	24 32 12 38 42	33 56 90 92 57
54 68 97 20 54	33 26 74 03 30	74 22 19 13 48	30 28 01 92 49	58 61 52 27 03
02 92 65 68 99	05 53 15 26 70	04 69 22 64 07	04 73 25 74 82	78 35 22 21 88
83 52 57 78 62	98 61 70 48 22	68 50 64 55 75	42 70 32 09 60	58 70 61 43 97
82 82 76 31 33	85 13 41 38 10	16 47 61 43 77	83 27 19 70 41	34 78 77 60 25
38 61 34 09 49	04 41 66 09 76	20 50 73 40 95	24 77 95 73 20	47 42 80 61 03
01 01 11 88 38	03 10 16 82 24	39 58 20 12 39	82 77 02 18 88	33 11 49 15 16
21 66 14 38 28	54 08 18 07 04	92 17 63 36 75	33 14 11 11 78	97 30 53 62 38
32 29 30 69 59	68 50 33 31 47	15 64 88 75 27	04 51 41 61 96	86 62 93 66 71
04 59 21 65 47	39 90 89 86 77	46 86 06 88 86	50 09 13 24 91	54 80 67 78 66
38 64 50 07 36	56 50 45 94 25	48 28 48 30 51	60 73 73 03 87	68 47 37 10 84
48 33 50 83 53	59 77 64 59 90	58 92 62 50 18	93 09 45 89 06	13 26 98 86 29
25 19 64 82 84	62 74 29 92 24	61 03 91 22 48	64 94 63 15 07	66 85 12 00 27
23 02 41 46 04	44 31 52 43 07	44 06 03 09 34	19 83 94 62 94	48 28 01 51 92
55 85 66 96 28	28 30 62 58 83	65 68 62 42 45	13 08 60 46 28	95 68 45 52 43
68 45 19 69 59	35 14 82 56 80	22 06 52 26 39	59 78 98 76 14	36 09 03 01 86
69 31 46 29 85	18 88 26 95 54	01 02 14 03 05	48 00 26 43 85	33 93 81 45 95

37 31 61 28 98	94 61 47 03 10	67 80 84 41 26	88 84 59 69 14	77 32 82 81 89
66 42 19 24 94	13 13 38 69 96	76 69 76 24 13	43 83 10 13 24	18 32 84 85 04
33 65 78 12 35	91 59 11 38 44	23 31 48 75 74	05 30 08 46 32	90 04 93 56 16
76 32 06 19 35	22 95 30 19 29	57 74 43 20 90	20 25 36 70 69	38 32 11 01 01
43 33 42 02 59	20 39 84 95 61	58 22 04 02 99	99 78 78 83 82	43 67 16 38 95
28 31 93 43 94	87 73 19 38 47	54 36 90 98 10	83 43 32 26 26	22 00 90 59 22
97 19 21 63 34	69 33 17 03 02	11 15 50 46 08	42 69 60 17 42	14 68 61 14 48
82 80 37 14 20	56 39 59 89 63	33 90 38 44 50	78 22 87 10 88	06 58 87 39 67
03 68 03 13 60	64 13 09 37 11	86 02 57 41 99	31 66 60 65 64	03 03 02 58 97
65 16 58 11 01	98 78 80 63 23	07 37 66 20 56	20 96 06 79 80	33 39 40 49 42
24 65 58 57 04	18 62 85 28 24	26 45 17 82 76	39 65 01 73 91	50 37 49 38 73
02 72 64 07 75	85 66 48 38 73	75 10 96 59 31	48 78 58 08 88	72 08 54 57 17
79 16 78 63 99	43 61 00 66 42	76 26 71 14 33	33 86 76 71 66	37 85 05 56 07
04 75 14 93 39	68 52 16 83 34	64 09 44 62 58	48 32 72 26 95	32 67 35 49 71
40 64 64 57 60	97 00 12 91 33	22 14 73 01 11	83 97 68 95 65	67 77 80 98 87
06 27 07 34 26	01 52 48 69 57	19 17 53 55 96	02 41 03 89 33	86 85 73 02 32
62 40 03 87 10	96 88 22 46 94	35 56 60 94 20	60 73 04 84 98	96 45 18 47 07
00 98 48 18 97	91 51 63 27 95	74 25 84 03 07	88 29 04 79 84	03 71 13 78 26
50 64 19 18 91	98 55 83 46 09	49 66 41 12 45	41 49 36 83 43	53 75 35 13 39
38 54 52 25 78	01 98 00 89 85	86 12 22 89 25	10 10 71 19 45	88 84 77 00 07
46 86 80 97 78	65 12 64 64 70	58 41 05 49 08	68 68 88 54 00	81 61 61 80 41
90 72 92 93 10	09 12 81 93 63	69 30 02 04 26	92 36 48 69 45	91 99 08 07 65
66 21 41 77 60	99 35 72 61 22	52 40 74 67 29	97 50 71 39 79	57 82 14 88 06
87 05 46 52 76	89 96 34 22 37	27 11 57 04 19	57 93 08 35 69	07 51 19 92 66
46 90 61 03 06	89 85 33 22 80	34 89 12 29 37	44 71 38 40 37	15 49 55 51 08
11 88 53 06 09	81 18 33 98 29	91 27 59 43 09	70 72 51 49 73	35 97 25 83 41
11 05 92 06 97	68 82 34 08 83	25 40 58 40 64	56 42 78 54 06	60 96 96 12 82
33 94 24 20 28	62 42 07 12 63	34 39 02 92 31	80 61 68 44 19	09 92 14 73 49
24 89 74 75 61	61 02 73 36 85	67 28 50 49 85	37 79 95 02 66	73 19 76 28 13

Continued on next page.

Table A.—Continued

15 19 74 67 23	16 38 93 73 68	76 23 15 58 20	35 36 82 82 59	01 33 48 17 66
05 64 12 70 88	80 85 35 06 88	73 48 27 39 43	43 40 13 35 45	55 10 54 38 50
57 49 36 44 06	74 93 55 39 26	27 70 98 76 68	78 36 26 24 06	43 24 56 40 80
77 82 96 96 97	60 42 17 18 48	16 34 92 19 52	98 84 48 42 92	83 19 00 15 42
24 10 70 06 51	59 62 37 95 42	53 67 14 95 29	84 65 46 07 30	77 54 00 15 42
50 00 07 78 23	49 54 36 85 14	18 50 54 18 82	23 79 80 71 37	60 62 95 40 30
44 37 76 21 96	37 03 08 98 64	90 85 59 43 64	17 79 96 52 35	21 05 22 59 30
90 57 55 17 47	53 26 79 20 38	69 90 58 64 03	33 48 32 91 54	68 44 90 24 25
50 74 64 67 42	95 28 12 73 23	32 54 98 64 94	82 17 18 17 14	55 10 61 64 29
44 04 70 22 02	84 31 64 64 08	52 55 04 24 29	91 95 43 81 14	66 13 18 47 44
32 74 61 64 73	21 46 51 44 77	72 48 92 00 05	83 59 89 65 06	53 76 70 58 78
75 73 51 70 49	12 53 67 51 54	38 10 11 67 73	22 32 61 43 75	31 61 22 21 11
76 18 36 16 34	16 28 25 82 98	64 26 70 54 87	49 48 55 11 39	94 25 20 80 85
00 17 37 71 81	64 21 91 15 82	81 04 14 52 11	39 07 30 60 77	39 18 27 85 68
54 95 57 55 04	12 77 40 70 14	79 86 61 57 50	52 49 41 73 46	05 63 34 92 33
69 99 95 54 63	44 37 33 53 17	38 06 58 37 93	47 10 62 31 28	63 59 40 40 32
13 13 92 66 99	47 24 49 57 74	32 25 43 62 17	10 97 11 69 84	99 63 22 32 98
33 78 80 87 15	38 30 06 38 21	14 47 07 17 26	54 31 58 50 08	11 39 03 96 25
04 31 17 21 56	63 73 99 19 87	26 72 39 27 67	53 77 57 68 93	60 61 97 22 61
61 06 98 03 91	87 14 77 43 96	43 00 65 98 50	45 60 33 01 07	98 99 46 50 47

SOURCE: Tables A, C, F, and G are taken from Tables III, VII, Viii, and XXXIII of Fisher and Yates: *Statistical Tables for Biological, Agricultural and Medical Research* published by Longman Group UK Ltd. London (previously published by Oliver and Boyd Ltd., Edinburgh) and by permission of the authors and publishers.

Table B. Proportional Areas of the Normal Distribution for Given z Values

(1) z	(2) Area	(3) Area	(1) z	(2) Area	(3) Area	(1) z	(2) Area	(3) Area
0.00	0.0000	0.5000	0.30	0.1179	0.3821	0.60	0.2257	0.2743
0.01	0.0040	0.4960	0.31	0.1217	0.3783	0.61	0.2291	0.2709
0.02	0.0080	0.4920	0.32	0.1255	0.3745	0.62	0.2324	0.2676
0.03	0.0120	0.4880	0.33	0.1293	0.3707	0.63	0.2357	0.2643
0.04	0.0160	0.4840	0.34	0.1331	0.3669	0.64	0.2389	0.2611
0.05	0.0199	0.4801	0.35	0.1368	0.3632	0.65	0.2422	0.2578
0.06	0.0239	0.4761	0.36	0.1406	0.3594	0.66	0.2454	0.2546
0.07	0.0279	0.4721	0.37	0.1443	0.3557	0.67	0.2486	0.2514
0.08	0.0319	0.4681	0.38	0.1480	0.3520	0.68	0.2517	0.2483
0.09	0.0359	0.4641	0.39	0.1517	0.3483	0.69	0.2549	0.2451
0.10	0.0398	0.4602	0.40	0.1554	0.3446	0.07	0.2580	0.2420
0.11	0.0438	0.4562	0.41	0.1591	0.3409	0.71	0.2611	0.2389
0.12	0.0478	0.4522	0.42	0.1628	0.3372	0.72	0.2642	0.2358
0.13	0.0517	0.4483	0.43	0.1664	0.3336	0.73	0.2673	0.2327
0.14	0.0557	0.4443	0.44	0.1700	0.3300	0.74	0.2704	0.2296
0.15	0.0596	0.4404	0.45	0.1736	0.3264	0.75	0.2734	0.2266
0.16	0.0636	0.4364	0.46	0.1772	0.3228	0.76	0.2764	0.2236
0.17	0.0675	0.4325	0.47	0.1808	0.3192	0.77	0.2794	0.2206
0.18	0.0714	0.4286	0.48	0.1844	0.3156	0.78	0.2823	0.2177
0.19	0.0753	0.4247	0.49	0.1879	0.3121	0.79	0.2852	0.2148
0.20	0.0793	0.4207	0.50	0.1915	0.3085	0.80	0.2881	0.2119
0.21	0.0832	0.4168	0.51	0.1950	0.3050	0.81	0.2910	0.2090
0.22	0.0871	0.4129	0.52	0.1985	0.3015	0.82	0.2939	0.2061
0.23	0.0910	0.4090	0.53	0.2019	0.2981	0.83	0.2967	0.2033

Continued on next page.

Table B.—Continued

(1) z	(2) Area	(3) Area	(1) z	(2) Area	(3) Area	(1) z	(2) Area	(3) Area
0.24	0.0948	0.4052	0.54	0.2054	0.2946	0.84	0.2995	0.2005
0.25	0.0987	0.4013	0.55	0.2088	0.2912	0.85	0.3023	0.1977
0.26	0.1026	0.3974	0.56	0.2123	0.2877	0.86	0.3051	0.1949
0.27	0.1064	0.3936	0.57	0.2157	0.2843	0.87	0.3078	0.1922
0.28	0.1103	0.3897	0.58	0.2190	0.2810	0.88	0.3106	0.1894
0.29	0.1141	0.3859	0.59	0.2224	0.2776	0.89	0.3133	0.1867
0.90	0.3159	0.1841	1.35	0.4115	0.0885	1.80	0.4641	0.0359
0.91	0.3186	0.1814	1.36	0.4131	0.0869	1.81	0.4649	0.0351
0.92	0.3212	0.1788	1.37	0.4147	0.0853	1.82	0.4656	0.0344
0.93	0.3238	0.1762	1.38	0.4162	0.0838	1.83	0.4664	0.0336
0.94	0.3264	0.1736	1.39	0.4177	0.0823	1.84	0.4671	0.0329
0.95	0.3289	0.1711	1.40	0.4192	0.0808	1.85	0.4678	0.0322
0.96	0.3315	0.1685	1.41	0.4207	0.0793	1.86	0.4686	0.0314
0.97	0.3340	0.1660	1.42	0.4222	0.0778	1.87	0.4693	0.0307
0.98	0.3365	0.1635	1.43	0.4236	0.0764	1.88	0.4699	0.0301
0.99	0.3389	0.1611	1.44	0.4251	0.0749	1.89	0.4706	0.0294
1.00	0.3413	0.1587	1.45	0.4265	0.0735	1.90	0.4713	0.0287
1.01	0.3438	0.1562	1.46	0.4279	0.0721	1.91	0.4719	0.0281
1.02	0.3461	0.1539	1.47	0.4292	0.0708	1.92	0.4726	0.0274
1.03	0.3485	0.1515	1.48	0.4306	0.0694	1.93	0.4732	0.0268
1.04	0.3508	0.1492	1.49	0.4319	0.0681	1.94	0.4738	0.0262
1.02	0.3531	0.1469	1.50	0.4332	0.0668	1.95	0.4744	0.0256
1.06	0.3554	0.1446	1.51	0.4345	0.0655	1.96	0.4750	0.0250
1.07	0.3577	0.1423	1.52	0.4357	0.0643	1.97	0.4756	0.0244
1.08	0.3599	0.1401	1.53	0.4370	0.0630	1.98	0.4761	0.0239

1.09	0.3621	0.1379	1.54	0.4382	0.0618	1.99	0.4767	0.0233
1.10	0.3643	0.1357	1.55	0.4394	0.0606	2.00	0.4772	0.0228
1.11	0.3665	0.1335	1.56	0.4406	0.0594	2.01	0.4778	0.0222
1.12	0.3686	0.1314	1.57	0.4418	0.0582	2.02	0.4783	0.0217
1.13	0.3708	0.1292	1.58	0.4429	0.0571	2.03	0.4788	0.0212
1.14	0.3729	0.1271	1.59	0.4441	0.0559	2.04	0.4793	0.0207
1.15	0.3749	0.1251	1.60	0.4452	0.0548	2.05	0.4798	0.0202
1.16	0.3770	0.1230	1.61	0.4463	0.0537	2.06	0.4803	0.0197
1.17	0.3790	0.1210	1.62	0.4474	0.0526	2.07	0.4808	0.0192
1.18	0.3810	0.1190	1.63	0.4484	0.0516	2.08	0.4812	0.0188
1.19	0.3830	0.1170	1.64	0.4495	0.0505	2.09	0.4817	0.0183
			1.645	0.4500	0.0500			
1.20	0.3849	0.1151	1.65	0.4505	0.0495	2.10	0.4821	0.0179
1.21	0.3869	0.1131	1.66	0.4515	0.0485	2.11	0.4826	0.0174
1.22	0.3888	0.1112	1.67	0.4525	0.0475	2.12	0.4830	0.0170
1.23	0.3907	0.1093	1.68	0.4535	0.0465	2.13	0.4834	0.0166
1.24	0.3925	0.1075	1.69	0.4545	0.0455	2.14	0.4838	0.0162
1.25	0.3944	0.1056	1.70	0.4554	0.0446	2.15	0.4842	0.0158
1.26	0.3962	0.1038	1.71	0.4564	0.0436	2.16	0.4846	0.0154
1.27	0.3980	0.1020	1.72	0.4573	0.0427	2.17	0.4850	0.0150
1.28	0.3997	0.1003	1.73	0.4582	0.0418	2.18	0.4854	0.0146
1.29	0.4015	0.0985	1.74	0.4591	0.0409	2.19	0.4857	0.0143
1.30	0.4032	0.0968	1.75	0.4599	0.0401	2.20	0.4861	0.0139
1.31	0.4049	0.0951	1.76	0.4608	0.0392	2.21	0.4864	0.0136
1.32	0.4066	0.0934	1.77	0.4616	0.0384	2.22	0.4868	0.0132
1.33	0.4082	0.0918	1.78	0.4625	0.0375	2.23	0.4871	0.0129
1.34	0.4099	0.0901	1.79	0.4633	0.0367	2.24	0.4875	0.0125
2.25	0.4878	0.0122	2.64	0.4959	0.0041	3.00	0.4987	0.0013
2.26	0.4881	0.0119	2.65	0.4960	0.0040	3.01	0.4987	0.0013

Continued on next page.

Table B.—Continued

(1) z	(2) Area	(3) Area	(1) z	(2) Area	(3) Area	(1) z	(2) Area	(3) Area
2.27	0.4884	0.0116	2.66	0.4961	0.0039	3.02	0.4987	0.0013
2.28	0.4887	0.0113	2.67	0.4962	0.0038	3.03	0.4988	0.0012
2.29	0.4890	0.0110	2.68	0.4963	0.0037	3.04	0.4988	0.0012
2.30	0.4893	0.0107	2.69	0.4964	0.0036	3.05	0.4989	0.0011
2.31	0.4896	0.0104	2.70	0.4965	0.0035	3.06	0.4989	0.0011
2.32	0.4898	0.0102	2.71	0.4966	0.0034	3.07	0.4989	0.0011
2.33	0.4901	0.0099	2.72	0.4967	0.0033	3.08	0.4990	0.0010
2.34	0.4904	0.0096	2.73	0.4968	0.0032	3.09	0.4990	0.0010
2.35	0.4906	0.0094	2.74	0.4969	0.0031	3.10	0.4990	0.0010
2.36	0.4909	0.0091	2.75	0.4970	0.0030	3.11	0.4991	0.0009
2.37	0.4911	0.0089	2.76	0.4971	0.0029	3.12	0.4991	0.0009
2.38	0.4913	0.0087	2.77	0.4972	0.0028	3.13	0.4991	0.0009
2.39	0.4916	0.0084	2.78	0.4973	0.0027	3.14	0.4992	0.0008
2.40	0.4918	0.0082	2.79	0.4974	0.0026	3.15	0.4992	0.0008
2.41	0.4920	0.0080	2.80	0.4974	0.0026	3.16	0.4992	0.0008
2.42	0.4922	0.0078	2.81	0.4975	0.0025	3.17	0.4992	0.0008
2.43	0.4925	0.0075	2.82	0.4976	0.0024	3.18	0.4993	0.0007

2.44	0.4927	0.0073	2.83	0.4977	0.0023	3.19	0.4993	0.0007
2.45	0.4929	0.0071	2.84	0.4977	0.0023	3.20	0.4993	0.0007
2.46	0.4931	0.0069	2.85	0.4978	0.0022	3.21	0.4993	0.0007
2.47	0.4932	0.0068	2.86	0.4979	0.0021	3.22	0.4994	0.0006
2.48	0.4934	0.0066	2.87	0.4979	0.0021	3.23	0.4994	0.0006
2.49	0.4936	0.0064	2.88	0.4980	0.0020	3.24	0.4994	0.0006
2.50	0.4938	0.0062	2.89	0.4981	0.0019	3.25	0.4994	0.0006
2.51	0.4940	0.0060	2.90	0.4981	0.0019	3.30	0.4995	0.0005
2.52	0.4941	0.0059	2.91	0.4982	0.0018	3.35	0.4996	0.0004
2.53	0.4943	0.0057	2.92	0.4982	0.0018	3.40	0.4997	0.0003
2.54	0.4945	0.0055	2.93	0.4983	0.0017	3.45	0.4997	0.0003
2.55	0.4946	0.0054	2.94	0.4984	0.0016	3.50	0.4998	0.0002
2.56	0.4948	0.0052	2.95	0.4984	0.0016	3.60	0.4998	0.0002
2.57	0.4949	0.0051	2.96	0.4985	0.0015	3.70	0.4999	0.0001
2.576	0.4950	0.0050	2.97	0.4985	0.0015	3.80	0.4999	0.0001
2.58	0.4951	0.0049	2.98	0.4986	0.0014	3.90	0.49995	0.00005
2.59	0.4952	0.0048	2.99	0.4986	0.0014	4.00	0.49997	0.00003
2.60	0.4953	0.0047						
2.61	0.4955	0.0045						
2.62	0.4956	0.0044						
2.63	0.4957	0.0043						

Table C. Critical Values of *t*

	Level of significance for one-tailed test					
	0.10	0.05	0.025	0.01	0.005	0.0005
	Level of significance for two-tailed test					
	0.20	**0.10**	**0.05**	**0.02**	**0.01**	**0.001**
1	3.078	6.314	12.706	31.821	63.657	636.619
2	1.886	2.920	4.303	6.965	9.925	31.598
3	1.638	2.353	3.182	4.541	5.841	12.941
4	1.533	2.132	2.776	3.747	4.604	8.610
5	1.476	2.015	2.571	3.365	4.032	6.859
6	1.440	1.943	2.447	3.143	3.707	5.959
7	1.415	1.895	2.365	2.998	3.499	5.405
8	1.397	1.860	2.306	2.896	3.355	5.041
9	1.383	1.833	2.262	2.821	3.250	4.781
10	1.372	1.812	2.228	2.764	3.169	4.587
11	1.363	1.796	2.201	2.718	3.106	4.437
12	1.356	1.782	2.179	2.681	3.055	4.318
13	1.350	1.771	2.160	2.650	3.012	4.221
14	1.345	1.761	2.145	2.624	2.977	4.140
15	1.341	1.753	2.131	2.602	2.947	4.073
16	1.337	1.746	2.120	2.583	2.921	4.015
17	1.333	1.740	2.110	2.567	2.898	3.965
18	1.330	1.734	2.101	2.552	2.878	3.922
19	1.328	1.729	2.093	2.539	2.861	3.883
20	1.325	1.725	2.086	2.528	2.845	3.850
21	1.323	1.721	2.080	2.518	2.831	3.819
22	1.321	1.717	2.074	2.508	2.819	3.792

Table C.—Continued

	Level of significance for one-tailed test					
	0.10	**0.05**	**0.025**	**0.01**	**0.005**	**0.0005**
	Level of significance for two-tailed test					
	0.20	**0.10**	**0.05**	**0.02**	**0.01**	**0.001**
23	1.319	1.714	2.069	2.500	2.807	3.767
24	1.318	1.711	2.064	2.492	2.797	3.745
25	1.316	1.708	2.060	2.485	2.787	3.725
26	1.315	1.706	2.056	2.479	2.779	3.707
27	1.314	1.703	2.052	2.473	2.771	3.690
28	1.313	1.701	2.048	2.467	2.763	3.674
29	1.311	1.699	2.045	2.462	2.756	3.659
30	1.310	1.697	2.042	2.457	2.750	3.646
35	1.306	1.690	2.030	2.438	2.724	3.591
40	1.303	1.684	2.021	2.423	2.704	3.551
50	1.299	1.676	2.009	2.403	2.678	3.496
60	1.296	1.671	2.000	2.390	2.660	3.460
70	1.294	1.667	1.994	2.381	2.648	3.435
80	1.292	1.664	1.990	2.374	2.639	3.416
90	1.291	1.662	1.987	2.368	2.632	3.402
100	1.290	1.660	1.984	2.364	2.626	3.390
120	1.289	1.658	1.980	2.358	2.617	3.373
∞	1.282	1.645	1.960	2.326	2.576	3.291

SOURCE: Tables A, C, F, and G are taken from Tables III, VII, Viii, and XXXIII of Fisher and Yates: *Statistical Tables for Biological, Agricultural and Medical Research* published by Longman Group UK Ltd. London (previously published by Oliver and Boyd Ltd., Edinburgh) and by permission of the authors and publishers.

Table D. 5 Percent (Lightface Type) and 1 Percent

f_2	f_1 Degrees of Freedom (for											
	1	2	3	4	5	6	7	8	9	10	11	12
1	161	200	216	225	230	234	237	239	241	242	243	244
	4,052	**4,999**	**5,403**	**5,625**	**5,764**	**5,859**	**5,928**	**5,981**	**6,022**	**6,056**	**6,082**	**6,106**
2	18.51	19.00	19.16	19.25	19.30	19.33	19.36	19.37	19.38	19.39	19.40	19.41
	98.49	**99.00**	**99.17**	**99.25**	**99.30**	**99.33**	**99.34**	**99.36**	**99.38**	**99.40**	**99.41**	**99.42**
3	10.13	9.55	9.28	9.12	9.01	8.94	8.88	8.84	8.81	8.78	8.76	8.74
	34.12	**30.82**	**29.46**	**28.71**	**28.24**	**27.91**	**29.67**	**27.49**	**27.34**	**27.23**	**27.13**	**27.05**
4	7.71	6.94	6.59	6.39	6.26	6.16	6.09	6.04	6.00	5.96	5.93	5.91
	21.20	**18.00**	**16.69**	**15.98**	**15.52**	**15.21**	**14.98**	**14.80**	**14.66**	**14.54**	**14.45**	**14.37**
5	6.61	5.79	5.41	5.19	5.05	4.95	4.88	4.82	4.78	4.74	4.70	4.68
	16.26	**13.27**	**12.06**	**11.39**	**10.97**	**10.67**	**10.45**	**10.27**	**10.15**	**10.05**	**9.96**	**9.89**
6	5.99	5.14	4.76	4.53	4.39	4.28	4.21	4.15	4.10	4.06	4.03	4.00
	13.74	**10.92**	**9.78**	**9.15**	**8.75**	**8.47**	**8.26**	**8.10**	**7.98**	**7.87**	**7.79**	**7.72**
7	5.59	4.74	4.35	4.12	3.97	3.87	3.79	3.73	3.68	3.63	3.60	3.57
	12.25	**9.55**	**8.45**	**7.85**	**7.46**	**7.19**	**7.00**	**6.84**	**6.71**	**6.62**	**6.54**	**6.47**
8	5.32	4.46	4.07	3.84	3.69	3.58	3.50	3.44	3.39	3.34	3.31	3.28
	11.26	**8.65**	**7.59**	**7.01**	**6.63**	**6.37**	**6.19**	**6.03**	**5.91**	**5.82**	**5.74**	**5.67**
9	5.12	4.26	3.86	3.63	3.48	3.37	3.29	3.23	3.18	3.13	3.10	3.07
	10.56	**8.02**	**6.99**	**6.42**	**6.06**	**5.80**	**5.62**	**5.47**	**5.35**	**5.26**	**5.18**	**5.11**
10	4.96	4.10	3.71	3.48	3.33	3.22	3.14	3.07	3.02	2.97	2.94	2.91
	10.04	**7.56**	**6.55**	**5.99**	**5.64**	**5.39**	**5.21**	**5.06**	**4.95**	**4.85**	**4.78**	**4.71**
11	4.84	3.98	3.59	3.36	3.20	3.09	3.01	2.95	2.90	2.86	2.82	2.79
	9.65	**7.20**	**6.22**	**5.67**	**5.32**	**5.07**	**4.88**	**4.74**	**4.63**	**4.54**	**4.46**	**4.40**
12	4.75	3.88	3.49	3.26	3.11	3.00	2.92	2.85	2.80	2.76	2.72	2.69
	9.33	**6.93**	**5.95**	**5.41**	**5.06**	**4.82**	**4.65**	**4.50**	**4.39**	**4.30**	**4.22**	**4.16**
13	4.67	3.80	3.41	3.18	3.02	2.92	2.84	2.72	2.77	2.63	2.63	2.60
	9.07	**6.70**	**5.74**	**5.20**	**4.86**	**4.62**	**4.44**	**4.30**	**4.19**	**4.10**	**4.02**	**3.96**
14	4.60	3.74	3.34	3.11	2.96	2.85	2.77	2.70	2.65	2.60	2.56	2.53
	8.86	**6.51**	**5.56**	**5.03**	**4.69**	**4.46**	**4.28**	**4.14**	**4.03**	**3.94**	**3.86**	**3.80**
15	4.54	3.68	3.29	3.06	2.90	2.79	2.70	2.64	2.59	2.55	2.51	2.48
	8.68	**6.36**	**5.42**	**4.89**	**4.56**	**4.32**	**4.14**	**4.00**	**3.89**	**3.80**	**3.73**	**3.67**
16	4.49	3.63	3.24	3.01	2.85	2.74	2.66	2.59	2.54	2.49	2.45	2.42
	8.53	**6.23**	**5.29**	**4.77**	**4.44**	**4.20**	**4.03**	**3.89**	**3.78**	**3.69**	**3.61**	**3.55**
17	4.45	3.59	3.20	2.96	2.81	2.70	2.62	2.55	2.50	2.45	2.41	2.38
	8.40	**6.11**	**5.18**	**4.67**	**4.34**	**4.10**	**3.93**	**3.79**	**3.68**	**3.59**	**3.52**	**3.45**
18	4.41	3.55	3.16	2.93	2.77	2.66	2.58	2.51	2.46	2.41	2.37	2.34
	8.28	**6.01**	**5.09**	**4.58**	**4.25**	**4.01**	**3.85**	**3.71**	**3.60**	**3.51**	**3.44**	**3.37**

(Boldface Type) Points for the Distribution of F

Greater Mean Square)

14	16	20	24	30	40	50	75	100	200	500	∞	f_2
245	246	248	249	250	251	252	253	253	254	254	254	1
6,142	**6,169**	**6,208**	**6,234**	**6,258**	**6,286**	**6,302**	**6,323**	**6,334**	**6,352**	**6,361**	**6,366**	
19.42	19.43	19.44	19.45	19.46	19.47	19.47	19.48	19.49	19.49	19.50	19.50	2
99.43	**99.44**	**99.45**	**99.46**	**99.47**	**99.48**	**99.48**	**99.49**	**99.49**	**99.49**	**99.50**	**99.50**	
8.71	8.69	8.66	8.64	8.62	8.60	8.58	8.57	8.56	8.54	8.54	8.53	3
26.92	**26.83**	**26.69**	**26.60**	**26.50**	**26.41**	**26.35**	**26.27**	**26.23**	**26.18**	**26.14**	**26.12**	
5.87	5.84	5.80	5.77	5.74	5.71	5.70	5.68	5.66	5.65	5.64	5.63	4
14.24	**14.15**	**14.02**	**13.93**	**13.83**	**13.74**	**13.69**	**13.61**	**13.57**	**13.52**	**13.48**	**13.46**	
4.64	4.60	4.56	4.53	4.50	4.46	4.44	4.42	4.40	4.38	4.37	4.36	5
9.77	**9.68**	**9.55**	**9.47**	**9.38**	**9.29**	**9.24**	**9.17**	**9.13**	**9.07**	**9.04**	**9.02**	
3.96	3.92	3.87	3.84	3.81	3.77	3.75	3.72	3.71	3.69	3.68	3.67	6
7.60	**7.52**	**7.39**	**7.31**	**7.23**	**7.14**	**7.09**	**7.02**	**6.99**	**6.94**	**6.90**	**6.88**	
3.52	3.49	3.44	3.41	3.38	3.34	3.32	3.29	3.28	3.25	3.24	3.23	7
6.35	**6.27**	**6.15**	**6.07**	**5.98**	**5.90**	**5.85**	**5.78**	**5.75**	**5.70**	**5.67**	**5.65**	
3.23	3.20	3.15	3.12	3.08	3.05	3.03	3.00	2.98	2.96	2.94	2.93	8
5.56	**5.48**	**5.36**	**5.28**	**5.20**	**5.11**	**5.06**	**5.00**	**4.96**	**4.91**	**4.88**	**4.86**	
3.02	2.98	2.93	2.90	2.86	2.82	2.80	2.77	2.76	2.73	2.72	2.71	9
5.00	**4.92**	**4.80**	**4.73**	**4.64**	**4.56**	**4.51**	**4.45**	**4.41**	**4.36**	**4.33**	**4.31**	
2.86	2.82	2.77	2.74	2.70	2.67	2.64	2.61	2.59	2.56	2.55	2.54	10
4.60	**4.52**	**4.41**	**4.33**	**4.25**	**4.17**	**4.12**	**4.05**	**4.01**	**3.96**	**3.93**	**3.91**	
2.74	2.70	2.65	2.61	2.57	2.53	2.50	2.47	2.45	2.42	2.41	2.40	11
4.29	**4.21**	**4.10**	**4.02**	**3.94**	**3.86**	**3.80**	**3.74**	**3.70**	**3.66**	**3.62**	**3.60**	
2.64	2.60	2.54	2.50	2.46	2.42	2.40	2.36	2.35	2.32	2.31	2.30	12
4.05	**3.98**	**3.86**	**3.78**	**3.70**	**3.61**	**3.56**	**3.49**	**3.46**	**3.41**	**3.38**	**3.36**	
2.55	2.51	2.46	2.42	2.38	2.34	2.32	2.28	2.26	2.24	2.22	2.21	13
3.85	**3.78**	**3.67**	**3.59**	**3.51**	**3.42**	**3.37**	**3.30**	**3.27**	**3.21**	**3.18**	**3.16**	
2.48	2.44	2.39	2.35	2.31	2.27	2.24	2.21	2.19	2.16	2.14	2.13	14
3.70	**3.62**	**3.51**	**3.43**	**3.34**	**3.26**	**3.21**	**3.14**	**3.11**	**3.06**	**3.02**	**3.00**	
2.43	2.39	2.33	2.29	2.25	2.21	2.18	2.15	2.12	2.10	2.08	2.07	15
3.56	**3.48**	**3.36**	**3.29**	**3.20**	**3.12**	**3.07**	**3.00**	**2.97**	**2.92**	**2.89**	**2.87**	
2.37	2.33	2.28	2.24	2.20	2.16	2.13	2.09	2.07	2.04	2.02	2.01	16
3.45	**3.37**	**3.25**	**3.18**	**3.10**	**3.01**	**2.96**	**2.89**	**2.86**	**2.80**	**2.77**	**2.75**	
2.33	2.29	2.23	2.19	2.15	2.11	2.08	2.04	2.02	1.99	1.97	1.96	17
3.35	**3.27**	**3.16**	**3.08**	**3.00**	**2.92**	**2.86**	**2.79**	**2.76**	**2.70**	**2.67**	**2.65**	
2.29	2.25	2.19	2.15	2.11	2.07	2.04	2.00	1.98	1.95	1.93	1.92	18
3.27	**3.19**	**3.07**	**3.00**	**2.91**	**2.83**	**2.78**	**2.71**	**2.68**	**2.62**	**2.59**	**2.57**	

Continued on next page.

Table D.—Continued

									f_1 Degrees of Freedom (for			
f_2	1	2	3	4	5	6	7	8	9	10	11	12
19	4.38	3.52	3.13	2.90	2.74	2.63	2.55	2.48	2.43	2.38	2.34	2.31
	8.18	**5.93**	**5.01**	**4.50**	**4.17**	**3.94**	**3.77**	**3.63**	**3.52**	**3.43**	**3.36**	**3.30**
20	4.35	3.49	3.10	2.87	2.71	2.60	2.52	2.45	2.40	2.35	2.31	2.28
	8.10	**5.85**	**4.94**	**4.43**	**4.10**	**3.87**	**3.71**	**3.56**	**3.45**	**3.37**	**3.30**	**3.23**
21	4.32	3.47	3.07	2.84	2.68	2.57	2.49	2.42	2.37	2.32	2.28	2.25
	8.02	**5.78**	**4.87**	**4.37**	**4.04**	**3.81**	**3.65**	**3.51**	**3.40**	**3.31**	**3.24**	**3.17**
22	4.30	3.44	3.05	2.82	2.66	2.55	2.47	2.40	2.35	2.30	2.26	2.23
	7.94	**5.72**	**4.82**	**4.31**	**3.99**	**3.76**	**3.59**	**3.45**	**3.35**	**3.26**	**3.18**	**3.12**
23	4.28	3.42	3.03	2.80	2.64	2.53	2.45	2.38	2.32	2.28	2.24	2.20
	7.88	**5.66**	**4.76**	**4.26**	**3.94**	**3.71**	**3.54**	**3.41**	**3.30**	**3.21**	**3.14**	**3.07**
24	4.26	3.40	3.01	2.78	2.62	2.51	2.43	2.36	2.30	2.26	2.22	2.18
	7.82	**5.61**	**4.72**	**4.22**	**3.90**	**3.67**	**3.50**	**3.36**	**3.25**	**3.17**	**3.09**	**3.03**
25	4.24	3.38	2.99	2.76	2.60	2.49	2.41	2.34	2.28	2.24	2.20	2.16
	7.77	**5.57**	**4.68**	**4.18**	**3.86**	**3.63**	**3.46**	**3.32**	**3.21**	**3.13**	**3.05**	**2.99**
26	4.22	3.37	2.98	2.74	2.59	2.47	2.39	2.32	2.27	2.22	2.18	2.15
	7.72	**5.53**	**4.64**	**4.14**	**3.82**	**3.59**	**3.42**	**3.29**	**3.17**	**3.09**	**3.02**	**2.96**
27	4.21	3.35	2.96	2.73	2.57	2.46	2.37	2.30	2.25	2.20	2.16	2.13
	7.68	**5.49**	**4.60**	**4.11**	**3.79**	**3.56**	**3.39**	**3.26**	**3.14**	**3.06**	**2.98**	**2.93**
28	4.20	3.34	2.95	2.71	2.56	2.44	2.36	2.29	2.24	2.19	2.15	2.12
	7.64	**5.45**	**4.57**	**4.07**	**3.76**	**3.53**	**3.36**	**3.23**	**3.11**	**3.03**	**2.95**	**2.90**
29	4.18	3.33	2.93	2.70	2.54	2.43	2.35	2.28	2.22	2.18	2.14	2.10
	7.60	**5.42**	**4.54**	**4.04**	**3.73**	**3.50**	**3.33**	**3.20**	**3.08**	**3.00**	**2.92**	**2.87**
30	4.17	3.32	2.92	2.69	2.53	2.42	2.34	2.27	2.21	2.16	2.12	2.09
	7.56	**5.39**	**4.51**	**4.02**	**3.70**	**3.47**	**3.30**	**3.17**	**3.06**	**2.98**	**2.90**	**2.84**
32	4.15	3.30	2.90	2.67	2.51	2.40	2.32	2.25	2.19	2.14	2.10	2.07
	7.50	**5.34**	**4.46**	**3.97**	**3.66**	**3.42**	**3.25**	**3.12**	**3.01**	**2.94**	**2.86**	**2.80**
34	4.13	3.28	2.88	2.65	2.49	2.38	2.30	2.23	2.17	2.12	2.08	2.05
	7.44	**5.29**	**4.42**	**3.93**	**3.61**	**3.38**	**3.21**	**3.08**	**2.97**	**2.89**	**2.82**	**2.76**
36	4.11	3.26	2.86	2.63	2.48	2.36	2.28	2.21	2.15	2.10	2.06	2.03
	7.39	**5.25**	**4.38**	**3.89**	**3.58**	**3.35**	**3.18**	**3.04**	**2.94**	**2.86**	**2.78**	**2.72**
38	4.10	3.25	2.85	2.62	2.46	2.35	2.26	2.19	2.14	2.09	2.05	2.02
	7.35	**5.21**	**4.34**	**3.86**	**3.54**	**3.32**	**3.15**	**3.02**	**2.91**	**2.82**	**2.75**	**2.69**
40	4.08	3.23	2.84	2.61	2.45	2.34	2.25	2.18	2.12	2.07	2.04	2.00
	7.31	**5.18**	**4.31**	**3.83**	**3.51**	**3.29**	**3.12**	**2.99**	**2.88**	**2.80**	**2.73**	**2.66**
42	4.07	3.22	2.83	2.59	2.44	2.32	2.24	2.17	2.11	2.06	2.02	1.99
	7.27	**5.15**	**4.29**	**3.80**	**3.49**	**3.26**	**3.10**	**2.96**	**2.86**	**2.77**	**2.70**	**2.64**
44	4.06	3.21	2.82	2.58	2.43	2.31	2.23	2.16	2.10	2.05	2.01	1.98
	7.24	**5.12**	**4.26**	**3.78**	**3.46**	**3.24**	**3.07**	**2.94**	**2.84**	**2.75**	**2.68**	**2.62**

Greater Mean Square)

14	16	20	24	30	40	50	75	100	200	500	∞	f_2
2.26	2.21	2.15	2.11	2.07	2.02	2.00	1.96	1.94	1.91	1.90	1.88	19
3.19	**3.12**	**3.00**	**2.92**	**2.84**	**2.76**	**2.70**	**2.63**	**2.60**	**2.54**	**2.51**	**2.49**	
2.23	2.18	2.12	2.08	2.04	1.99	1.96	1.92	1.90	1.87	1.85	1.84	20
3.13	**3.05**	**2.94**	**2.86**	**2.77**	**2.69**	**2.63**	**2.56**	**2.53**	**2.47**	**2.44**	**2.42**	
2.20	2.15	2.09	2.05	2.00	1.96	1.93	1.89	1.87	1.84	1.82	1.81	21
3.07	**2.99**	**2.88**	**2.80**	**2.72**	**2.63**	**2.58**	**2.51**	**2.47**	**2.42**	**2.38**	**2.36**	
2.18	2.13	2.07	2.03	1.98	1.93	1.91	1.87	1.84	1.81	1.80	1.78	22
3.02	**2.94**	**2.83**	**2.75**	**2.67**	**2.58**	**2.53**	**2.46**	**2.42**	**2.37**	**2.33**	**2.31**	
2.14	2.10	2.04	2.00	1.96	1.91	1.88	1.84	1.82	1.79	1.77	1.76	23
2.97	**2.89**	**2.78**	**2.70**	**2.62**	**2.53**	**2.48**	**2.41**	**2.37**	**2.32**	**2.28**	**2.26**	
2.13	2.09	2.02	1.98	1.94	1.89	1.86	1.82	1.80	1.76	1.74	1.73	24
2.93	**2.85**	**2.74**	**2.66**	**2.58**	**2.49**	**2.44**	**2.36**	**2.33**	**2.27**	**2.23**	**2.21**	
2.11	2.06	2.00	1.96	1.92	1.87	1.84	1.80	1.77	1.74	1.72	1.71	25
2.89	**2.81**	**2.70**	**2.62**	**2.54**	**2.45**	**2.40**	**2.32**	**2.29**	**2.23**	**2.19**	**2.17**	
2.10	2.05	1.99	1.95	1.90	1.85	1.82	1.78	1.76	1.72	1.70	1.69	26
2.86	**2.77**	**2.66**	**2.58**	**2.50**	**2.41**	**2.36**	**2.28**	**2.25**	**2.19**	**2.15**	**2.13**	
2.08	2.03	1.97	1.93	1.88	1.84	1.80	1.76	1.74	1.71	1.68	1.67	27
2.83	**2.74**	**2.63**	**2.55**	**2.47**	**2.38**	**2.33**	**2.25**	**2.21**	**2.16**	**2.12**	**2.10**	
2.06	2.02	1.96	1.91	1.87	1.81	1.78	1.75	1.72	1.69	1.67	1.65	28
2.80	**2.71**	**2.60**	**2.52**	**2.44**	**2.35**	**2.30**	**2.22**	**2.18**	**2.13**	**2.09**	**2.06**	
2.05	2.00	1.94	1.90	1.85	1.80	1.77	1.73	1.71	1.68	1.65	1.64	29
2.77	**2.68**	**2.57**	**2.49**	**2.41**	**2.32**	**2.27**	**2.19**	**2.15**	**2.10**	**2.06**	**2.03**	
2.04	1.99	1.93	1.89	1.84	1.79	1.76	1.72	1.69	1.66	1.64	1.62	30
2.74	**2.66**	**2.55**	**2.47**	**2.38**	**2.29**	**2.24**	**2.16**	**2.13**	**2.07**	**2.03**	**2.01**	
2.02	1.97	1.91	1.86	1.82	1.76	1.74	1.69	1.67	1.64	1.61	1.59	32
2.70	**2.62**	**2.51**	**2.42**	**2.34**	**2.25**	**2.20**	**2.12**	**2.08**	**2.02**	**1.98**	**1.96**	
2.00	1.95	1.89	1.84	1.80	1.74	1.71	1.67	1.64	1.61	1.59	1.57	34
2.66	**2.58**	**2.47**	**2.38**	**2.30**	**2.21**	**2.15**	**2.08**	**2.04**	**1.98**	**1.94**	**1.91**	
1.98	1.93	1.87	1.82	1.78	1.72	1.69	1.65	1.62	1.59	1.56	1.55	36
2.62	**2.54**	**2.43**	**2.35**	**2.26**	**2.17**	**2.12**	**2.04**	**2.00**	**1.94**	**1.90**	**1.87**	
1.96	1.92	1.85	1.80	1.76	1.71	1.67	1.63	1.60	1.57	1.54	1.53	38
2.59	**2.51**	**2.40**	**2.32**	**2.22**	**2.14**	**2.08**	**2.00**	**1.97**	**1.90**	**1.86**	**1.84**	
1.95	1.90	1.84	1.79	1.74	1.69	1.66	1.61	1.59	1.55	1.53	1.51	40
2.56	**2.49**	**2.37**	**2.29**	**2.20**	**2.11**	**2.05**	**1.97**	**1.94**	**1.88**	**1.84**	**1.81**	
1.94	1.89	1.82	1.78	1.73	1.68	1.64	1.60	1.57	1.54	1.51	1.49	42
2.54	**2.46**	**2.35**	**2.26**	**2.17**	**2.08**	**2.02**	**1.94**	**1.91**	**1.85**	**1.80**	**1.78**	
1.92	1.88	1.81	1.76	1.72	1.66	1.63	1.58	1.56	1.52	1.50	1.48	44
2.52	**2.44**	**2.32**	**2.24**	**2.15**	**2.06**	**2.00**	**1.92**	**1.88**	**1.82**	**1.78**	**1.75**	

Continued on next page.

Table D.—Continued

	f_1 Degrees of Freedom (for											
f_2	1	2	3	4	5	6	7	8	9	10	11	12
46	4.05	3.20	2.81	2.57	2.42	2.30	2.22	2.14	2.09	2.04	2.00	1.97
	7.21	**5.10**	**4.24**	**3.76**	**3.44**	**3.22**	**3.05**	**2.92**	**2.82**	**2.73**	**2.66**	**2.60**
48	4.04	3.19	2.80	2.56	2.41	2.30	2.21	2.14	2.08	2.03	1.99	1.96
	7.19	**5.08**	**4.22**	**3.74**	**3.42**	**3.20**	**3.04**	**2.90**	**2.80**	**2.71**	**2.64**	**2.58**
50	4.03	3.18	2.79	2.56	2.40	2.29	2.20	2.13	2.07	2.02	1.98	1.95
	7.17	**5.06**	**4.20**	**3.72**	**3.41**	**3.18**	**3.02**	**2.88**	**2.78**	**2.70**	**2.62**	**2.56**
55	4.02	3.17	2.78	2.54	2.38	2.27	2.18	2.11	2.05	2.00	1.97	1.93
	7.12	**5.01**	**4.16**	**3.68**	**3.37**	**3.15**	**2.98**	**2.85**	**2.75**	**2.66**	**2.59**	**2.53**
60	4.00	3.15	2.76	2.52	2.37	2.25	2.17	2.10	2.04	1.99	1.95	1.92
	7.08	**4.98**	**4.13**	**3.65**	**3.34**	**3.12**	**2.95**	**2.82**	**2.72**	**2.63**	**2.56**	**2.50**
65	3.99	3.14	2.75	2.51	2.36	2.24	2.15	2.08	2.02	1.98	1.94	1.90
	7.04	**4.95**	**4.10**	**3.62**	**3.31**	**3.09**	**2.93**	**2.79**	**2.70**	**2.61**	**2.54**	**2.47**
70	3.98	3.13	2.74	2.50	2.35	2.23	2.14	2.07	2.01	1.97	1.93	1.89
	7.01	**4.92**	**4.08**	**3.60**	**3.29**	**3.07**	**2.91**	**2.77**	**2.67**	**2.59**	**2.51**	**2.45**
80	3.96	3.11	2.72	2.48	2.33	2.21	2.12	2.05	1.99	1.95	1.91	1.88
	6.96	**4.88**	**4.04**	**3.56**	**3.25**	**3.04**	**2.87**	**2.74**	**2.64**	**2.55**	**2.48**	**2.41**
100	3.94	3.09	2.70	2.46	2.30	2.19	2.10	2.03	1.97	1.92	1.88	1.85
	6.90	**4.82**	**3.98**	**3.51**	**3.20**	**2.99**	**2.82**	**2.69**	**2.59**	**2.51**	**2.43**	**2.36**
125	3.92	3.07	2.68	2.44	2.29	2.17	2.08	2.01	1.95	1.90	1.86	1.83
	6.84	**4.78**	**3.94**	**3.47**	**3.17**	**2.95**	**2.79**	**2.65**	**2.56**	**2.47**	**2.40**	**2.33**
150	3.91	3.06	2.67	2.43	2.27	2.16	2.07	2.00	1.94	1.89	1.85	1.82
	6.81	**4.75**	**3.91**	**3.44**	**3.14**	**2.92**	**2.76**	**2.62**	**2.53**	**2.44**	**2.37**	**2.30**
200	3.89	3.04	2.65	2.41	2.26	2.14	2.05	1.98	1.92	1.87	1.83	1.80
	6.76	**4.71**	**3.88**	**3.41**	**3.11**	**2.90**	**2.73**	**2.60**	**2.50**	**2.41**	**2.34**	**2.28**
400	3.86	3.02	2.62	2.39	2.23	2.12	2.03	1.96	1.90	1.85	1.81	1.78
	6.70	**4.66**	**3.83**	**3.36**	**3.06**	**2.85**	**2.69**	**2.55**	**2.46**	**2.37**	**2.29**	**2.23**
1000	3.85	3.00	2.61	2.38	2.22	2.10	2.02	1.95	1.89	1.84	1.80	1.76
	6.66	**4.62**	**3.80**	**3.34**	**3.04**	**2.82**	**2.66**	**2.53**	**2.43**	**2.34**	**2.26**	**2.20**
∞	3.84	2.99	2.60	2.37	2.21	2.09	2.01	1.94	1.88	1.83	1.79	1.75
	6.64	**4.60**	**3.78**	**3.32**	**3.02**	**2.80**	**2.64**	**2.51**	**2.41**	**2.32**	**2.24**	**2.18**

Greater Mean Square)

14	16	20	24	30	40	50	75	100	200	500	∞	f_2
1.91	1.87	1.80	1.75	1.71	1.65	1.62	1.57	1.54	1.51	1.48	1.46	46
2.50	**2.42**	**2.30**	**2.22**	**2.13**	**2.04**	**1.98**	**1.90**	**1.86**	**1.80**	**1.76**	**1.72**	
1.90	1.86	1.79	1.74	1.70	1.64	1.61	1.56	1.53	1.50	1.47	1.45	48
2.48	**2.40**	**2.28**	**2.20**	**2.11**	**2.02**	**1.96**	**1.88**	**1.84**	**1.78**	**1.73**	**1.70**	
1.90	1.85	1.78	1.74	1.69	1.63	1.60	1.55	1.52	1.48	1.46	1.44	50
2.46	**2.39**	**2.26**	**2.18**	**2.10**	**2.00**	**1.94**	**1.86**	**1.82**	**1.76**	**1.71**	**1.68**	
1.88	1.83	1.76	1.72	1.67	1.61	1.58	1.52	1.50	1.46	1.43	1.41	55
2.43	**2.35**	**2.23**	**2.15**	**2.06**	**1.96**	**1.90**	**1.82**	**1.78**	**1.71**	**1.66**	**1.64**	
1.86	1.81	1.75	1.70	1.65	1.59	1.56	1.50	1.48	1.44	1.41	1.39	60
2.40	**2.32**	**2.20**	**2.12**	**2.03**	**1.93**	**1.87**	**1.79**	**1.74**	**1.68**	**1.63**	**1.60**	
1.85	1.80	1.73	1.68	1.63	1.57	1.54	1.49	1.46	1.42	1.39	1.37	65
2.37	**2.30**	**2.18**	**2.09**	**2.00**	**1.90**	**1.84**	**1.76**	**1.71**	**1.64**	**1.60**	**1.56**	
1.84	1.79	1.72	1.67	1.62	1.56	1.53	1.47	1.45	1.40	1.37	1.35	70
2.35	**2.28**	**2.15**	**2.07**	**1.98**	**1.88**	**1.82**	**1.74**	**1.69**	**1.62**	**1.56**	**1.53**	
1.82	1.77	1.70	1.65	1.60	1.54	1.51	1.45	1.42	1.38	1.35	1.32	80
2.32	**2.24**	**2.11**	**2.03**	**1.94**	**1.84**	**1.78**	**1.70**	**1.65**	**1.57**	**1.52**	**1.49**	
1.79	1.75	1.68	1.63	1.57	1.51	1.48	1.42	1.39	1.34	1.30	1.28	100
2.26	**2.19**	**2.06**	**1.98**	**1.89**	**1.79**	**1.73**	**1.64**	**1.59**	**1.51**	**1.46**	**1.43**	
1.77	1.72	1.65	1.60	1.55	1.49	1.45	1.39	1.36	1.31	1.27	1.25	125
2.23	**2.15**	**2.03**	**1.94**	**1.85**	**1.75**	**1.68**	**1.59**	**1.54**	**1.46**	**1.40**	**1.37**	
1.76	1.71	1.64	1.59	1.54	1.47	1.44	1.37	1.34	1.29	1.25	1.22	150
2.20	**2.12**	**2.00**	**1.91**	**1.83**	**1.72**	**1.66**	**1.56**	**1.51**	**1.43**	**1.37**	**1.33**	
1.74	1.69	1.62	1.57	1.52	1.45	1.42	1.35	1.32	1.26	1.22	1.19	200
2.17	**2.09**	**1.97**	**1.88**	**1.79**	**1.69**	**1.62**	**1.53**	**1.48**	**1.39**	**1.33**	**1.28**	
1.72	1.67	1.60	1.54	1.49	1.42	1.38	1.32	1.28	1.22	1.16	1.13	400
2.12	**2.04**	**1.92**	**1.84**	**1.74**	**1.64**	**1.57**	**1.47**	**1.42**	**1.32**	**1.24**	**1.19**	
1.70	1.65	1.58	1.53	1.47	1.41	1.36	1.30	1.26	1.19	1.13	1.08	1000
2.09	**2.01**	**1.89**	**1.81**	**1.71**	**1.61**	**1.54**	**1.44**	**1.38**	**1.28**	**1.19**	**1.11**	
1.69	1.64	1.57	1.52	1.46	1.40	1.35	1.28	1.24	1.17	1.11	1.00	∞
2.07	**1.99**	**1.87**	**1.79**	**1.69**	**1.59**	**1.52**	**1.41**	**1.36**	**1.25**	**1.15**	**1.00**	

Table E. Critical Values of Chi-Square

df	$\alpha = 0.05$	$\alpha = 0.01$	df	$\alpha = 0.05$	$\alpha = 0.01$	df	$\alpha = 0.05$	$\alpha = 0.01$
1	3.841	6.635	46	62.830	71.201	91	114.268	125.289
2	5.991	9.210	47	64.001	72.443	92	115.390	126.462
3	7.815	11.345	48	65.171	73.683	93	116.511	127.633
4	9.488	13.277	49	66.339	74.919	94	117.632	128.803
5	11.071	15.086	50	67.505	76.154	95	118.752	129.973
6	12.592	16.812	51	68.669	77.386	96	119.871	131.141
7	14.067	18.475	52	69.832	78.616	97	120.990	132.309
8	15.507	20.090	53	70.993	79.843	98	122.108	133.476
9	16.919	21.666	54	72.153	81.069	99	123.225	134.642
10	18.307	23.209	55	73.311	82.292	100	124.342	135.807
11	19.675	24.725	56	74.468	83.513	102	126.574	138.134
12	21.026	26.217	57	75.624	84.733	104	128.804	140.459
13	22.362	27.688	58	76.778	85.950	106	131.031	142.780
14	23.685	29.141	59	77.931	87.166	108	133.257	145.099
15	24.996	30.578	60	79.082	88.379	110	135.480	147.414
16	26.296	32.000	61	80.232	89.591	112	137.701	149.727
17	27.587	33.409	62	81.381	90.802	114	139.921	152.037
18	28.869	34.805	63	82.529	92.010	116	142.138	154.344
19	30.144	36.191	64	83.675	93.217	118	144.354	156.648
20	31.410	37.566	65	84.821	94.422	120	146.567	158.950
21	32.671	38.932	66	85.965	95.626	122	148.799	161.250

22	33.924	40.289	67	87.108	96.828	124	150.989	163.546
23	35.172	41.638	68	88.250	98.028	126	153.198	165.841
24	36.415	42.980	69	89.391	99.228	128	155.405	168.133
25	37.652	44.314	70	90.531	100.425	130	157.610	170.423
26	38.885	45.642	71	91.670	101.621	132	159.814	172.711
27	40.113	46.963	72	92.808	102.816	134	162.016	174.996
28	41.337	48.278	73	93.945	104.010	136	164.216	177.280
29	42.557	49.588	74	95.081	105.202	138	166.415	179.561
30	43.773	50.892	75	96.217	106.393	140	168.613	181.840
31	44.985	52.191	76	97.351	107.583	142	170.809	184.118
32	46.194	53.486	77	98.484	108.771	144	173.004	186.393
33	47.400	54.776	78	99.617	109.958	146	175.198	188.666
34	48.602	56.061	79	100.749	111.114	148	177.390	190.938
35	49.802	57.342	80	101.879	112.329	150	179.581	193.208
36	50.998	58.619	81	103.010	113.512	200	233.994	249.445
37	52.192	59.892	82	104.139	114.695	250	287.882	304.940
38	53.384	61.162	83	105.267	115.876	300	341.395	359.906
39	54.572	62.428	84	106.395	117.057	400	447.632	468.724
40	55.758	63.691	85	107.522	118.236	500	553.127	576.493
41	56.942	64.950	86	108.648	119.414	600	658.094	683.516
42	58.124	66.206	87	109.773	120.591	700	762.661	789.974
43	59.304	67.459	88	110.898	121.767	800	866.911	895.984
44	60.481	68.710	89	112.022	122.942	900	970.904	1001.630
45	61.656	69.957	90	113.145	124.116	1000	1074.679	1106.969

Table F. Critical Values of the Pearson r

	Level of Significance for a One-Tailed Test			
	0.05	0.025	0.01	0.005
	Level of Significance for a Two-Tailed Test			
df = $n - 2$	**0.10**	**0.05**	**0.02**	**0.01**
1	0.988	0.997	0.9995	0.9999
2	0.900	0.950	0.980	0.990
3	0.805	0.378	0.934	0.959
4	0.729	0.811	0.882	0.917
5	0.699	0.754	0.833	0.874
6	0.622	0.707	0.789	0.834
7	0.582	0.666	0.750	0.798
8	0.549	0.632	0.716	0.765
9	0.521	0.602	0.685	0.735
10	0.497	0.576	0.658	0.708
11	0.476	0.553	0.634	0.684
12	0.458	0.532	0.612	0.661
13	0.441	0.514	0.592	0.641
14	0.426	0.497	0.574	0.623
15	0.412	0.482	0.558	0.606
16	0.400	0.468	0.542	0.590
17	0.389	0.456	0.528	0.575
18	0.378	0.444	0.516	0.561
19	0.369	0.433	0.503	0.549
20	0.360	0.423	0.492	0.537
21	0.352	0.413	0.482	0.526
22	0.344	0.404	0.472	0.515
23	0.337	0.396	0.462	0.505
24	0.330	0.388	0.453	0.496
25	0.323	0.381	0.445	0.487

Table F.—Continued

	Level of Significance for a One-Tailed Test			
	0.05	0.025	0.01	0.005
	Level of Significance for a Two-Tailed Test			
df = $n - 2$	0.10	0.05	0.02	0.01
26	0.317	0.374	0.437	0.479
27	0.311	0.367	0.430	0.471
28	0.306	0.361	0.423	0.463
29	0.301	0.355	0.416	0.456
30	0.296	0.349	0.409	0.449
35	0.275	0.325	0.381	0.418
40	0.257	0.304	0.358	0.393
45	0.243	0.288	0.338	0.372
50	0.231	0.273	0.322	0.354
60	0.211	0.250	0.295	0.325
70	0.195	0.232	0.274	0.302
80	0.183	0.217	0.256	0.283
90	0.173	0.205	0.242	0.267
100	0.164	0.195	0.230	0.254
120	0.150	0.178	0.210	0.232
150	0.134	0.159	0.189	0.208
200	0.116	0.138	0.164	0.181
300	0.095	0.113	0.134	0.148
400	0.082	0.098	0.116	0.128
500	0.073	0.088	0.104	0.115

SOURCE: Tables A, C, F, and G are taken from Tables III, VII, Viii, and XXXIII of Fisher and Yates: *Statistical Tables for Biological, Agricultural and Medical Research* published by Longman Group UK Ltd. London (previously published by Oliver and Boyd Ltd., Edinburgh) and by permission of the authors and publishers.

Table G. Conversion of a Pearson r into a Fisher z Coefficient

r	z	r	z	r	z	r	z	r	z
0.000	0.000	0.200	0.203	0.400	0.424	0.600	0.693	0.800	1.099
0.005	0.005	0.205	0.208	0.405	0.430	0.605	0.701	0.805	1.113
0.010	0.010	0.210	0.213	0.410	0.436	0.610	0.709	0.810	1.127
0.015	0.015	0.215	0.218	0.415	0.442	0.615	0.717	0.815	1.142
0.020	0.020	0.220	0.224	0.420	0.448	0.620	0.725	0.820	1.157
0.025	0.025	0.225	0.229	0.425	0.454	0.625	0.733	0.825	1.172
0.030	0.030	0.230	0.234	0.430	0.460	0.630	0.741	0.830	1.118
0.035	0.035	0.235	0.239	0.435	0.466	0.635	0.750	0.835	1.204
0.040	0.040	0.240	0.245	0.440	0.472	0.640	0.758	0.840	1.221
0.045	0.045	0.245	0.250	0.445	0.478	0.645	0.767	0.845	1.238
0.050	0.050	0.250	0.255	0.450	0.485	0.650	0.775	0.850	1.256
0.055	0.055	0.255	0.261	0.455	0.491	0.655	0.784	0.855	1.274
0.060	0.060	0.260	0.266	0.460	0.497	0.660	0.793	0.860	1.293
0.065	0.065	0.265	0.271	0.465	0.504	0.665	0.802	0.865	1.313
0.070	0.070	0.270	0.277	0.470	0.510	0.670	0.811	0.870	1.333
0.075	0.075	0.275	0.282	0.475	0.517	0.675	0.820	0.875	1.354
0.080	0.080	0.280	0.288	0.480	0.523	0.680	0.829	0.880	1.376
0.085	0.085	0.285	0.293	0.485	0.530	0.685	0.838	0.885	1.398
0.090	0.090	0.290	0.299	0.490	0.536	0.690	0.848	0.890	1.422
0.095	0.095	0.295	0.304	0.495	0.543	0.695	0.858	0.895	1.447

0.100	0.100	0.300	0.310	0.500	0.549	0.700	0.867	0.900	1.472
0.105	0.105	0.305	0.315	0.505	0.556	0.705	0.877	0.905	1.499
0.110	0.110	0.310	0.321	0.510	0.563	0.710	0.887	0.910	1.528
0.115	0.116	0.315	0.326	0.515	0.570	0.715	0.897	0.915	1.557
0.120	0.121	0.320	0.332	0.520	0.576	0.720	0.908	0.920	1.589
0.125	0.126	0.325	0.337	0.525	0.583	0.725	0.918	0.925	1.623
0.130	0.131	0.330	0.343	0.530	0.590	0.730	0.929	0.930	1.658
0.135	0.136	0.335	0.348	0.535	0.597	0.735	0.940	0.935	1.697
0.140	0.141	0.340	0.354	0.540	0.604	0.740	0.950	0.940	1.738
0.145	0.146	0.345	0.360	0.545	0.611	0.745	0.962	0.945	1.783
0.150	0.151	0.350	0.365	0.550	0.618	0.750	0.973	0.950	1.832
0.155	0.156	0.355	0.371	0.555	0.626	0.755	0.984	0.955	1.886
0.160	0.167	0.360	0.377	0.560	0.633	0.760	0.996	0.960	1.946
0.165	0.167	0.365	0.383	0.565	0.640	0.765	1.008	0.965	2.014
0.170	0.172	0.370	0.388	0.570	0.648	0.770	1.020	0.970	2.092
0.175	0.177	0.375	0.394	0.575	0.655	0.755	1.033	0.975	2.185
0.180	0.182	0.380	0.400	0.580	0.662	0.780	1.045	0.980	2.298
0.185	0.187	0.385	0.406	0.585	0.670	0.785	1.058	0.985	2.443
0.190	0.192	0.390	0.412	0.590	0.678	0.790	1.071	0.990	2.647
0.195	0.198	0.395	0.418	0.595	0.685	0.795	1.085	0.995	2.994

SOURCE: Tables A, C, F, and G are taken from Tables III, VII, Viii, and XXXIII of Fisher and Yates: *Statistical Tables for Biological, Agricultural and Medical Research* published by Longman Group UK Ltd. London (previously published by Oliver and Boyd Ltd., Edinburgh) and by permission of the authors and publishers.

Table H. Critical Values of Spearman Rank Correlation Coefficient

	Level of Significance for a One-Tailed Test			
	0.05	0.025	0.01	0.005
Number of Pairs = *n*	**Level of Significance for a Two-Tailed Test**			
	0.10	**0.05**	**0.02**	**0.01**
5	0.900	1.000	1.000	
6	0.829	0.886	0.943	1.000
7	0.714	0.786	0.893	0.929
8	0.643	0.738	0.833	0.881
9	0.606	0.696	0.783	0.833
10	0.564	0.648	0.746	0.794
11	0.536	0.619	0.709	0.764
12	0.503	0.587	0.678	0.734
13	0.484	0.560	0.648	0.703
14	0.464	0.538	0.626	0.679
15	0.446	0.521	0.604	0.657
16	0.429	0.504	0.585	0.635
17	0.414	0.488	0.566	0.618
18	0.401	0.474	0.550	0.600
19	0.391	0.460	0.535	0.584
20	0.380	0.447	0.522	0.570
21	0.370	0.436	0.509	0.556
22	0.361	0.425	0.497	0.544
23	0.353	0.416	0.486	0.532
24	0.344	0.407	0.476	0.521
25	0.337	0.398	0.466	0.511
26	0.331	0.390	0.475	0.501
27	0.324	0.383	0.449	0.492
28	0.318	0.375	0.441	0.483
29	0.312	0.369	0.433	0.475
30	0.306	0.362	0.426	0.467

SOURCE: From *Elementary Statistics,* 2nd Ed. by Roger Kirk.

Table I. Critical Values of the Studentized Range Statistic

df		***k*** **= number of means or steps between ordered means**								
		2	**3**	**4**	**5**	**6**	**7**	**8**	**9**	**10**
1	.05	18.0	27.0	32.8	37.1	40.4	43.1	45.4	47.4	49.1
	.01	90.0	135.0	164.0	186.0	202.0	216.0	227.0	237.0	246.0
2	.05	6.09	8.3	9.8	10.9	11.7	12.4	13.0	13.5	14.0
	.01	14.0	19.0	22.3	24.7	26.6	28.2	29.5	30.7	31.7
3	.05	4.50	5.91	6.82	7.50	8.04	8.48	8.85	9.18	9.46
	.01	8.26	10.60	12.20	13.30	14.20	15.00	15.60	16.20	16.70
4	.05	3.93	5.04	5.76	6.29	6.71	7.05	7.35	7.60	7.83
	.01	6.51	8.12	9.12	9.96	10.60	11.10	11.50	11.90	12.30
5	.05	3.64	4.60	5.22	5.67	6.03	6.33	6.58	6.80	6.99
	.01	5.70	6.97	7.80	8.42	8.91	9.32	9.67	9.97	10.20
6	.05	3.46	4.34	4.90	5.31	5.63	5.89	6.12	6.32	6.49
	.01	5.24	6.33	7.03	7.56	7.97	8.32	8.61	8.87	9.10
7	.05	3.34	4.16	4.69	5.06	5.36	5.61	5.82	6.00	6.16
	.01	4.95	5.92	6.54	7.01	7.37	7.68	7.94	8.17	8.37
8	.05	3.26	4.04	4.53	4.89	5.17	5.40	5.60	5.77	5.92
	.01	4.74	5.63	6.20	6.63	6.96	7.24	7.47	7.68	7.87

Continued on next page.

Table I.—Continued

df		k = number of means or steps between ordered means								
		2	3	4	5	6	7	8	9	10
9	.05	3.20	3.95	4.42	4.76	5.02	5.24	5.43	5.60	5.74
	.01	4.60	5.43	5.96	6.35	6.66	6.91	7.13	7.32	7.49
10	.05	3.15	3.88	4.33	4.65	4.91	5.12	5.30	5.46	5.60
	.01	4.48	5.27	5.77	6.14	6.43	6.67	6.87	7.05	7.21
11	.05	3.11	3.82	4.26	4.57	4.82	5.03	5.20	5.35	5.49
	.01	4.39	5.14	5.62	5.97	6.25	6.48	6.67	6.84	6.99
12	.05	3.08	3.77	4.20	4.51	4.75	4.95	5.12	5.27	5.40
	.01	4.32	5.04	5.50	5.84	6.10	6.32	6.51	6.67	6.81
13	.05	3.06	3.73	4.15	4.45	4.69	4.88	5.05	5.19	5.32
	.01	4.26	4.96	5.40	5.73	5.98	6.19	6.37	6.53	6.67
14	.05	3.03	3.70	4.11	4.41	4.64	4.83	4.99	5.13	5.25
	.01	4.21	4.89	5.32	5.63	5.88	6.08	6.26	6.41	6.54
16	.05	3.00	3.65	4.05	4.33	4.56	4.74	4.90	5.03	5.15
	.01	4.13	4.78	5.19	5.49	5.72	5.92	6.08	6.22	6.35

18	.05	2.97	3.61	4.00	4.28	4.49	4.67	4.82	4.96	5.07
	.01	4.07	4.70	5.09	5.38	5.60	5.79	5.94	6.08	6.20
20	.05	2.95	3.58	3.96	4.23	4.45	4.62	4.77	4.90	5.01
	.01	4.02	4.64	5.02	5.29	5.51	5.69	5.84	5.97	6.09
24	.05	2.92	3.53	3.90	4.17	4.37	4.45	4.68	4.81	4.92
	.01	3.96	4.54	4.91	5.17	5.37	5.54	5.69	5.81	5.92
30	.05	2.89	3.49	3.84	4.10	4.30	4.46	4.60	4.72	4.83
	.01	3.89	4.45	4.80	5.05	5.24	5.40	5.54	5.56	5.76
40	.05	2.86	3.44	3.79	4.04	4.23	4.39	4.52	4.63	4.74
	.01	3.82	4.37	4.70	4.93	5.11	5.27	5.39	5.50	5.60
60	.05	2.83	3.40	3.74	3.98	4.16	4.31	4.44	4.55	4.65
	.01	3.76	4.28	4.60	4.82	4.99	5.13	5.25	5.36	5.45
120	.05	2.80	3.36	3.69	3.92	4.10	4.24	4.36	4.48	4.56
	.01	3.70	4.20	4.50	4.71	4.87	5.01	5.12	5.21	5.30
∞	.05	2.77	3.31	3.63	3.86	4.03	4.17	4.29	4.39	4.47
	.01	3.64	4.12	4.40	4.60	4.76	4.88	4.99	5.08	5.16

Table J. Critical Values of the Dunnett Test for Comparing Treatment Means with a Control

One-Tailed Comparisons

df		k = number of treatment means, including control								
		2	3	4	5	6	7	8	9	10
5	.05	2.02	2.44	2.68	2.85	2.98	3.08	3.16	3.24	3.30
	.01	3.37	3.90	4.21	4.43	4.60	4.73	4.85	4.94	5.03
6	.05	1.94	2.34	2.56	2.71	2.83	2.92	3.00	3.07	3.12
	.01	3.14	3.61	3.88	4.07	4.21	4.33	4.43	4.51	4.59
7	.05	1.89	2.27	2.48	2.62	2.73	2.82	2.89	2.95	3.01
	.01	3.00	3.42	3.66	3.83	3.96	4.07	4.15	4.23	4.30
8	.05	1.86	2.22	2.42	2.55	2.66	2.74	2.81	2.87	2.92
	.01	2.90	3.29	3.51	3.67	3.79	3.88	3.96	4.03	4.09
9	.05	1.83	2.18	2.37	2.50	2.60	2.68	2.75	2.81	2.86
	.01	2.82	3.19	3.40	3.55	3.66	3.75	3.82	3.89	3.94
10	.05	1.81	2.15	2.34	2.47	2.56	2.64	2.70	2.76	2.81
	.01	2.76	3.11	3.31	3.45	3.56	3.64	3.71	3.78	3.83
11	.05	1.80	2.13	2.31	2.44	2.53	2.60	2.67	2.72	2.77
	.01	2.72	3.06	3.25	3.38	3.48	3.56	3.63	3.69	3.74
12	.05	1.78	2.11	2.29	2.41	2.50	2.58	2.64	2.69	2.74
	.01	2.68	3.01	3.19	3.32	3.42	3.50	3.56	3.62	3.67

13	.05	1.77	2.09	2.27	2.39	2.48	2.55	2.61	2.66	2.71
	.01	2.65	2.97	3.15	3.27	3.37	3.44	3.51	3.56	3.61
14	.05	1.76	2.08	2.25	2.37	2.46	2.53	2.59	2.64	2.69
	.01	2.62	2.94	3.11	3.23	3.32	3.40	3.46	3.51	3.56
15	.05	1.75	2.07	2.24	2.36	2.44	2.51	2.57	2.62	2.67
	.01	2.60	2.91	3.08	3.20	3.29	3.36	3.42	3.47	3.52
16	.05	1.75	2.06	2.23	2.34	2.43	2.50	2.56	2.61	2.65
	.01	2.58	2.88	3.05	3.17	3.26	3.33	3.39	3.44	3.48
17	.05	1.74	2.05	2.22	2.33	2.42	2.49	2.54	2.59	2.64
	.01	2.57	2.86	3.03	3.14	3.23	3.30	3.36	3.41	3.45
18	.05	1.73	2.04	2.21	2.32	2.41	2.48	2.53	2.58	2.62
	.01	2.55	2.84	3.01	3.12	3.21	3.27	3.33	3.38	3.42
19	.05	1.73	2.03	2.20	2.31	2.40	2.47	2.52	2.57	2.61
	.01	2.54	2.83	2.99	3.10	3.18	3.25	3.31	3.36	3.40
20	.05	1.72	2.03	2.19	2.30	2.39	2.46	2.51	2.56	2.60
	.01	2.53	2.81	2.97	3.08	3.17	3.23	3.29	3.34	3.38
24	.05	1.71	2.01	2.17	2.28	2.36	2.43	2.48	2.53	2.57
	.01	2.49	2.77	2.92	3.03	3.11	3.17	3.22	3.27	3.31
30	.05	1.70	1.99	2.15	2.25	2.33	2.40	2.45	2.50	2.54
	.01	2.46	2.72	2.87	2.97	3.05	3.11	3.16	3.21	3.24

Continued on next page.

Table J.—Continued

One-Tailed Comparisons
***k* = number of treatment means, including control**

df		2	3	4	5	6	7	8	9	10
40	.05	1.68	1.97	2.13	2.23	2.31	2.37	2.42	2.47	2.51
	.01	2.42	2.68	2.82	2.92	2.99	3.05	3.10	3.14	3.18
60	.05	1.67	1.95	2.10	2.21	2.28	2.35	2.39	2.44	2.48
	.01	2.39	2.64	2.78	2.87	2.94	3.00	3.04	3.08	3.12
120	.05	1.66	1.93	2.08	2.18	2.26	2.32	2.37	2.41	2.45
	.01	2.36	2.60	2.73	2.82	2.89	2.94	2.99	3.03	3.06
∞	.05	1.64	1.92	2.06	2.16	2.23	2.29	2.34	2.38	2.42
	.01	2.33	2.56	2.68	2.77	2.84	2.89	2.93	2.97	3.00

Two-Tailed Comparisons
***k* = number of treatment means, including control**

df		2	3	4	5	6	7	8	9	10
5	.05	2.57	3.03	3.29	3.48	3.62	3.73	3.82	3.90	3.97
	.01	4.03	4.63	4.98	5.22	5.41	5.56	5.69	5.80	5.89
6	.05	2.45	2.86	3.10	3.26	3.39	3.49	3.57	3.64	3.71
	.01	3.71	4.21	4.51	4.71	4.87	5.00	5.10	5.20	5.28

7	.05	2.36	2.75	2.97	3.12	3.24	3.33	3.41	3.47	3.53
	.01	3.50	3.95	4.21	4.39	4.53	4.64	4.74	4.82	4.89
8	.05	2.31	2.67	2.88	3.02	3.13	3.22	3.29	3.35	3.41
	.01	3.36	3.77	4.00	4.17	4.29	4.40	4.48	4.56	4.62
9	.05	2.26	2.61	2.81	2.95	3.05	3.14	3.20	3.26	3.32
	.01	3.25	3.63	3.85	4.01	4.12	4.22	4.30	4.37	4.43
10	.05	2.23	2.57	2.76	2.89	2.99	3.07	3.14	3.19	3.24
	.01	3.17	3.53	3.74	3.88	3.99	4.08	4.16	4.22	4.28
11	.05	2.20	2.53	2.72	2.84	2.94	3.02	3.08	3.14	3.19
	.01	3.11	3.45	3.65	3.79	3.89	3.98	4.05	4.11	4.16
12	.05	2.18	2.50	2.68	2.81	2.90	2.98	3.04	3.09	3.14
	.01	3.05	3.39	3.58	3.71	3.81	3.89	3.96	4.02	4.07
13	.05	2.16	2.48	2.65	2.78	2.87	2.94	3.00	3.06	3.10
	.01	3.01	3.33	3.52	3.65	3.74	3.82	3.89	3.94	3.99
14	.05	2.14	2.46	2.63	2.75	2.84	2.91	2.97	3.02	3.07
	.01	2.98	3.29	3.47	3.59	3.69	3.76	3.83	3.88	3.93
15	.05	2.13	2.44	2.61	2.73	2.82	2.89	2.95	3.00	3.04
	.01	2.95	3.25	3.43	3.55	3.64	3.71	3.78	3.83	3.88
16	.05	2.12	2.42	2.59	2.71	2.80	2.87	2.92	2.97	3.02
	.01	2.92	3.22	3.39	3.51	3.60	3.67	3.73	3.78	3.83

Continued on next page.

Table J.—Continued

df		Two-Tailed Comparisons k = number of treatment means, including control								
		2	3	4	5	6	7	8	9	10
17	.05	2.11	2.41	2.58	2.69	2.78	2.85	2.90	2.95	3.00
	.01	2.90	3.19	3.36	3.47	3.56	3.63	3.69	3.74	3.79
18	.05	2.10	2.40	2.56	2.68	2.76	2.83	2.89	2.94	2.98
	.01	2.88	3.17	3.33	3.44	3.53	3.60	3.66	3.71	3.75
19	.05	2.09	2.39	2.55	2.66	2.75	2.81	2.87	2.92	2.96
	.01	2.86	3.15	3.31	3.42	3.50	3.57	3.63	3.68	3.72
20	.05	2.09	2.38	2.54	2.65	2.73	2.80	2.86	2.90	2.95
	.01	2.85	3.13	3.29	3.40	3.48	3.55	3.60	3.65	3.69

24	.05	2.06	2.35	2.51	2.61	2.70	2.76	2.81	2.86	2.90
	.01	2.80	3.07	3.22	3.32	3.40	3.47	3.52	3.57	3.61
30	.05	2.04	2.32	2.47	2.58	2.66	2.72	2.77	2.82	2.86
	.01	2.75	3.01	3.15	3.25	3.33	3.39	3.44	3.49	3.52
40	.05	2.02	2.29	2.44	2.54	2.62	2.68	2.73	2.77	2.81
	.01	2.70	2.95	3.09	3.19	3.26	3.32	3.37	3.41	3.44
60	.05	2.00	2.27	2.41	2.51	2.58	2.64	2.69	2.73	2.77
	.01	2.66	2.90	3.03	3.12	3.19	3.25	3.29	3.33	3.37
120	.05	1.98	2.24	2.38	2.47	2.55	2.60	2.65	2.69	2.73
	.01	2.62	2.85	2.97	3.06	3.12	3.18	3.22	3.26	3.29
∞	.05	1.96	2.21	2.35	2.44	2.51	2.57	2.61	2.65	2.69
	.01	2.58	2.79	2.92	3.00	3.06	3.11	3.15	3.19	3.22

SOURCE: Reproduced with permission from Dunnett, C. W. *J. Am. Stat. Assoc. 55*, 1096–1121, 1955. Copyright 1955, JASA.

Table K. Coefficients of Orthogonal Polynomials

k	Polynomial	Coefficients						Σc^2
3	Linear	−1	0	1				2
	Quadratic	1	−2	1				6
	Linear	3	−1	1	3			20
4	Quadratic	1	−1	−1	1			4
	Cubic	−1	3	−3	1			20
	Linear	−2	−1	0	1	2		10
5	Quadratic	2	−1	−2	−1	2		14
	Cubic	−1	2	0	−2	1		10
	Quartic	1	−4	6	−4	1		70
	Linear	−5	−3	−1	1	3	5	70
6	Quadratic	5	−1	−4	−4	−1	5	84
	Cubic	−5	7	4	−4	−7	5	180
	Quartic	1	−3	2	2	−3	1	28

7	Linear	−3	−2	−1	0	1	2	3				28
	Quadratic	5	0	−3	−4	−3	0	5				84
	Cubic	−1	1	1	0	−1	−1	1				6
	Quartic	3	−7	1	6	1	−7	3				154
8	Linear	−7	−5	−3	−1	1	3	5	7			168
	Quadratic	7	1	−3	−5	−5	−3	1	7			168
	Cubic	−7	5	7	3	−3	−7	−5	7			264
	Quartic	7	−13	−3	9	9	−3	−13	7			616
	Quintic	−7	23	−17	−15	15	17	−23	7			2184
9	Linear	−4	−3	−2	−1	0	1	2	3	4		60
	Quadratic	28	7	−8	−17	−20	−17	−8	7	28		2772
	Cubic	−14	7	13	9	0	−9	−13	−7	14		990
	Quartic	14	−21	−11	9	18	9	−11	−21	14		2002
	Quintic	−4	11	−4	−9	0	9	4	−11	4		468
10	Linear	−9	−7	−5	−3	−1	1	3	5	7	9	330
	Quadratic	6	2	−1	−3	−4	−4	−3	−1	2	6	132
	Cubic	−42	14	35	31	12	−12	−31	−35	−14	42	8580
	Quartic	18	−22	−17	3	18	18	3	−17	−22	18	2860
	Quintic	−6	14	−1	−11	−6	6	11	1	−14	6	780

Table L. Sample Sizes Required to Detect Prescribed Differences Between Means for Two-Tailed Tests

The table entry is the sample size (n) required to detect with probability $1 - \beta$ that the mean for a group differs from a hypothesized mean or the mean of another group. The standardized difference (d) is expressed as a z value.

$\alpha = 0.01$

$1-\beta$ / d	0.50	0.60	0.70	0.80	0.90	0.95	0.99
0.1	664	801	962	1168	1488	1782	2404
0.2	166	201	241	292	372	446	601
0.4	42	51	61	73	93	112	151
0.6	19	23	27	33	42	50	67
0.8	11	13	16	19	24	28	38
1.0	7	9	10	12	15	18	25
1.2	5	6	7	9	11	13	17
1.4	4	5	5	6	8	10	13
1.6	3	4	4	5	6	7	10
1.8	3	3	3	4	5	6	8
2.0	2	3	3	3	4	5	7
3.0	1	1	2	2	2	2	3

If σ is estimated and the t test is used, add 4 to the tabulated values.

$\alpha = 0.05$

$1-\beta$ / d	0.50	0.60	0.70	0.80	0.90	0.95	0.99
0.1	385	490	618	785	1051	1300	1838
0.2	97	123	155	197	263	325	460
0.4	25	31	39	50	66	82	115
0.6	11	14	18	22	30	37	52
0.8	7	8	10	13	17	21	29
1.0	4	5	7	8	11	13	19
1.2	3	4	5	6	8	10	13
1.4	2	3	4	5	6	7	10
1.6	2	2	3	4	5	6	8
1.8	2	2	2	3	4	5	6
2.0	1	2	2	2	3	4	5
3.0	1	1	1	1	2	2	3

If α is estimated and the t test is used, add 2 to the tabulated values.

SOURCE: Reproduced from *Experimental Statistics*, Mary G. Natrella, National Bureau of Standards Handbook 91, U.S. Government Printing Office, Washington, DC, August 1, 1963.

Table M. Sample Sizes Required to Detect Prescribed Differences Between Means for One-Tailed Tests

The table entry is the sample size (*n*) required to detect with probability $1 - \beta$ that the mean of one group exceeds a hypothesized mean, or that the mean of one group exceeds the mean of another group. The standardized difference (*d*) is expressed as a *z* value.

$\alpha = 0.01$

$1 - \beta$ / d	0.50	0.60	0.70	0.80	0.90	0.95	0.99
0.1	542	666	813	1004	1302	1578	2165
0.2	136	167	204	251	326	395	542
0.4	34	42	51	63	82	99	136
0.6	16	19	23	28	37	44	61
0.8	9	11	13	16	21	25	34
1.0	6	7	9	11	14	16	22
1.2	4	5	6	7	10	11	16
1.4	3	4	5	6	7	9	12
1.6	3	3	4	4	6	7	9
1.8	2	3	3	4	5	5	7
2.0	2	2	3	3	4	4	6
3.0	1	1	1	2	2	2	3

If σ is estimated and the *t* test is used, add 3to the tabulated values.

Continued on next page.

Table M.—Continued

	$\alpha = 0.05$						
$1 - \beta$ / d	**0.50**	**0.60**	**0.70**	**0.80**	**0.90**	**0.95**	**0.99**
0.1	271	361	471	619	857	1083	1578
0.2	68	91	118	155	215	271	395
0.4	17	23	30	39	54	68	99
0.6	8	11	14	18	24	31	44
0.8	5	6	8	10	14	17	25
1.0	3	4	5	7	9	11	16
1.2	2	3	4	5	6	8	11
1.4	2	2	3	4	5	6	9
1.6	2	2	2	3	4	5	7
1.8	1	2	2	2	3	4	5
2.0	1	1	2	2	3	3	4
3.0	1	1	1	1	1	2	2

If α is estimated and the t test is used, add 2 to the tabulated values.

SOURCE: Reproduced from *Experimental Statistics*, Mary G. Natrella, National Bureau of Standards Handbook 91, U.S. Government Printing Office, Washington, DC, August 1, 1963.

Glossary of Terms

Addition theorem The rule used to determine the probability of occurrence of several mutually exclusive events, which is the sum of their separate probabilities.

Additivity A property of numbers that indicates that they can be added together in a meaningful way so that the results are consistent with the rational number system.

Aliases In an incomplete factorial design, the two or more designations given to the same sum of squares.

Alpha error *See* Type I error.

Alpha (α) level *See* level of significance.

Alternative hypothesis The hypothesis that remaines tenable if the null hypothesis is rejected, and is usually consistent with the scientific hypothesis, symbolized as H_1.

Analysis of variance (ANOVA) A statistical method to test the null hypothesis involving two or more means by partitioning the total variance into between groups variance and within group variance.

A posteriori tests Test statistics designed to evaluate all possible comparisons among means computed to determine which comparisons are significant following an overall test which is significant.

A priori tests A set of hypotheses that a researcher plans to test as an experiment is being planned; usually refers to comparisons among means.

1453–0/88/0451$06.00/1

Arithmetic mean The sum of the values in a data set divided by the number of values included.

Average Refers to one of the measures at the central location point in a distribution—mean, median, or mode.

Backward stepping A technique used in stepwise multiple regression to remove variables from the equation according to specified criteria.

Beta error *See* Type II error.

Between-groups variance In ANOVA, the estimate of the population variance based on the deviations of subgroup means from the grand mean.

Bimodal A frequency distribution of values that has two modes or peaks.

Binomial expansion Equation for determining the probabilities of dichotomous events, the general expression of which is $(p + q)^n$, where n is the number of trials.

Bivariate frequency distributions A joint distribution of two variables, for which individual values are paired in some logical way.

Block A homogeneous group of elements that are similar on some criterion.The elements are then assigned to different treatments in a randomized-blocks design.

Blocking A method of planned grouping for an experimental study accomplished by grouping elements into relatively homogeneous subgroups as one way to reduce experimental error.

Canonical correlation A multivariate statistical method used to determine the correlation between two sets of variables.

Cartesian coordinate system A graph constructed by using two number scales placed at right angles to each other with the zero point in the middle where the two lines meet.

Categorical data Data that is tabulated into unique categories by using a nominal scale.

Cells The compartments in a matrix or table.

Central limit theorem A theorem that states that if a population has a finite variance and mean, then the distribution of sample means taken from random samples of n independent observations approaches a normal distribution with the mean equal to the population mean and the standard deviation equal to the population standard deviation divided by the square root of n.

Central location Refers to the value at or near the center of a frequency distribution.

Chain block design A special type of incomplete block design used when observations are expensive and the experimental error is small.

Charts Visual representation of data, usually in tabular form.

Chi-square A nonparametric method to test the null hypothesis for qualitative data expressed as frequencies.

Class intervals The interval between the highest and lowest value for each interval or class of the original values in a grouped frequency distribution.

Cluster analysis A multivariate statistical procedure that can be used to separate objects or things into relatively homogeneous subgroups.

Cluster sampling A sampling procedure in which groups of elements, called clusters, are taken as the sample units rather than individual elements.

Coefficient of alienation An index number that indicates the degree of lack of relationship between two variables, which is computed as the square root of one minus the square of the correlation coefficient.

Coefficient of determination An index number that indicates the proportion of variance two variables have in common, which is computed as the square of the correlation coefficient.

Coefficient of nondetermination An index number that indicates the proportion of variance that is unique to each variable or that is not common between the two variables, usually designated by K^2, and is computed as one minus the square of the correlation coefficient.

Confidence coefficient The probability that the confidence interval will bracket the population parameter.

Confidence interval A range of values which provides the lower and upper limits in which some designated population parameter has a specified probability of being included, if all possible samples are considered.

Confounding of effects In incomplete factorial designs, the same computed sum of squares has two or more interpretations. Thus, certain effects cannot be distinguished from other effects.

Consistent estimator An estimator of a population parameter is consistent if it approaches the parameter more closely as the sample size increases, which is the principle underlying the law of large numbers.

Constant A property of an entity that does not vary from one unit to another, or a number that has a fixed value in an equation.

Construct validity This type of validity deals with how conditions, variables, and practices were defined and carried out in a research study.

Contingency coefficient An index number that indicates the degree of relationship between two categorical attributes, each of which consist of two or more categories.

Contingency table A table showing the joint frequency distribution for two nominal variables paired in a logical way.

Correlation A measure of the degree of linear relationship between two variables.

Correlation coefficient An index number that indicates the degree of linear relationship between two variables.

Correlation ratio (η) An index number indicating the degree of relationship between two variables which are curvilinearly related.

Critical region The area of the sampling distribution that covers the values of the statistical test for rejecting the null hypothesis.

Critical values The value of the statistical test that cuts off the critical region of the sampling distribution.

Cross tabulations A table showing the joint frequency distribution for two or more nominal variables related in a logical way.

Crossed treatments In a factorial design, an experimental arrangement in which all possible combinations of levels of two or more treatments occur together.

Cumulative frequency Indicates how many observations fall at and below each value or interval in a frequency distribution.

Cumulative frequency polygon A line graph of a cumulative frequency distribution formed by placing dots above the upper real limit of each class interval at a height representing the cumulative frequency of that interval, and then connecting the dots with straight lines.

Curvilinear A nonlinear relationship between two variables that indicates a curved trend rather than a straight-line trend.

Data reduction A statistical procedure that is used to reduce a large number of variables to a smaller number of dimensions called factors, e.g., factor analysis.

Data set A collection of measures obtained from observations or measuring instruments for a group of objects or things, which is the basic data used for statistical analyses.

Degrees of freedom The number of values that are free to vary. In statistical analyses, depends upon the mathematical restrictions on the entire set of values, and is usually the number of observations minus the number of necessary statistics used in the computations to compute that statistic.

Dependent variable The variable in an experiment that is affected by manipulation of the independent variable, also referred to as the criterion variable.

Descriptive statistics The branch of statistics concerned with organizing, summarizing, and describing data.

Deviation method A computational procedure using deviation values rather than the actual observed values.

Directional hypothesis A hypothesis which is usually consistent with the scientific hypothesis in that the direction of the relationship or difference is specified.

Discriminant function A multivariate statistical procedure used to classify objects or things into subgroups on the basis of how those groups differ on a given set of variables.

Disordinal interaction In factorial designs, an interaction in which the effects at one level of one factor reverse themselves at different levels of the other factors.

Element In sampling theory, the unit selected from which information is collected. One of the necessary values on which statistics are calculated.

Estimator A rule, usually in the form of an equation, that is used to calculate an estimate of a population parameter using sample data.

Experimental error The effects of extraneous variables or factors on the dependent variable which includes all uncontrolled sources of variance.

Experimentwise error rate The number of experiments that could be run with at least one difference falsely identified as being significant divided by the total number of experiments, which results in an increase in Type I errors.

External validity The extent to which the conclusions from an experiment or a research study can be generalized to other cases, situations, and/or times.

Factor analysis A multivariate stastistical analysis that is used to reduce a large number of variables to a smaller number of dimensions called factors.

Factorial design An experimental design that makes it possible for an investigator to evaluate the individual and combined effects of two or more treatments in a single experiment.

Finite statistical model A statistical model based upon the assumption that each value in all subgroups can be expressed as a deviation from the overall grand mean for those subgroups, and that this deviation can be partitioned into additive components.

Fixed-effects model A factorial design in which the investigator specifies or fixes the particular levels of each factor and these remain set from one study to another.

Forecasting efficiency An index number indicating the percentage reduction in errors of estimation due to the correlation between the two variables, usually symbolized as E.

Forward stepping A technique used in stepwise multiple regression to enter variables into the equation according to specified criteria.

Fourfold point correlation Another designation for the phi coefficient.

Fractional factorial design An incomplete factorial design which includes only a fraction of the total possible treatment combinations.

Frequency distribution The organization of observations in a systematic order that shows the frequencies of occurence of each observation.

Frequency polygon A graph used to display the frequency distribution of a continuous variable, constructed by placing a dot above each value at a height representing the frequency of observations for that value and then connecting those dots with straight lines.

Geometric mean The nth root of the product of the included values in a data set, usually used when dealing with data in the form of ratios.

Graph Visual representation of data, usually in the form of bar, line, or circle graphs.

Greco-Latin-square design A Latin-square design expanded to control for three sources of error.

Grouped frequency distributions Frequency distributions which use intervals or classes of the original values or units of measurements, each of which contains several scale values.

Half-life The time required for half of something to change or to undergo a process such as to become extinct.

Harmonic mean The reciprocal of the arithmetic mean of the reciprocals of the individual values in a data set, usually used when dealing with rates of work or distances covered per time span.

Heterogeneous distribution A frequency distribution in which there are large differences in the magnitude of the values. The values vary considerably from low to highand thus result in high variability.

Hierarchal design An experimental design in which it is not possible to have all possible combinations of the levels of the different factors in the study, resulting in an arrangement in whichd one factor is uniquely defined at each of the levels of another factor or combination of factors.

Histogram A bar graph used to represent the frequency or relative frequency of a continuous interval variable, with the bars erected contiguously over the real limits of the values or the class intervals.

Homogeneity of variance One of the assumptions for the analylsis of variance, which is that the variances among groups are equivalent.

Homogeneous distribution A frequency distribution in which most of the values are similar in magnitude and cluster together, resulting in low variability.

Honestly significant difference (HSD) *See* Tukey test.

Hyper-Greco-Latin-square design A Latin-square design expanded to control for four sources of error.

Hypothesis A tentative statement about an expected relationship between two or more parameters for a given population.

Hypothesized effect size (HES) A ratio computed by dividing the minimum difference that a researcher wants to detect by the standard deviation of that variable, similar to a z ratio.

Hypothetical population A statistical population that may have no real existence but is hypothesized to exist based upon sample data or is imagined by theory.

Identity A property of a number that identifies an object or thing as a discrete category and is usually qualitatitively described.

Independent events Events that are not influence by the probability of occurrence of any other events.

Independent variable The variable in an experiment that is selected and manipulated by the ivestigator so that its effect can be determined.

Inferential statistics The branch of statistics concerned with inferences about the characteristics of population parameters from the characteristics of samples.

Interaction In factorial designs, factors interact if values obtained under levels of one factor are different for different levels of the other factors, which result from differential effects of the independent variables.

Internal validity The extent to which conclusions about cause and effect relationships within an experiment are accurate.

Interquartile range The range of values in a frequency distribution between the 75th percentile and the 25th percentile that cuts off the middle 50% of the values.

Interval estimation A method which uses a range of values as the estimate of a population parameter constructed so that, with a specified degree of confidence, it would bracket that population parameter.

Interval measurement A scale of measurement in which objects or things are measured on a scale with equal differences between numbers which represent equal differences in what is being measured.

Judgment sampling A sampling procedure in which elements are selected on the basis of the researcher's knowledge or opinion.

Kurtosis One of the properties of a frequency distribution that refers to the shape of the distribution of values as being relatively flat or peaked.

Latin-square design An experimental design in which two extraneous factors are controlled by grouping elements or cases into homogeneous cells within a matrix.

Law of large numbers The concept that, other factors remaining constant, the larger the sample size, the more probable it is that the sample statistic will represent the population paremater that it represents.

Leptokurtic A frequency distribution that has a shape more peaked than the normal distribution.

Level of significance The probability which is set by the investigator that a true null hypothesis will be rejected, designated by the Greek letter alpha (α) and which specifies the probability of making a Type I error.

Linear regression A relationship between two variables that follows a pattern such that a straight line can be fitted to the points of the bivariate frequency distribution.

Main effects In factorial designs, the effect of any individual factor over all levels of other factors.

Mean Usually refers to the arithmetic mean.

Measurement The assigning of numbers to objects or events according to specified rules.

Measurement scaling The development of a measurement scale according to specified rules.

Median The point in a frequency distribution that divides the distribution of values into two equal halves, so that half of the values are below this point and half are above it.

Mesokurtic Refers to the shape of the normal frequency distribution, literally means an intermediate distribution that falls between leptokurtic and platykurtic distributions.

Mixed-effects model A factorial design in which the levels of some factors are fixed and the levels of other factors are chosen at random.

Mode The value of the most frequently occurring observation in a frequency distribution.

Multimodal A frequency distribution of values that has more than two modes or peaks.

Multiple correlation A correlation method that relates several predictor variables to one predicted variable.

Multiplication theorem The rule used to determine the probability of two or more events occurring together, which is the product of their separate probabilities.

Multivariate analysis Usually refers to a statistical analysis in which there is more than one independent and dependent variables.

Multivariate analysis of variance (MANOVA) A multivariate statistical analysis designed to analyzed results of studies which involve both multiple independent and dependent variables.

Multivariate statistics *See* multivariate analysis.

Mutually exclusive and exhaustive A classification procedure that provides a set of unique categories of data into which each object or thing can be classified into one and only one category.

Negative correlation An inverse relationship in which as the trend in one variable increases the trend in another variable decreases.

Negative skew A frequency distribution of values in which most of the values are at the high end of the distribution with a few low values pulling the curve out to the left.

Nested factor A factor is said to be nested if levels of that factor occur only at one level of another factor.

Nesting An experimental design in which all levels of one factor occur at only one level of another factor.

Nominal measurement A scale of measurement in which objects or things are assigned to categories that have no inherent order or quantitative value.

Nominal scale *See* nominal measurement.

Nondirectional hypothesis A hypothesis which is concerned with a relationship or difference regardless of the direction of that relationship or difference.

Nonparametrics Hypothesis-testing procedures that do not make stringent assumptions about population parameters and usually do not involve the estimation of a specific population parameter.

Normal distribution A frequency distribution of values that is symmetrical, which is usually bell-shaped, and whose curve is described mathematically by a general equation. Sometimes referred to as the Gaussian curve.

Null hypothesis A hypothesis that specifies that there is no relationship or difference between or among parameters, and is the hypothesis that is tested in the hypothesis testing model, usually symbolized as H_0.

Ordinal measurement A scale of measurement in which objects or things are rank ordered, resulting in a set of numbers which indicate the rank order of the objects or things being measured.

Ogive An S-shaped cumulative frequency polygon which results when a symmetrical distribution is graphed.

One-tailed test A statistical test of the directional hypothesis in which the critical region lies in only one tail of the sampling distribution.

Ordinal interaction In factorial designs, an interaction which indicates a differential effect for the levels of one factor compared to the levels of other factors, but where the means for all levels of that factor are equal to or higher than the means for the other factors.

Orthogonal comparison Independent or uncorrelated comparisons. For a priori comparisons, two comparisons are said to be orthogonal if the products of their corresponding coefficients sum to zero.

Orthogonal designs Factorial designs in which the sample sizes in each cell are either equal or proportional to each other.

Outliers Extreme values in a data set that deviate markedly from the majority of the other values.

Parallel-forms reliability A method to determine the reliability of a procedure or instrument by taking two independent measures of the same objects or things at the same time.

Pascal's triangle A table showing the expected frequencies of binomials for the special case where $p = q = ½$.

Parameter A characteristic or measure of a population, often symbolized by a Greek letter.

Partial correlation A statistical method that is used to determine the correlation between two variables while eliminating the effects of a third or a number of other variables.

Pearson correlation coefficient The most commonly computed correlation coefficient, which was developed by Karl Pearson.

Percentile A point in a frequency distribution below which a specified percentage of the values fall.

Percentile rank A transformed value that indicates the percentage of the values in a frequency distribution falling at or below a given value.

Per-comparison error rate The number of comparisons between means falsely declared significant divided by the total number of comparisons, which results in an increase in Type I errors.

Periodicity The arrangements of elements in a population so that there are series of repeated values or trends.

Phi coefficient (ϕ) An index number that indicates the degree of relationship between two dichotomous variables.

Pictorial charts A type of bar graph in which pictures of the item of interest are used to represent the frequencies of each value.

Platykurtic A frequency distribution that has a shape less peaked or flatter than the normal distribution.

Point estimation A method that uses a single value to estimate the population parameter being estimated.

Polygon *See* frequency polygon.

Population The specified aggregation of all elements that possess a specified set of common characteristics that define it, and in inferential statistics, the group to which inferences are made.

Positive skew A frequency distribution of values in which most of the values are at the low end of the distribution with a few high values pulling the curve out to the right.

Post hoc tests *See* a posteriori tests.

Power ($1 - \beta$) The probability that the null hypothesis will be rejected when in actuality it is false and should be rejected.

Probability A measure of the likelihood that a specified event will occur, usually symbolized as p.

Probability distribution A distribution of events that indicates the probability that each event will occur.

Qualitative data *See* categorical data.

Quartile deviation *See* semiinterquartile range.

Quota sampling A sampling procedure in which those doing the selecting are instructed to obtain a specified number of elements. Often the most accessible elements are selected, and they might not be representative of the population.

Range Usually defined as the difference between the highest and lowest value in a data set plus one.

Random-effects model A factorial design in which the levels of each factor are chosen at random and these levels will probably vary from one study to another.

Random sample *See* simple random sample.

Randomized-blocks design An experimental design in which an extraneous factor is controlled by grouping elements into homogeneous blocks.

Rank order A property of a number that indicates the rank order from smallest to largest, with the smallest rank assigned a 1, the next a 2, etc.

Raw score approach A computational procedure using the actual observed values in the equations.

Real limits The inclusive distance along a continuum that each value represents, usually one-half a unit below to one-half a unit above the value limits for each interval.

Regression A statistical method that deals with estimating or predicting values on one variable from values on a second variable. Originally referred to a phenomenon that extreme values in a distribution will regress towards the mean of the general population when measured again or when successive generations are measured.

Regression line The straight line of best fit for a set of bivariate data computed from the equation $bX + a$, and is used to predict values of one variable from known values of another variable.

Relative frequency The frequency of a value reported as a proportion of the total number of observations.

Relative standard deviation A measure of variability that is independent of the unit of measurement and which takes into account the averages of each data set, often used when comparing the variability of several data sets that use different scaled values. Is defined as the standard deviation divided by the mean.

Reliability A measure of the consistency of a measurement instrument or procedure.

Repeatability A measure of the degree of closeness of agreement between individual results obtained under the same conditions.

Reproducibility A measure of the degree of closeness of agreement between individual results obtained under different conditions, related to generalization of the results of a study to other situations.

Research hypothesis A tentative statement about phenomena, usually based on theory or experiential evidence, that is considered to account for observed relationships, and serves to guide the formulation of a research study.

Robustness The property of an inferential statistic to resist the effects of violations of assumptions and to yield the same results regardless of whether the assumptions for that test are met.

Sample A subset of the population usually taken in a way as to make it similar to the population with regard to specified characteristics.

Sample population The population from which a sample is taken.

Sampling distribution A distribution of all possible values of a statistic computed from all possible samples of a given size selected from a given population, usually used to estimate the sample-to-sample variation expected among samples.

Scheffe′ test An a posteriori test developed by Scheffe′ for comparing all possible pairs of means.

Semiinterquartile range Half of the interquartile range, which in a symmetrical distribution is the difference between the value at the 50th percentile and the value at either the 75th percentile or the 25th percentile.

Sensitivity The extent to which a measure is able to sense minute differences between objects or things.

Significance level *See* level of significance.

Simple main effects In factorial designs, if interaction effects are present, the comparison of means for the levels of one factor at each level of another factor.

Simple random sampling (SRS) A sample selected in a way so that all elements have an equal probability of being selected.

Simple, simple main effects In a three-way factorial design, the comparison of means collapsed across all other factors used when interaction effects are present.

Spearman rank-order correlation (ρ) A statistical method to determine the degree of relationship between two variables, both of which are measured by at least an ordinal scale.

Standard deviation A standard measure of variability that indicates the average amount of variability of observed values from the mean of those values, which is computed as the square root of the variance.

Standard error The standard deviation of the sampling distribution of a statistic.

Standard error of estimate The standard deviation of the differences between predicted and actual values in a regression analysis, which gives an indication of the accuracy of prediction.

Standard error of measurement The standard deviation of measurements taken on an object or thing one would expect to obtain from a large number of measures using the same instrument, which indicates the extent to which measurements will vary due to lack of reliability.

Standard score approach A computational procedure using values that have been transformed into standardized values such as z, Z, or T values.

Standard square A Latin-square design arranged so that the letters occur in their natural order in the first row and the first column, also referred to as standard form.

Statistic A measure computed from observations in a sample, usually designated by a Latin letter.

Statistics A branch of mathematics concerned with the collection, analysis, presentation, and interpretation of data.

Statistical hypothesis A tentative statement about one or more parameters of a population. Null and alternative hypotheses are two types of statistical hypotheses.

Stratified sampling A sampling procedure in which the population is divided into subpopulations, called strata, and then elements are randomly selected from each strata.

Sum of squares The sum of the squared deviations from the mean, a value used in the computation of most descriptive and inferential statistics.

Systematic sampling A sampling procedure in which a consistent method is used, such as selecting every kth element or one element from every kth time period.

Target population The population to which inferences are to be made.

Test-retest form of reliability A method to determine the reliability of a procedure or an instrument by taking two measures of the same objects or things using the same procedure or instrument within a reasonable time period.

Test statistic The statistical method selected to test the null hypothesis.

Tests of main effects. *See* Scheffe′ and Tukey tests.

Treatment-by-levels design *See* randomized-blocks design.

Trend analysis An a priori test to determine if there is a relationship between the independent and dependent variables and if there is, what type of trend there is.

Tukey A test An a posteriori test developed by Tukey for comparing all possible pairs of means.

Two-tailed test A statistical test of the nondirectional hypothesis in which the critical region lies in both tails of the sampling distribution.

Type I error A decision made to reject a null hypothesis when in actuality it is true and should not have been rejected.

Type II error A decision made not to reject a null hypothesis when in actuality it is false and should have been rejected.

Type III error A decision to correctly reject a false null hypothesis but incorrectly interpreting the direction of the difference.

Unbiased estimate An estimate of a population parameter is unbiased if, for a large number of random samples of the same size, the mean of the estimates equals the parameter it estimates.

Ungrouped frequency distributions Frequency distributions which use the original units of measurements for each interval.

Unimodal A frequency distribution of values that has one mode or peak.

Univariate analysis A statistical analysis in which there is only one dependent variable.

Unweighted means analysis A procedure for computing a factorial design for a matrix with cells of unequal and nonproportional sample sizes.

Validity The extent to which a measurement instrument measures what it is intended to measure.

Value limits Highest and lowest values for each class interval expressed in terms of the original units of measurements.

Variability Refers to differences among the values in a data set.

Variable Any observable or measurable property of an entity that varies from one unit to another, usually synonymous with the terms factor or characteristic.

Variance Measure of variability that is based on the mean squared deviations from the mean, symbolized σ^2 or S^2.

Variation Refers to the extent to which values scatter from the central location in a frequency distribution.

Venn diagram A scheme useful for illustrating a data set, usually picturing a set as all points contained within a circle or other closed geometrical figure.

Within-groups variance In ANOVA, the estimate of the population variance computed by averaging the variance estimates within each subgroup, also referred to as error variance.

Youden square A special Latin-square analysis developed by W. J. Youden in which one must have the same number of rows and treatments, but a choice in the number of columns is possible.

Index

A

B

D

F

I

N

O

P

Q

R

T

U

V

Copy editing and indexing by Keith B. Belton
Production by Barbara Libengood.
Book design by Carla L. Clemens

Typeset by Techna Type, York, PA
Printed and bound by Maple Press Company, York, PA